AI and the Future of Smart Cooking

Ankit Shukla
Rajamangala University of Technology, Thanyaburi, Thailand

Nagendra Yadav
Welcomgroup Graduate School of Hotel Administration, Manipal Academy of Higher Education, India

Partho Pratim Seal
Manipal Academy of Higher Education, India

Abhishek Tiwari
Burapha University International College, Thailand

IGI Global
Scientific Publishing
Publishing Tomorrow's Research Today

Vice President of Editorial	Melissa Wagner
Director of Acquisitions	Mikaela Felty
Director of Book Development	Jocelynn Hessler
Production Manager	Mike Brehm
Cover Design	Jose Rosado

Published in the United States of America by
IGI Global Scientific Publishing
701 East Chocolate Avenue
Hershey, PA, 17033, USA
Tel: 717-533-8845 | Fax: 717-533-7115
Website: https://www.igi-global.com E-mail: cust@igi-global.com

Library of Congress Cataloging-in-Publication Data

Names: Shukla, Ankit, 1992- editor
Title: AI and the future of smart cooking / edited by Ankit Shukla, Nagendra Yadav, Partho Pratim Seal, Abhishek Tiwari.
Description: Hershey, PA : IGI Global Scientific Publishing, 2026. | Includes bibliographical references and index. | Summary: "This book aims to critically explore and document the transformative role of Artificial Intelligence in the evolution of smart cooking systems and technologies"-- Provided by publisher.
Identifiers: LCCN 2025022159 (print) | LCCN 2025022160 (ebook) | ISBN 9798337340425 hardcover | ISBN 9798337340432 paperback | ISBN 9798337340449 ebook
Subjects: LCSH: Food service--Data processing | Cooking--Data processing | Artificial intelligence
Classification: LCC TX911.3.E4 A526 2025 (print) | LCC TX911.3.E4 (ebook) | DDC 647.950285/63--dc23/eng/20250702
LC record available at https://lccn.loc.gov/2025022159
LC ebook record available at https://lccn.loc.gov/2025022160

British Cataloguing in Publication Data
A Cataloguing in Publication record for this book is available from the British Library.

Table of Contents

Detailed Table of Contents

transformative role of artificial intelligence (AI) in hotel kitchens. Food preparation, inventory control, and menu customization are being revolutionized by AI-driven solutions such as robotic cooking assistants, sensor-based safety systems, and predictive analytics. By automating tedious tasks and enabling data-driven decision-making, AI fosters chefs' creativity rather than replacing them. AI's effects on operations, customer interaction, and profitability are demonstrated through case studies of global hospitality chains. The balance between human knowledge and intelligent systems offers a scalable and sustainable path forward, despite obstacles like cost, cultural, and ethical concerns. By ensuring that innovation and tradition coexist peacefully in kitchens of the future, this convergence redefines culinary greatness.

The emergence of Ghost kitchens, called cloud kitchens, is causing a radical change in the food service sector. These off-site locations avoid the conventional dine-in restaurant concept by specializing only in meal preparation for delivery. Ghost kitchens provide some advantages, such as increased speed and convenience, cost-effectiveness, and scalability, by using developments in automation, artificial intelligence, and innovative kitchen appliances. They also support sustainability by reducing food waste and streamlining delivery routes. Despite quality control, brand impression, and regulatory compliance obstacles, Ghost kitchens have a bright future. New developments like virtual branding, communal kitchen areas, and connection with smart city infrastructure point to future expansion and creativity. Ghost kitchens have the potential to completely change the food industry by giving customers efficient and customized dining experiences and giving food service operators adaptable and affordable business options.

Ghost kitchens, sometimes called cloud or dark kitchens, are a fast growing subset of the food service industry that focusses solely on delivery and does not have any dining areas. These facilities offer a cost-effective and scalable strategy to meet the growing demand for online meal delivery. However, they face substantial challenges, such as the need for stringent quality control and sanitary measures, intense operating pressure, and a lack of available workers. In order to tackle these difficulties, automation—especially through robotics—is becoming an essential tool for creativity and efficiency. In this chapter, we look at how robotics is changing the face of ghost kitchen operations by automating mundane but necessary tasks like chopping, cooking, plating, packaging, and cleaning. Autonomous cooking systems, robotic arms, and AI-driven inventory and order management technologies are all investigated. This chapter delves into how modern technologies are changing the nature of the workforce, looking at the pros and cons of job loss and new skill acquisition.

The chapter discusses the AI revolution in the operations of restaurants and the enhancement of eating experiences. Robotic prep systems, predictive analytics, recommendation engines fueled by consumer data, and other artificial intelligence (AI) technology automate the time-consuming tasks, including food preparation and inventory management, making them more productive and reducing waste. AI additionally enables the development of personalized menus based on analyzing nutritional requirements, cultural contexts, and consumer preferences, which will facilitate inclusion and enhance customer joy. The practical implications could be observed in the case studies of restaurants applying AI to personalized menus, dynamic pricing, and sustainability initiatives. The chapter also addresses challenges such as training of employees, data protection issues and the fact that it is easy to lose the human element of eating.

The way food is prepared, cooked, and consumed is changing as a result of the culinary industry's adoption of artificial intelligence (AI). This essay examines how artificial intelligence (AI) is improving cooking techniques, from automating meal preparation to customizing culinary experiences. By increasing productivity, encouraging creativity, and promoting sustainability, AI-powered appliances like smart ovens, robotic chefs, and meal planning assistants are transforming kitchens in both homes and businesses. Applications for computer vision and machine learning make it possible to identify ingredients, automate cooking procedures, and create novel flavour combinations based on personal preferences. Additionally, by decreasing food waste, maximizing energy use, and enhancing cooking efficiency, AI-driven tools provide substantial sustainability benefits. With systems made to meet dietary and health needs, this study also emphasizes AI's expanding impact on personalized nutrition.

This study explores how cultural values, traditions, and attitudes influence the adoption and perception of AI-driven smart cooking technologies (e.g., smart kitchen appliances, AI recipe assistants, automated cooking robots) across different cultural contexts. While some cultures may embrace technology for its convenience and innovation, others may view it as a threat to traditional culinary practices, family roles, or food identity. The research aims to compare user acceptance, usage patterns, and cultural resistance to AI in the kitchen across diverse regions or demographic groups.

The rapid advancement of technology is reshaping the food and hospitality industry, with automation and intelligent systems driving efficiency, safety, and sustainability. Smart kitchens integrate artificial intelligence (AI), robotics, the Internet of Things (IoT), and real-time analytics to manage food preparation, storage, and service with precision. These systems ensure compliance with hygiene standards, control temperatures, prevent cross-contamination, and manage allergens, thereby reducing human error while meeting regulatory requirements. Predictive analytics and machine learning enable proactive risk detection, while resource optimization and energy efficiency reduce waste and support sustainability. Yet challenges remain, including high costs, workforce adaptation, reliance on digital systems, and ethical issues concerning transparency and accountability. The future of food safety will depend on human–machine collaboration, where automation secures consistency and chefs provide creativity and cultural value, together redefining trust and guest satisfaction.

The increasing demand for personalized dietary experiences and sustainable living has led to the emergence of smart kitchens enhanced with artificial intelligence (AI). This chapter explores the integration of AI and data analytics in creating adaptive cooking systems that cater to individual health profiles, flavor preferences, and ecological concerns. We present an in-depth analysis of the technologies enabling real-time decision-making in meal planning and preparation, including machine learning, natural language processing, and computer vision. Emphasis is placed on how such systems balance nutrition, taste, and environmental sustainability. Through case studies and practical frameworks, this chapter proposes future-ready solutions for AI-powered culinary personalization.

Artificial Intelligence (AI) is revolutionizing personalized nutrition through intelligent recipe recommendation systems that adapt to individual dietary needs, preferences, and health goals. These systems utilize machine learning algorithms, natural language processing, and user-generated data to curate meal suggestions that align with personal health profiles, such as allergies, fitness objectives, medical conditions, or cultural preferences. By analyzing nutritional content, consumption patterns, and user feedback, AI-driven platforms enhance user engagement and foster healthier eating habits. Moreover, integration with wearable devices and health apps allows for real-time dietary tracking and adaptive recommendations. Despite promising outcomes, challenges such as data privacy, algorithm transparency, and user trust remain critical. This paper explores the mechanisms, benefits, and limitations of AI-powered recipe recommendation systems and emphasizes their potential to transform modern dietary practices into personalized, data-driven nutrition models.

This chapter explores the intersection of personalized nutrition and artificial intelligence (AI) in transforming dietary planning for optimal health. Personalized nutrition focuses on tailoring dietary recommendations based on an individual's unique genetic makeup, lifestyle, and health conditions. AI plays a pivotal role in this field by processing vast amounts of data to offer precise, dynamic, and scalable dietary plans. The chapter delves into AI-driven tools, such as machine learning algorithms and predictive analytics, which enable real-time, data-driven dietary advice. Furthermore, it examines how AI enhances food tracking, nutrient analysis, and the creation of personalized meal plans, leading to better health outcomes, weight management, and chronic disease prevention. Through these innovations, AI offers a promising future for individualized health optimization and wellness.

The convergence of artificial intelligence (AI) and health data is revolutionising how consumers approach nutrition, particularly in the e-grocery sector. Groceries are sold in India through websites and apps designed to engender a unique customer experience. This chapter explores the transformative potential of AI-driven personalised nutrition solutions in e-grocery shopping platforms. Based on recent research and data gleaned from the grocery industry, the chapter examines how machine learning and behavioural analytics can be integrated with dietary preferences and real-time consumption data to deliver hyper-personalised food recommendations to Indians. Findings suggest that AI not only enhances consumer health outcomes but also drives business value through increased engagement and loyalty. However, scalability depends on overcoming barriers like data standardisation, algorithmic transparency, and equitable access.

The rise of AI-powered large language models (LLMs) has reshaped digital learning, especially for Generation Z seeking personalized and interactive experiences. This study examines how informativeness and trendiness influence the continuance intention to use AI tools for cuisine learning. Grounded in Uses and Gratifications Theory (UGT), it positions trendiness as a mediator between perceived informativeness and continued usage. Data were collected from 119 Gen Z users in Northern India with prior experience using AI tools like ChatGPT for cooking tasks. Structural equation modeling revealed that informativeness significantly affects both trendiness and continuance intention, with trendiness playing a strong mediating role. The findings expand UGT by incorporating socially driven gratifications and offer practical insights for enhancing AI-based learning tools aligned with Gen Z's preferences.

Augmented reality gamified experiences may entertain and engage visitors. A hotel might implement an augmented reality treasure hunt that directs guests to various locations inside the facility. Product visualization Augmented Reality (AR) is a technology that overlays computer-generated visuals over a user's perception of the actual environment, so creating a composite vision. Room customizations - Augmented reality applications for room customization enable customers to digitally alter the décor of their hotel accommodations, including furniture, artwork, and lighting. Augmented reality (AR) is transforming travel and our engagement with the world. This research seeks to examine the transformational effects of technological breakthroughs in the hospitality sector which reflects the technology is transforming the visitor experience and the operational framework of hotels, resorts, and other hospitality venues. Augmented reality enhances comprehension of how technology innovations are transforming guest experiences and operational methodologies in the hotel industry.

This study examines TripAdvisor reviews from three wineries in Europe to analyze customer perceptions and preferences in wine tourism. The investigation aims to understand the impact of online reviews on winery reputations and customer satisfaction in the context of wine tourism experiences. The study utilizes an advanced data analytics tool, namely Power BI, to analyze the development and demographics of the reviews and Orange Data Mining to conduct sentiment analysis and text mining. By examining reviews from wineries in France, Portugal, and Italy, this work identifies key themes, sentiments, and preferences expressed by customers, providing actionable insights for wineries to enhance visitor experiences and strategic decision-making in the wine tourism industry using an artificial intelligence approach.

Preface

INTRODUCTION

The rapid advancement of Artificial Intelligence (AI) is reshaping nearly every aspect of our lives—from how we communicate and work to how we learn and care for our health. Among the most intriguing and dynamic areas of change is the kitchen, a traditionally analog space now at the forefront of digital innovation. As AI integrates more seamlessly into domestic environments, smart cooking has emerged as a vibrant frontier where technology meets culture, creativity, and everyday necessity.

AI and the Future of Smart Cooking was conceived in response to the growing interest, complexity, and interdisciplinary nature of this transformation. This edited volume brings together researchers, practitioners, technologists, and visionaries who are working at the intersection of culinary practice and computational intelligence. The book offers a timely and critical examination of how AI is not only automating tasks in the kitchen but also enabling entirely new forms of interaction, personalization, and problem-solving.

We have structured this volume to explore three core objectives:

1. To examine current and emerging applications of AI in smart kitchens—from voice-guided meal preparation and nutritional analytics to robotic cooking and adaptive user interfaces.
2. To interrogate the broader implications of these technologies—ethical, cultural, environmental, and social—as they become increasingly embedded in our food systems and daily lives.
3. To identify research gaps, forecast future trends, and surface practical opportunities for innovation across disciplines.

This book is designed to serve as both a reference and a springboard. For academic researchers, it offers new theoretical frameworks and case studies. For developers and designers, it provides user-centered insights into evolving consumer needs and expectations. For policymakers, it raises critical questions about responsibility, equity, and sustainability. And for chefs, nutritionists, and culinary educators, it opens up exciting possibilities for creativity, health, and inclusion.

Our contributors represent a diverse array of fields—from artificial intelligence, robotics, and human-computer interaction to food science, hospitality, design, and ethics. This multidisciplinary approach is not just deliberate—it is necessary. The future of smart cooking lies not in siloed innovation, but in the convergence of perspectives, technologies, and values.

We extend our deepest gratitude to the authors who have contributed their expertise to this volume, and to the readers who continue to push the boundaries of what it means to cook, eat, and live in a world shaped by intelligent systems.

ORGANIZATION OF THE BOOK

This volume begins by laying the conceptual groundwork for understanding how AI is reshaping modern gastronomy. In Chapter 1, *AI and Smart Cooking: Innovation in Gastronomy*, the focus is on the integration of AI, data science, and machine learning into commercial kitchens, particularly within the hospitality sector. Drawing from interviews with chefs and an extensive review of existing literature, the chapter highlights the transformative potential of AI in enhancing food safety, quality, operational efficiency, and sustainability. It underscores how smart cooking technologies offer more than convenience—they are redefining the future of culinary practice.

Chapter 2, *The Digital Chef: Future-Proofing Hotel Kitchens with AI*, expands on this theme by introducing the "Digital Chef" model, a framework that fuses intelligent automation with human creativity in hotel kitchens. Through global case studies, the chapter illustrates how AI tools like predictive analytics, robotic assistants, and smart inventory systems can increase profitability while maintaining a personal touch in guest experiences. The narrative strikes a balance between technological progress and the preservation of culinary heritage, addressing ethical, cultural, and financial barriers along the way.

The evolution of food service is further explored in Chapter 3, *Ghost Kitchens and the Transformation of Future Food Services*. Here, we turn our attention to the rise of ghost (or cloud) kitchens—delivery-only operations that represent a fundamental shift in food production and consumption. The chapter explores how these tech-driven environments leverage AI and automation to optimize efficiency, scalability, and sustainability. Despite regulatory and quality control challenges, ghost kitchens offer a compelling model for the future of urban food logistics and digital branding.

Building on this, Chapter 4, *Automation in Ghost Kitchens: Exploring the Role of Robotics in Revolutionizing Food Service Operations*, delves deeper into the operational backbone of these delivery-focused facilities. Through a detailed examination of robotic technologies—ranging from automated cooking systems to intelligent packaging and cleaning solutions—the chapter reveals how automation is addressing labor shortages and increasing productivity. It also raises important questions about workforce transformation, job displacement, and the new skills required in AI-enabled kitchens.

Chapter 5, *AI in Restaurants: Creating Smarter Kitchens and More Customized Dining Experiences*, shifts the focus back to traditional restaurant environments and explores how AI is enhancing both back-of-house operations and front-of-house guest experiences. Technologies such as recommendation engines, robotic prep stations, and dynamic pricing models are examined alongside real-world examples. The chapter also critically addresses concerns around data privacy, employee training, and the preservation of human connection in dining.

In Chapter 6, *AI-Powered Culinary Innovation: Exploring the Intersection of Technology and Gastronomy in Smart Cooking*, we examine how AI is enabling creativity and personalization in unprecedented ways. From smart ovens and AI-generated recipes to personalized nutrition planning and sustainable meal design, the chapter showcases a range of applications that highlight AI's capacity to enhance both efficiency and imagination. This intersection of culinary artistry and computational intelligence is a defining feature of next-generation smart kitchens.

Chapter 7, *From Kitchens to Code: Cross-Cultural Perspectives on AI and Smart Cooking Adoption*, introduces a much-needed global and cultural lens. It explores how traditions, values, and societal attitudes shape the acceptance—or resistance—toward smart cooking technologies. By comparing adoption patterns across regions and demographics, the chapter provides insights into the socio-cultural dynamics that influence technological innovation in domestic and commercial kitchens alike.

Food safety and compliance are the focus of Chapter 8, *Food Safety Meets Automation: Smart Kitchens of the Future*. Here, the convergence of AI, IoT, and robotics is positioned as a powerful tool for minimizing human error, ensuring hygiene, and optimizing resource use. The chapter discusses how predictive analytics and real-time monitoring systems can help meet increasingly complex regulatory standards while advancing sustainability goals. Yet, it also warns of the challenges posed by high implementation costs and the ethical implications of digital dependency.

Chapter 9, *AI-Powered Personalization in Smart Kitchens: Adaptive Cooking Systems for Health, Taste, and Sustainability* synthesizes many of the book's key themes—personalization, sustainability, and technological adaptability—into a comprehensive exploration of how AI can deliver deeply customized culinary experiences. Focusing on adaptive systems that adjust recipes and ingredients in real time based on health profiles and ecological goals, it presents a forward-looking framework for AI-enabled cooking that is both human-centric and planet-conscious.

Chapter 10, *AI-Powered Recipe Recommendation Systems and their Role in Personalized Nutrition*, begins this exploration by highlighting how machine learning and user-centric data models are enabling recipe recommendations tailored to individual dietary needs and health goals. These intelligent systems leverage personal health profiles, real-time inputs, and even data from wearable devices to guide nutritional behavior, while also raising vital questions around privacy and algorithmic transparency. The chapter makes a compelling case for how AI can act as both a coach and a curator in modern dietary practices.

Building on this, Chapter 11, *Personalized Nutrition and AI-Driven Dietary Planning: A Technological Leap Toward Health Optimization*, broadens the scope to examine how AI transforms the very foundations of nutritional planning. By harnessing genomics, lifestyle data, and predictive modeling, AI-powered tools offer unprecedented levels of precision in dietary recommendations. The chapter explores how these technologies support not only wellness but also preventive healthcare through chronic disease management and real-time health optimization.

Chapter 12, *Customer-Centric E-Grocery Services in India: Leveraging AI for Personalized Nutrition Solutions*, grounds these innovations within the context of India's rapidly evolving e-grocery market. It explores how digital platforms integrate AI, consumer behavior analytics, and health data to deliver hyper-personalized food recommendations. This chapter provides insights into scalability, equity, and infrastructure challenges while showcasing how AI can serve both public health goals and commercial growth in emerging markets.

Generational shifts in digital behavior are addressed in Chapter 13, *Influence of Informativeness and Trendiness on Gen Z's Continued Use of AI for Culinary Learning*. Focused on the rising use of AI platforms like ChatGPT among Gen Z learners, this study applies Uses and Gratifications Theory to explore how perceived informativeness and social relevance drive continued engagement. The findings provide key insights for designers of AI-powered educational platforms and underscore the role of trendiness in shaping user adoption among digital-native generations.

Sustainability remains a central concern throughout this volume, and Chapter 14, *Sustainable Cooking Solutions Powered by AI and Smart Technologies*, directly addresses this intersection. Focusing on global sustainability goals, this chapter investigates how AI and smart systems—from IoT appliances to energy-efficient meal planning—are redefining eco-conscious cooking in households, commercial kitchens, and community settings. Through compelling case studies, the chapter advances a vision of AI-enabled sustainability that is inclusive, ethical, and practical.

Shifting toward the broader hospitality sector, Chapter 15, *AI-Driven Transformation in the Hospitality Industry: Enhancing Guest Services and Operational Performance*, reviews how AI tools are being adopted to optimize guest satisfaction and streamline operations. The chapter evaluates AI's potential to resolve long-standing industry challenges—such as security, financial management, and service personalization—while critically examining the costs, accountability concerns, and structural changes required to realize its full promise.

Chapter 16, *Assessing the Role of Augmented Reality in Shaping the Future of the Hotel Industry*, complements the previous chapter by showcasing how augmented reality (AR) is transforming the guest experience. From immersive entertainment to room customization and enhanced product visualization, AR is redefining hospitality environments. This chapter situates AR as a bridge between physical and digital spaces, offering a dynamic tool for guest engagement and operational innovation within hotels and resorts.

The book concludes with Chapter 17, *Exploring Consumer Experiences in Wine Tourism through Sentiment Analysis*, which demonstrates how AI, specifically sentiment analysis and data visualization tools, can be applied to tourism and experiential marketing. Focusing on wineries in France, Portugal, and Italy, the chapter examines customer reviews to extract meaningful insights into visitor satisfaction and brand perception. By applying tools like Power BI and Orange Data Mining, the study reveals how wineries—and the broader tourism sector—can leverage AI to refine services and elevate consumer experiences.

Together, these chapters offer a sweeping and multifaceted view of how AI is shaping not just cooking, but the broader food and hospitality ecosystems. From personalized nutrition to AR-driven hotel experiences, this volume illustrates the far-reaching and interdisciplinary impact of intelligent systems. As editors, we hope this book not only informs but also inspires future exploration, innovation, and dialogue across academia, industry, and policy.

CONCLUSION

As we conclude this volume, it is clear that artificial intelligence is not merely an accessory to modern cooking—it is rapidly becoming a transformative force across the entire culinary and hospitality spectrum. From the automation of commercial kitchens and personalized nutrition planning to the reinvention of hospitality experiences and sustainable food practices, AI is challenging long-held assumptions about how we prepare, consume, and engage with food.

What emerges across these chapters is a shared recognition: the future of cooking will be not only more intelligent, but also more interconnected, inclusive, and responsive to both human and planetary needs. AI offers the tools to personalize nutrition, optimize operations, reduce waste, and democratize access to culinary knowledge. Yet with these advances come critical responsibilities—to ensure transparency in algorithms, protect user data, preserve cultural foodways, and design systems that enhance, rather than replace, human creativity and care.

This book does not claim to offer all the answers. Instead, it aims to open up new lines of inquiry and spark dialogue across disciplines and sectors. By bringing together diverse voices—from chefs and engineers to designers, researchers, and policy thinkers—we hope to offer a dynamic and forward-looking framework for understanding the possibilities and pitfalls of AI in the culinary domain.

We invite readers to approach these chapters not only as a snapshot of current developments, but as a call to action. The future of smart cooking is being shaped right now—by the questions we ask, the technologies we build, and the values we embed in them. Whether you are a scholar, a technologist, a policymaker, or a food enthusiast, we hope this volume equips and inspires you to take part in that shaping.

Acknowledgement

I would like to express my sincere gratitude to my fellow editors — Nagendra Yadav, Partho Seal, and Abhishek Tiwari — whose collaboration, expertise, and commitment have been instrumental in shaping AI and the Future of Smart Cooking. Working alongside such dedicated colleagues has been both an enriching and inspiring experience.

On a more personal note, I am deeply indebted to my wife, Dipika Sharma, and my son, Anvit, for their unconditional love, patience, and constant encouragement. Their belief in me has been my greatest motivation throughout this journey. I also wish to extend my heartfelt appreciation to my family, whose unwavering support has given me the strength to balance both professional and personal responsibilities.

Finally, I would like to thank all the contributors for their valuable insights, research, and creativity. Each chapter reflects the dedication and passion of its author, and without their collective effort, this book would not have been possible.

Chapter 1
AI and Smart Cooking: Innovation in Gastronomy

Partho Pratim Seal
https://orcid.org/0000-0001-6867-8713
Welcomgroup Graduate School of Hotel Administration, Manipal Academy of Higher Education, India

Nagendra Yadav
https://orcid.org/0009-0008-9743-9250
Welcomgroup Graduate School of Hotel Administration, Manipal Academy of Higher Education, India

Ankit Shukla
https://orcid.org/0000-0002-7646-3517
Rajamangala University of Technology, Thanyburi, Thailand

Abhishek Tiwari
https://orcid.org/0000-0003-2970-2092
Burapha University International College, Thailand

ABSTRACT

Cooking is a basic requirement of human beings, as food is one of the necessities to meet human needs. Technological concepts like artificial intelligence, data science, and machine learning are being used to advance the smart kitchen in hotels and pave the way for the smart kitchen revolution. Smart cooking, along with artificial intelligence, can bring in a lot of automation in cooking. The chapter explores how innovation in the culinary field is changing the commercial kitchen with a focus on food safety, food quality, efficiency, and sustainability. The usage of AI in the culinary domain is restricted not only to a better dining experience for the customer but also to enhancing kitchen operations. The study follows a qualitative method of interviews with chefs to examine the academic literature on the chosen concepts by

DOI: 10.4018/979-8-3373-4042-5.ch001

examining the literature in a wide range of contexts. The findings of the study state that smart cooking does provide convenience in working life, and technology will be a blessing in the future.

INTRODUCTION

Cooking is an art that combines and processes ingredients creatively to make a delicious dish, which is magical (Pollan, 2013). Humans have evolved the art of cooking, and the creative process has led to rich culinary legacies that define the culture globally (Freeman, 2009). Gastronomy is artistic, and with the increasing usage of structured data and various computational methods, it has changed the artistic approach of gastronomy vividly (Bagler & Goel, 2024). Over the centuries, gastronomy has evolved as an intersection of art, science, and culture. It includes the continuous development of innovative cooking techniques and kitchen equipment. The smart cooking technology allows the kitchen masters to showcase their creativity, making their cooking process more efficient and accurate, along with innovation.

Artificial Intelligence has gone through incredible growth and innovation, with the potential for growth in various fields, including the food industry. The food industry is implementing AI technology in various forms, such as recipe generation, meal planning, food waste management, and nutritional analysis. The AI algorithm can process large data and provide personalized recommendations, and streamline decision-making processes, which can increase efficiency and improve outcomes. As AI becomes more accessible and user-friendly, it now has a significant role in the food industry and also in our daily lives. The usage of smart cooking tools and technology is adding new life to the domain of culinary arts. The usage of smart ovens, sensors, and AI-powered devices helps chefs to perfect the dishes, maintaining the standards, along with sustainability and efficiency.

SMART CUISINE

Smart Cuisine is an innovative system that features an ingredient calculator, an AI-generated recipe, and an OpenAI completion chat service that provides informed decisions about food choices and reducing food wastage. The ingredient calculator will calculate the exact amount of cost of ingredients, along with the proportion key, which will help to plan meals accordingly and save money. The function also ensures that the time, energy, and money a consumer spends lead to an appropriate meal. The AI-generated recipes will enable users to make recipes based on the availability of ingredients, allowing them to be more creative and reduce leftover ingredients. The OpenAI completion chat service provides users with expert advice on food, nutrition, and health-related areas, providing informed decisions about dietary choices. Smart Cuisine will also promote healthier eating habits, save time, money, reduce food wastage, and help in creating a sustainable future.

COMPUTATIONAL GASTRONOMY

A data-driven approach for food that will transform the food landscape. A data science that will transform the artistry of food and cooking. Integrating food, data, and computation will transform the landscape with data innovation. It will help to create a recipe data structure that will provide new opportunities for food and cooking. The process, if combined with robotics, can help accelerate kitchen innovations. With the understanding of the molecular correlation of taste and odor of dishes, ingredients, and the molecules, it will help to have a flavor that generates molecular connotations of sensory profile. The precise nutrition knowledge on food and empirical understanding of health can create a digital diet coach that recommends personalized diets appropriate for their health parameters. Net-zero recipes will be possible through the fine-grained data of carbon footprints of the ingredients, which is traceable and a blockchain-enabled supply chain.

DIGITAL GASTRONOMY

Digital Gastronomy (DG) used digital technologies in the kitchen to enhance traditional cooking with new computational capabilities (Mizrahi et al., 2022). Digital gastronomy is in line with human-computer interaction, which includes digital manufacturing and fabrication, which currently have an important role in the kitchen setting and the culinary field, which includes 2D, 3D, and 4D food printing.

Integrating computers in the kitchen helps optimize food and minimize waste (Zoran, 2019). One of the best ways to reduce food waste is to rely on dried substances like powders, which have a longer shelf life than fresh products and can be easily transported and stored. With the powdered ingredients being homogeneous, it is easier to make recipes and personalize the nutritional value of the final dish.

Digital gastronomy enhances cooking with new digital capabilities. The integration of computers with cooking allows the cooks to individually address users' needs and desires, which calculates the nutritional content of ingredients that affect the cooking-related chemical reactions, which fit personal preferences. It also allows extending traditional recipes into wider opportunities that present themselves with the relationship and dependence of cooking elements, which a chef defines. It can also help in designing flavor patterns and structures in dishes.

ROBOTS IN FOOD PREPARATION

Applying robotics in the culinary field is a key to digital gastronomy. Robots powered by artificial intelligence do assist in executing various tasks, including slicing and cooking, presenting dishes, and enhancing the efficiency of food preparation (Bollini et al, 2013). Technological advancements can transform operations in a commercial kitchen, streamline processes, and reduce errors. It also assists in augmenting the dining experience through sensory enhancements. With the usage of augmented reality (AR) and virtual reality (VR), it is possible to have virtual eating scenarios where ambience, scents, and tastes can be modified digitally (Narumi et al, 2012). The innovation offers chefs a multisensory dimension, making it an unparalleled level of engagement. With artificial intelligence becoming accessible and economical, it is forecasted that it will influence food and nutrition in our regular lives (Miyazava et al, 2022). Sustainability has a major role in the current scenario, with solutions like cultured meats, algae-based ingredients, which provide eco-friendly alternatives to food sources, including resource depletion (Post, 2012).

SMART COOKING

Combining computer technology with the cooking process, the chefs can prepare meals meeting the specific tastes and requirements of the users, calculating the nutritional values and controlling cooking-induced chemical reactions, enabling customization of dishes to align with individual preferences digitally (Mizrahi et al., 2022). Smart cooking technologies do offer innovative solutions that change the kitchen experience. Smart ovens cook meals automatically and make necessary changes so users can make changes on their own and use their time efficiently. Smart pots make kitchen work less laborious by optimizing the cooking process, which can be controlled remotely. Smart stoves ensure that meals are cooked perfectly as they can adjust both heat and cooking time. App-based cooking assistants allow you to prepare delicious meals, presenting recipes through mobile devices that guide the cooking process stepwise.

Smart Oven

An indispensable kitchen device designed to enhance culinary efficiency is equipped with advanced sensors that monitor a variety of cooking conditions. This tool not only displays the ideal cooking temperature and duration tailored to different types of food but also incorporates smart technology to ensure precision. As the food cooks, if it reaches a temperature above the preset threshold, the device issues an immediate warning notification and automatically halts the cooking process to prevent overcooking or burning. Equipped with sophisticated sensors, the device accurately assesses the size, condition, and quantity of the ingredients being prepared, allowing it to optimize meal preparation based on these variables. Moreover, it continuously learns and adapts to the cooking habits and preferences of its users, ensuring a personalized cooking experience. This innovative solution is revolutionizing the way we prepare meals, making cooking not only more convenient but also more reliable and consistent (Rational, 2024).

Smart Pans

Smart pans have been designed with sensors placed on the bottom of the pan and connected to the smartphone to prevent energy loss due to overcooking. The application also sends warnings when new ingredients are added, including turning, mixing and adjusting the heat settings, which creates an advantage by saving time, and the staff does not have to wait from the beginning of product preparation (Keşkekci & Gençer, 2023). Although it looks like a tilting pan but it has features of boiling, deep-frying, pan-frying, and pressure cooking. iVario Pro replaces all traditional appliances, which combines power with precision as it is four times faster than traditional cooking devices, consumes 40% less energy, and also saves space (Rational,2024).

FOOD SAFETY

IOT technology has a crucial role in enhancing food safety and compliance standards in commercial kitchens. The temperature monitoring sensors, which are embedded in the refrigerator, ensure that all perishable food is stored at a safe temperature, preventing the spoilage of food and its contamination. An automated temperature logging and reporting system generates digital records of temperature data, thereby simplifying regulatory compliance. IoT-enabled food safety solutions provide real-time alerts and notifications so that kitchen staff can create immediate corrective actions and mitigate the risks to food quality.

FOOD INVENTORY MANAGEMENT

AI plays a vital role in the effective management of food inventory by employing sophisticated data analysis techniques. This technology not only predicts when specific ingredients are running low but also provides insights into usage patterns and trends, allowing kitchens to restock supplies well before they reach critical levels. By continuously monitoring inventory status in real-time, staff are kept fully informed about stock availability, reducing the risk of running out of essential items during peak operation times.

Moreover, AI streamlines the ordering process through automated purchasing systems that generate purchase orders based on accurate forecasts of future needs. This feature considers factors such as seasonal demand fluctuations, historical usage data, and current inventory levels, making the ordering process more efficient and tailored to the kitchen's actual requirements.

As a result, commercial kitchens can ensure they always have the right ingredients on hand, minimize food waste by aligning purchases with actual consumption, and maintain a consistent workflow, ultimately leading to improved food quality and customer satisfaction. This intelligent inventory management system not only enhances operational efficiency but also contributes to better financial performance by optimizing resource utilization.

PREDICTIVE MAINTENANCE AND EQUIPMENT MONITORING

The benefits of IOT are in commercial kitchens, which can implement predictive maintenance and equipment. With the installation of sensors on equipment like refrigerators, freezers, and dishwashers, the establishment can monitor its performance in real time and detect potential malfunctions before they happen. A predictive maintenance algorithm analyses data patterns and equipment usage to identify maintenance needs and schedule repairs proactively to avoid costly equipment failure later. Remote monitoring capabilities allow kitchen managers to track energy consumption, optimize equipment settings, and reduce operating costs.

Currently, there is a need to develop a user-centric approach to Smart kitchen appliances and AI, which are recognized by both the industry and academia. As perceived value is the key driver for smart cooking appliances or AI-enabled cooking equipment, it leads an organization to adapt to rapid digital transformation. The study adopts smart kitchen appliances value of adopting

Usefulness

Usefulness is defined as the degree of improvement in the use of the product's performance in using a smart cooking device or an AI cooking appliance. Improving performance states that the degree of matching between the capabilities of the user and the functionalities offered by the technology affects the perceived value of the smart device (Jiang et al., 2020)

Flexibility

Flexibility is a system designed to address the evolving needs of users by adapting and reconfiguring equipment to align with specific customer requirements (Moura Leite et al., 2020). This adaptability allows for a more personalized user experience, ensuring that equipment can be modified to better suit individual preferences and varying contexts of use.

Researchers Zhen et al. (2020) further elaborate on this concept by discussing the role of smart cooking in embracing open innovation at multiple levels, including products, services, and systems. This approach fosters collaboration and creativity, empowering users to express their unique culinary needs and preferences. By facilitating user input and feedback, smart cooking technologies provide tailored solutions that enhance convenience, efficiency, and overall satisfaction in the kitchen.

Reliability

Reliability in the context of smart cooking devices pertains to both their accuracy and safety during use. These devices play a pivotal role in consumer empowerment by offering real-time feedback on various aspects, such as functionality, ingredient status, and cooking progress. This feedback helps users make informed decisions regarding their cooking processes.

However, it is essential to acknowledge that smart cooking and AI-enabled devices are not without their vulnerabilities. Issues can arise from network security breaches, leading to unauthorized access, data theft, or manipulation of settings. Additionally, equipment failures such as malfunctions in heating elements or sensors can compromise the cooking process. System errors, which may result from software bugs or compatibility problems, can further hinder device performance. Such challenges not only undermine the reliability of these appliances but can also pose significant health risks, such as undercooked food or appliance fires (Riel et al., 2017). Thus, ensuring robust security measures and regular maintenance is critical to safeguarding the user's experience and well-being.

Technicality

Technicality is the difficulty of using relevant products/services caused by technologies. The ICT leads new interaction for smart cooking, enabling the user to complete tasks independently. Although a new type of interaction requires learning and adaptations by the user, these technologies may have no clear gains in usage but allow some efforts to the user (Kim et al., 2007; Wang & Wang, 2010). So we can state that the more technological effort the users put into use, the less likely they are to get a high value perception.

The usage of AI in Hospitality & Culinary is quite impressive, with multiple uses including

AI AND DIGITAL TRANSFORMATION IN HOSPITALITY

Hospitality has been impacted by the digital revolution, with automation, machine learning, and data analytics, which have changed hospitality operations (Ivanov & Webster, 2019). With AI applications in hospitality, the usage of chatbots for hotel reservations and robotic concierge has become common (Tussyadiah & Park, 2018). However, the usage of technology has gone beyond customer service into the culinary environment, where AI is enabling the chefs and managers to enhance their culinary experience (Lu et al., 2021).

Emergence of Smart Kitchen

A concept of a smart kitchen is about digitally connected appliances, sensors, and AI systems that collaborate to assist in automating cooking processes. Commercial kitchens and smart systems do support operational efficiency by reducing errors, minimizing food wastage, and ensuring compliance with the hygiene regulations (Marvin et al., 2016). The innovation does align with the regulators by providing guests with a safe, consistent, and sustainable dining experience

AI and Culinary Creativity

The usage of AI applications helps in creativity, which includes the recipe generation algorithm created for ex, IBM's Chef Watson, which suggests the usage of novel ingredient combinations that are based on molecular flavor pairings and nutritional databases. There is a debate among chefs about whether to consider the machine-driven creativity (Amabile, 1996).

Sustainability in Smart Cooking

A challenge for the modern food industry is the sustainability of food. AI helps in sustainable practices by forecasting demand, reducing food wastage, and optimizing energy usage. Past research studies on sustainability by Filimonau (2020) state that smart systems assist in aligning them with the UN Sustainable Development Goals (SDGs).

ETHICS AND CULTURAL

The cultural practices surrounding food are deeply intertwined with local traditions and identities, reflecting the unique histories and social dynamics of various communities. However, the rise of automation in food production and preparation can lead to the commodification of food, transforming it into a standardized product that often lacks cultural significance and depth (Richards, 2018). To address this challenge, it is crucial to find a harmonious balance between the adoption of modern technologies and the preservation of the rich cultural essence that food represents. Furthermore, ethical considerations play a significant role in this evolving landscape, particularly as smart kitchens increasingly collect and analyze data regarding customer preferences, ingredient sourcing, and operational efficiencies. This data gathering raises important questions about privacy, transparency, and the responsibility involved in managing such information. Ensuring that this data is handled with care and integrity is vital to maintaining consumer trust and safeguarding cultural values in the food industry.

There are notable gaps in the current body of literature regarding the application of artificial intelligence (AI) in culinary settings. While a significant number of studies delve into the integration of AI within customer service operations, there is a pronounced scarcity of research pertaining to its role within the kitchen environment. Specifically, further investigation is needed to understand how AI impacts chefs' creativity and efficiency in meal preparation and menu development. Additionally, it is crucial to explore how chefs leverage these technological advancements while preserving the culinary traditions and techniques that have been passed down through generations, ensuring that the essence of their craft remains intact.

The application of artificial intelligence (AI) in gastronomy encompasses much more than simply improving operational efficiency; it significantly contributes to enhancing culinary creativity and the preservation of culinary traditions across cultures. Advanced smart cooking technologies introduce automation and consistency to food preparation processes, yet they raise important questions regarding the authenticity of traditional food practices and the protection of creative expression in the culinary arts. This complex interplay between innovation and tradition creates a fertile ground for both exciting opportunities and critical debates regarding the influence of AI in the gastronomic field (Bessière & Tibere, 2013).

At present, smart cooking systems play a pivotal role in the culinary landscape. AI technologies can facilitate precise inventory tracking, enabling kitchens to manage ingredients more effectively and reduce waste. Additionally, these systems can analyze consumer trends to forecast demand, allowing restaurants to adjust their offerings in real-time and maintain optimal supply levels. Furthermore, AI can provide sustainable ingredient substitutes, helping to promote environmentally friendly practices within the food industry.

The automation offered by AI not only minimizes human error in cooking processes but also plays a crucial role in ensuring compliance with stringent food safety regulations. By maintaining precise control over cooking temperatures, ingredient quality, and preparation methods, AI can enhance food safety while supporting chefs in delivering high-quality culinary experiences. Overall, the intersection of AI and gastronomy presents a dynamic arena filled with both promise and challenges.

As the application of artificial intelligence (AI) in the hospitality industry expands, the bulk of existing research has concentrated on specific technologies such as service robots and chatbots (Tussyadiah & Park, 2018). Despite the promising advancements these technologies offer, there remains a significant gap in understanding how chefs perceive and interact with AI-driven culinary systems. This lack of focus on the culinary professionals themselves limits our comprehension of how technology affects culinary creativity and innovation in the kitchen.

Moreover, there is a conspicuous deficiency of qualitative research investigating the viewpoints of chefs, who play a crucial role in preserving and evolving cultural identity through food. Understanding their perspectives is vital to discerning whether they see AI as an asset that enhances their craft or as a potential threat that undermines traditional culinary techniques and artistry. Engaging with chefs to explore their experiences and insights regarding AI's role in the culinary landscape can offer a more nuanced understanding of these technologies' integration into kitchens worldwide.

This study aims to investigate the transformative impact of artificial intelligence (AI) and smart cooking technologies on kitchen operations in commercial settings. It will delve into chefs' perspectives regarding how these innovations influence critical factors such as creativity, operational efficiency, and sustainability in culinary practices. Furthermore, the research will identify the specific challenges that chefs and kitchen managers face when integrating smart cooking solutions into their workflows, as well as the opportunities that arise from adopting these advanced technologies. By shedding light on these aspects, the study seeks to provide a comprehensive understanding of the evolving landscape of modern kitchens.

This study seeks to address these gaps by combining the literature with semi-structured interviews with professional chefs. The objectives are as follows:

1. To explore the role of AI and smart cooking technologies in transforming kitchen operations.
2. To examine chefs' perspectives on the impact of AI on creativity, efficiency, sustainability, and cultural heritage.
3. To identify the opportunities and challenges of adopting smart cooking technologies in commercial kitchens.
4. To contribute to the discourse on the future of gastronomy by integrating theoretical insights with practitioner experiences.

METHODOLOGY

The study adopts a qualitative approach with a document review of academic literature, industry reports on AI in gastronomy and smart cooking. The qualitative approach is considered as it helps to obtain the chef's perspective about their experience, attitude, and professional insights, which are many a times overlooked (Creswell & Poth, 2018). The study also helps to find the chef's expectations, their challenges, and ethical concerns. The data was collected through semi-structured interviews with 25 chefs from India, mostly in hotels and some fine dining restaurants.

The goal of qualitative research is to understand the study phenomenon (Polit & Beck, 2010). Researchers Kvale (1996) state that an interview is a mode of human interaction and we know people learn from experiences, feelings, and hopes, as we understand the world we live in. The semi-structured interview is a method commonly followed in social science. Researchers Megaldi and Berler (2020) state that semi-structured interviews enable researchers to go deep for discovery and are preferred by researchers as they go deep for a discovery. The semi-structured interview is preferred in social science. It is also stated that interviewees' viewpoints are more likely to be expressed in an openly designed manner rather than standardized conversation with the help of a questionnaire (Flick, 2002). A semi-structured interview is flexible and allows new questions to be brought up during the interview, considering what the interviewee has said. The semi-structured interview provides a framework for the theme (Rubin & Rubin, 2005) and a balance between the main questions, follow-ups, and probes. The semi-structured interviews are chosen to obtain consistency and flexibility. Although common questions were asked to all the chefs, the chefs were free to elaborate and add on their experiences and perspectives. To ensure that the interview is conducted only for the target users, the potential respondents were reminded not to participate in the interview if they do not have prior experience of using a smart large cooking device or an AI-powered cooking equipment.

To ensure that important aspects were not overlooked and that the questions asked during the interview were accurate, doctoral researchers and two professionals in smart cooking and AI in cooking reviewed the questions.

The study adopts a qualitative research design with semi-structured interviews to discover how the chefs' views and experiences on the integration of AI and smart cooking in their cooking process. The study used a purposive sampling strategy to select chefs with the requisite experience. The chefs were mostly from the luxury hotels serving food in fine dining restaurants. A sample size of 25 chefs was considered, primarily from India, with some of them having international exposure also. The inclusion criteria were a minimum of 5 years of experience in the culinary domain with exposure to smart kitchens. The sample size was considered appropriate for achieving the theoretical saturation necessary to obtain feasibility (Guest, Bunce, & Johnson, 2005).

Researchers suggest that it is beneficial for interviewers to have an interview guide made, which includes questions that the interviewer would ask different participants (Lindlof & Taylor,2002). It also allows interviewers to provide questions in the context, situation, and people being interviewed. After all the data is obtained, the data must be cleaned so that the raw data is in a form that can be easily worked with. Managing the data is important to ensure that the original information is maintained without distortion, and the researcher is familiar with the entire dataset. After the entire data is obtained, the next step is analysis, which requires the interviewer to try various ways of categorizing and reorganizing the prepared data, which investigates the patterns that are linked with. After this, the data is categorized and coded as per the research questions.

The interviews were conducted with the chefs through MS Teams, which lasted between 35-45 minutes. With the prior consent of the participants, the interviews were recorded and later transcribed verbatim. The data was collected over a period of three months, March- May 2025. The ethical rigor was maintained throughout the study as required for the qualitative research protocol (Bryman, 2016). Before the interview, all the participants were briefed about its purpose and were also told that the entire process would be confidential and the chef's identity would not be disclosed. All the chefs were assigned pseudonyms in the transcripts.

The data obtained from the interview were analyzed thematically (Braun & Clarke, 2006) as per the framework of the Technology Acceptance Model (Davis, 1989) and creativity in organizational settings (Amabile, 1996). NVivo software was used to support coding and organization of qualitative data. The triangulation helps to obtain richer insights, blending academic theory with the practitioner's perspective. The study contributes to the body of knowledge on digital transformation in hospitality by adopting smart kitchen appliances or AI in culinary operations by expanding its focus beyond service robots to include the views of chefs, thereby bridging the gap between academic theory and industry practice. It offers insights to both educators and policymakers.

RESULTS

The factors leading to the usage of Smart Cooking devices

Usefulness of Smart Kitchen

A chef from a five-star hotel stated, "Usage of smart ovens and smart pans has improved our performance by preparing our dishes on time". Another chef mentioned that smart ovens, "Culinary creativity, efficiency, and precision in making dishes have been better than before using them."

Smart Cuisine Systems

The smart systems help in making recipes, as a chef mentioned, "With the customization of the recipes and our equipment, the consistency of our products has been extremely good as it prevents overcooking, which is a huge relief in the rush." A chef stated that "AI-generated recipes allow me to be creative with ingredients that are left over, which otherwise would have been wasted. The usage of smart cooking appliances is allowing chefs to be more competent with their work and also making them more efficient.

Flexibility

With the usage of smart kitchen appliances such as smart ovens and smart pans, a chef working in a restaurant said, "I like how the smart stove lets me adjust heat and timing remotely, so I can multitask without worrying about the food being spoiled". The smart devices are also programmed differently so that they can cater to the requirements of various guests' needs. A chef who caters to a wider array of clientele said, "These systems adapt well when we switch between vegetarian, vegan, and meat-based menus—it's very flexible for different customer needs." As the equipment is designed and programmed to meet different needs, the temperature, time, and product specification can be varied.

Reliability

The chefs are accustomed to using different equipment in the kitchen. With the advancement of technology, new smart cooking devices are being installed in five-star hotels and in high-end restaurants. A chef in a specialty restaurant in western India said, "I am a bit worried that reliance on these systems could be risky if they malfunction during a high-pressure output period". A chef who has been working for more than twenty years in the industry was a little skeptical about the usage of the smart cooking equipment and said, "Reliability leads to trust building; if the system is consistent, I'll use it regularly. If not, I'll go back to traditional methods, which I used for so many years". From the statements, it could be concluded that chefs, though, have accepted the usage of smart cooking devices, are still not 100% accepting them, have a fear that they may fail during high output time, and are also concerned about the machine not being consistent, which may take them back to the usage of conventional cooking equipment.

Technicality

Although the smart devices are mostly user-friendly and are provided with demonstrations and manuals, there is still resistance by chefs to adopt smart cooking devices. A chef working in a multi-specialty restaurant stated, "Some of the smart devices are too technical—my junior chefs struggle with setting them up." This states that although smart ovens and pans are user-friendly, acceptance of them by the kitchen staff is crucial. A chef of a five-star hotel who had prior experience working in the kitchen said, "At first, when I was told to operate the smart oven, I felt confused, but once I learned the app interface, it became part of my routine." Initially, although there were some difficulties but most of the chefs were able to adapt to the smart devices, which contradicts the opinions of the chefs.

The Factors Leading to the Usage of AI in the Culinary

AI as a Tool for Efficiency and Precision

An executive chef of a hotel in North India emphasized, "AI helps to make a recipe standardized to ensure that whoever cooks the final product remains the same." Another chef stated that portion control is one of the major issues faced by chefs. "Use of AI in portion control has reduced the wastage and improved consistency in "Useoning of the dishes at various outlets". A chef from Eastern India commented that AI, along with predictive systems, has transformed his job, stating, "The inventory management powered by AI is helping me save hours of work, as it predicts and helps me to understand what ingredients I need for a period even before the order is placed."

AI and Culinary Creativity

The chefs state that AI is also a source of inspiration rather than just a tool. A chef from a restaurant said, "AI helps me in suggesting ingredients I would have never thought of, as it pushes my creativity." Another chef from Mumbai Restaurant stated, "I used AI to adapt traditional recipes for vegan guests so as not to lose the authentic flavors of the dish." He also went on to mention that "Usage of AI makes me feel as if I am collaborating with another chef who has access to thousands of recipes". A chef working in a high-end restaurant with great personalization and catering to a limited guests daily stated, "AI helps in personalizing dishes of regular guests, which are based upon the past choice of the guests". The statements of the chefs do suggest that AI is not only replacing the artistry of chefs but is also expanding the horizon of culinary creativity.

Sustainability

Currently, sustainability is of great concern for chefs, and AI helps to provide tangible benefits. A Chef of a hotel shared, "AI helps me to predict how much food to prepare, considering the bookings, which helps in reducing the food wastage". Another chef from southern India mentioned how AI is helping to regulate energy usage: "The system regulates energy usage by turning off the equipment when not operational". A chef even said that AI also helps in making environmentally responsible choices, like "AI helps in tracking carbon food prints as it analyses the suppliers and suggests greener alternatives". A chef also said that AI has helped them in sustainability, stating, "Sustainability is no more guesswork as it is driven by AI". The experiences shared by the chefs state that technology helps them not only in enhancing efficiency but also makes them use ingredients sustainably.

Skills, Culture, and Ethics

The chefs have stated the implications of AI, which are not positive. Some of the chefs have said that they fear that over-reliance on technology should be avoided. A chef from a restaurant mentioned, "I am a bit worried as young chefs are too dependent on AI and not on their instincts". Another chef said, "Since cooking is both science and art, and if AI takes over it entirely, then we may lose the cultural authenticity of our dishes." Another senior chef from a hotel mentioned that "Some young chefs are afraid that automation may replace their jobs in kitchens of hotels in large chain properties". Other than the technical skills, the chefs were also concerned about the human element and the interface, which cannot be replaced or sidelined. A chef from an international chain property said "The guests come back to hotels for their known human touch, which a machine cannot replicate, and also the emotional bonding the food creates." Regarding the ethics, some chefs did mention that too much data collection and surveillance may be a cause of concern and said, "At most carefulness has to be maintained for data collection from the kitchen and continuous surveillance can make the chefs feel that they are machines and robots". The statements from the chefs do mention that although acceptance of technology and innovation is accepted, there is a concern regarding cultural, ethical, and human dimensions in food, which cannot be ignored.

DISCUSSIONS AND IMPLICATIONS

The findings from the study do conclude that the chefs have a viewpoint that Artificial Intelligence (AI) can be game-changing for the culinary field. The statements mention the relationship between technological efficiency and preservation of human artistry, mentioning hospitality research and practice. The four themes of efficiency and precision, creativity, sustainability, skills, culture, and ethics are part of existing scholarship

The findings of the study on the usage of AI confirm their role in enhancing accuracy and consistency. The features of the predictive system and portion control with AI enhance the operational efficiency of the food establishments. The previous findings by researchers Lu et al. (2021) stated that AI helps to minimize variations and increase the scalability of hospitality operations. The hospitality and restaurant industry faces a labor shortage, and with AI taking over mechanical aspects, it can be a help to reduce pressure on staff. Although the efficiency of the usage of AI should not take over the decision-making role of chefs. AI can recommend a chef for improvisation of the dishes, but cannot replace a chef in decision-making or judgment-making.

Recent research studies have stated that chefs who use AI consider it to be a creative partner with algorithms used for ingredient pairing and the usage of different cooking techniques (Ahn et al., 2021). With the usage of AI, chefs can now be able to adapt and personalize the menu and provide a great dining experience (Batat, 2021). Past studies have also stated the importance of maintaining human control in culinary, although AI helps in co-creativity by enhancing rather than replacing humans (Daugherty & Wilson, 2018). In the future, it is expected that there will be a collaborative platform where chefs with AI will enable innovation that is grounded and does not overlook the artistic aspect of gastronomy.

The chefs have stated that AI helps reduce food wastage, places orders according to bookings, and optimizes energy consumption in the kitchen, which aligns with a past study by researchers. The use of AI in portioning and waste management supports the UN Sustainable Development Goal 12 on responsible consumption and production. As AI provides data-driven insights, it helps to turn sustainability into measurable outcomes. So, a culinary enterprise adopting AI not only gains operational efficiencies but also enhances its environmental credibility, a factor that increases guest preference. Although AI is quite beneficial, its accessibility to small and medium-sized establishments is limited due to a lack of financial resources. The chefs are concerned about the over-reliance on AI, which may lead to cultural dilution, and are concerned about the risks related to technological determinism, as found in previous studies (Ivanov & Webster, 2019). It is also feared that young chefs may adopt AI, which may overlook the traditional training for automated guidance, which may be a challenge to culinary heritage. A chef also stated that cooking is an art form that is connected to its culture, aligning with past studies (Richards, 2018) on gastronomy's role in cultural identity.

Ethical concerns have been raised about the surveillance of data in the workplace (Ball, 2010). Although AI provides valuable data to chefs, excessive monitoring may remove the trust and autonomy of the staff. AI also creates a fear of job displacement that hinders its adoption in operation. The industry should also be transparent about how AI is implemented in its adoption in the hospitality industry.

Implications

The study's findings reveal that artificial intelligence (AI) plays a multifaceted role in the culinary industry, functioning as a technical enabler, a creative partner, and an ethical disruptor. Rather than viewing AI as a potential replacement for chefs, it should be regarded as an advanced tool that enhances their skills and expertise. To prepare future culinary professionals adequately, educational institutions and policymakers must integrate AI technologies and smart cooking applications into their curricula, alongside traditional cooking techniques. This integration will equip aspiring chefs with a comprehensive understanding of both the opportunities and risks that AI presents in hotel kitchens and beyond. Further exploration is essential to gauge guests' perceptions of AI in their dining experiences. Insights from such research will shed light on how AI can heighten customer satisfaction and engagement. Moreover, investigating the role of AI in culinary education can help clarify how it may foster acceptance among both students and seasoned chefs.

Additionally, long-term studies are warranted to examine the potential for AI to transform kitchen culture and to ensure that culinary heritage is preserved amid technological advancements. Many chefs view AI as an invaluable asset in the kitchen, enhancing efficiency, fostering creativity, and promoting sustainability. However, there are valid concerns about the risks of deskilling and the erosion of diverse culinary traditions. For AI to be successfully integrated into the culinary landscape, it must serve as a supportive ally for chefs, empowering their creativity and skills rather than diminishing them. By adopting a balanced approach, the hospitality industry can harness the transformative potential of AI while safeguarding the essential human elements that define the culinary arts. This approach will not only enhance operational effectiveness but also enrich the overall dining experience, preserving the unique flavors and cultural narratives that are integral to gastronomy.

REFERENCES

Ahn, D., Almaatouq, A., Gulabani, M., & Hosanagar, K. (2021). Will we trust what we don't understand? Impact of model interpretability and outcome feedback on trust in AI. SSRN Electronic Journal. DOI: 10.2139/ssrn.3964332

Amabile, T. M. (1996). *Creativity in context.* Westview Press.

Bagler, G., & Goel, M. (2024). Computational gastronomy: Capturing culinary creativity by making food computable. *NPJ Systems Biology and Applications, 10*(1), 72. Advance online publication. DOI: 10.1038/s41540-024-00399-5

Ball, K. (2010). Workplace surveillance: An overview. *Labor History, 51*(1), 87–106. DOI: 10.1080/00236561003654776

Batat, W. (2021). How Michelin-Starred Chefs Are Being Transformed into Social Bricoleurs? An Online Qualitative Study of Luxury Foodservice during the Pandemic Crisis. *Journal of Service Management, 32*(1), 87–99. DOI: 10.1108/JOSM-05-2020-0142

Bessiere, J., & Tibere, L. (2013). Traditional food and tourism: French tourist experience and food heritage. *Journal of the Science of Food and Agriculture, 93*(14), 3420–3425. DOI: 10.1002/jsfa.6284

Bollini, M., Tellex, S., Thompson, T., Roy, N., & Rus, D. (2013). Interpreting and Executing Recipes with a Cooking Robot. In Springer tracts in advanced robotics (pp. 481–495). https://doi.org/DOI: 10.1007/978-3-319-00065-7_33

Braun, V., & Clarke, V. (2006). Using Thematic Analysis in Psychology. *Qualitative Research in Psychology, 3*(2), 77–101. DOI: 10.1191/1478088706qp063oa

Bryman, A. (2016). *Social Research Methods* (5th ed.). Oxford University Press.

Creswell, J. W., & Poth, C. N. (2018). *Qualitative Inquiry and Research Design Choosing among Five Approaches* (4th ed.). SAGE Publications, Inc.

Davis, F. D. (1989). Perceived usefulness, perceived ease of use, and user acceptance of information technology. *Management Information Systems Quarterly, 13*(3), 319. DOI: 10.2307/249008

Filimonau, V. (2020). The prospects of waste management in the hospitality sector post COVID-19. *Resources, Conservation and Recycling, 168*, 105272. DOI: 10.1016/j.resconrec.2020.105272

Flick, U. (2002). *An Introduction to Quality Research* (2nd ed.). Sage Publications.

Freeman, C. P. (2009). Michael Pollan, The Omnivore's Dilemma: A Natural History of four Meals. *Humanimalia, 1*(1), 63–69. DOI: 10.52537/humanimalia.10117

Guest, G., Bunce, A., & Johnson, L. (2005). How many interviews are enough? *Field Methods, 18*(1), 59–82. DOI: 10.1177/1525822X05279903

Ivanov, S., & Webster, C. (2019). Robots in tourism and hospitality: A research agenda. *Annals of Tourism Research, 78*, 102757. DOI: 10.1016/j.annals.2019.102757

Keskekci, D., & Gençer, K. (2023). Digital transformation in food and beverage businesses. *Journal of Travel and Tourism Research, 22*(22), 116–130.

Kvale, S. (1996). *Interview Views: An Introduction to Qualitative Research Interviewing*. Sage Publications.

Lu, Y., Papagiannidis, S., & Alamanos, E. (2021). Artificial intelligence in business: A systematic literature review and research agenda. *Technological Forecasting and Social Change, 165*, 120522. DOI: 10.1016/j.techfore.2020.120522

Marvin, H. J. P., Janssen, E. M., Bouzembrak, Y., Hendriksen, P. J. M., & Staats, M. (2016). Big data in food safety: An overview. *Critical Reviews in Food Science and Nutrition, 57*(11), 2286–2295. DOI: 10.1080/10408398.2016.1257481

Mizrahi, A. B., & Lachnish, A. (2022). Digital gastronomy testcase: A complete pipeline of robotic induced dish variations. *International Journal of Gastronomy and Food Science, 31*, 100625. DOI: 10.1016/j.ijgfs.2022.100625

Narumi, T., Ban, Y., Kajinami, T., Tanikawa, T., & Hirose, M. (2012). Augmented perception of satiety: Controlling food consumption by changing apparent size of food with augmented reality. *Proceeding of the SIGCHI Conference on Human Factors in Computing Systems*. 109-118. DOI: 10.1145/2207676.2207693

Polit, D. F., & Beck, C. T. (2010). Generalization in quantitative and qualitative research: Myths and strategies. *International Journal of Nursing Studies, 47*(11), 1451–1458. DOI: 10.1016/j.ijnurstu.2010.06.004

Pollan, M. (2013). Cooked: A natural history of transformation. *Choice (Chicago, Ill.), 51*(03), 51–1453. DOI: 10.5860/CHOICE.51-1453

Post, M. J. (2012). Cultured meat from stem cells: Challenges and prospects. *Meat Science, 92*(3), 297–301. DOI: 10.1016/j.meatsci.2012.04.008

Richards, G. (2018). Cultural tourism: A review of recent research and trends. *Journal of Hospitality and Tourism Management, 36*, 12–21. DOI: 10.1016/j.jhtm.2018.03.005

Rubin, H. J., & Rubin, I. S. (2005). *Qualitative Interviewing: The Art of Hearing Data* (2nd ed.). Sage. DOI: 10.4135/9781452226651

Tussyadiah, I., & Park, S. (2018). Consumer evaluation of AI service robots in hospitality. *International Journal of Hospitality Management, 72*, 93–106. DOI: 10.1016/j.ijhm.2018.06.010

Wilson, H. J., & Daugherty, P. R. (2018). *Human + Machine: Reimagining Work in the Age of AI*. Harvard Business Review Press.

Zoran, A. (2019). Cooking with Computers: The Vision of Digital Gastronomy [Point of View]. *Proceedings of the IEEE, 107*(8), 1467–1473. DOI: 10.1109/JPROC.2019.2925262

Chapter 2

The Digital Chef: Future-Proofing Hotel Kitchens With AI

Vikesh Kashyap
https://orcid.org/0000-0001-8532-2022
Bahra University, India

Vaibhav Verma
https://orcid.org/0000-0002-9041-7111
Bahra University, India

Monika Baltoo Verma
https://orcid.org/0009-0007-3317-7995
APG University, India

Ankit Shukla
https://orcid.org/0000-0002-7646-3517
Rajamangala University of Technology, Thanyburi, Thailand

ABSTRACT

Using the "Digital Chef" model, which integrates intelligent technologies for enhanced efficiency, sustainability, and guest personalization, this study explores the transformative role of artificial intelligence (AI) in hotel kitchens. Food preparation, inventory control, and menu customization are being revolutionized by AI-driven solutions such as robotic cooking assistants, sensor-based safety systems, and predictive analytics. By automating tedious tasks and enabling data-driven decision-making, AI fosters chefs' creativity rather than replacing them. AI's effects on operations, customer interaction, and profitability are demonstrated through case studies of global hospitality chains. The balance between human knowledge and intelligent

DOI: 10.4018/979-8-3373-4042-5.ch002

systems offers a scalable and sustainable path forward, despite obstacles like cost, cultural, and ethical concerns. By ensuring that innovation and tradition coexist peacefully in kitchens of the future, this convergence redefines culinary greatness.

INTRODUCTION

The modern hospitality landscape is undergoing a paradigm shift, driven by rapid advancements in digital technology and artificial intelligence (AI). One of the most significant transformations is occurring in hotel kitchens, where AI is being leveraged to optimize culinary operations, streamline workflows, and enhance guest satisfaction. This digital evolution, often conceptualized as "The Digital Chef," represents the integration of intelligent systems that support food preparation, inventory control, hygiene compliance, and personalized service delivery (M. G. A. Bautista, 2023).

AI adoption in hospitality kitchens is becoming a trend and strategic response to contemporary challenges such as food safety and hygiene demands, waste management, energy efficiency, sustainability and the growing expectation for customized dining experiences. With AI-driven tools like smart ovens, robotic cooking assistants, smart appliance, Voice Assistance, Robotic Sous chefs and predictive analytics for menu planning, hotel kitchens are becoming smarter, more sustainable, and resilient to operational disruptions. AI is increasingly becoming instrumental in enhancing menu designs and improving profitability by utilizing data-driven insights," states Jean-Georges Vongerichten, the chef and owner of more than 50 restaurants worldwide. He defines the way AI tools been used in the hotel kitchens to enhance both topline revenue and total profitability with entering simple data and getting the details for future perditions.

The case studies presented in this chapter illustrate the variety of possibilities for automation: from automating separate tasks to completely automating the back of the house (Spyce robotic kitchen, Miso Robotics, PreciTaste, Rational, Unox, Suvie and Chefling are best examples). The companies like Miso Robotics have developed automated cooking systems capable of preparing high-volume meals with minimal supervision, while AI-powered platforms like PreciTaste enable real-time kitchen monitoring and decision-making through data analysis. Although, the restaurant industry has adopted chatbots; voice and biometric technology; robots used in hosting, food running, cooking, and bartending; table ordering; conveyer belt; and robotic food delivery (Stephens, 2025). Furthermore, sensor-based systems are also adopted by the now strict adherence to hygiene and food safety protocols by monitoring temperature, humidity, and contamination risk (WHO). Simultaneously, AI's ability to analyze guest preferences, dietary restrictions, and consumption patterns empowers hotels to create personalized and health-conscious menus, thus enhancing the overall guest experience.

Artificial intelligence (AI) is increasingly making its mark in the culinary and hospitality industries, boosting efficiency and promoting sustainability. It's changing the game in inventory management by reducing waste, optimizing energy consumption, and providing predictive analytics for smarter resource allocation and demand forecasting. In hotel kitchens, AI is enhancing productivity, ensuring food safety, and improving guest experiences, all while aligning with the UN's Sustainable Development Goals (SDGs), particularly Goals 2, 3, 9, and 12. The future of sustainable gastronomy doesn't mean compromising on tradition or flavor; instead, it's about equipping staff and chefs with scalable, user-friendly tools that foster ethical practices. These technologies are not designed to replace chefs but to augment human capabilities by automating repetitive tasks, improving consistency, and allowing culinary professionals to focus on creativity and innovation. The human hands and intuition crafted meals, a new paradigm is emerging: human-chef collaborations with intelligent machines. These partnerships aren't about replacing chefs but enhancing their capabilities, improving consistency, and unlocking new dimensions of creativity. This chapter explores how hotel kitchens around the world are integrating AI systems, from menu design to food preparation, to balance efficiency with culinary tradition (Jobstreibizer et al., 2025).The world of haute cuisine has long been a realm of tradition, artistry, and meticulous craftsmanship. Yet, as technology advances, hotel kitchens-where precision, speed, and guest satisfaction are paramount—are embracing artificial intelligence (AI) to enhance, rather than replace, the human touch. AI is transforming high-end hotel kitchens through real-world case studies, demonstrating how the synergy between chefs and machines is redefining culinary excellence.

THE RISE OF AI IN HOSPITALITY

The evolution of technology in hotel kitchens reflects broader advancements in hospitality management and culinary arts. Historically, hotel kitchens were highly labour-intensive, relying on manual techniques and basic tools. Over time, technological innovations have transformed these kitchens into sophisticated operations that enhance efficiency, food safety, and guest satisfaction. Hotel kitchens have always been a vital part of hospitality operations, evolving in parallel with advances in technology to meet growing demands for efficiency, quality, and safety. From the earliest days of manual cooking methods to today's AI-powered systems, technological innovations have continuously transformed culinary workflows and guest experiences (Tuomi, 2020) (Dimitrios & Rosanna, 2018).

In the late 19th and early 20th centuries, the introduction of gas and electric stoves revolutionized cooking by providing more precise heat control compared to traditional open fires. This period also saw the emergence of refrigeration technology, which allowed for longer storage of perishable ingredients and improved food quality. Refrigeration was a game-changer for large-scale food service operations, enabling hotels to expand menu offerings and ensure consistent freshness (Sojecka, 2024).

The mid-20th century marked the rise of automation with the adoption of appliances such as dishwashers, mixers, and conveyor ovens, which reduced manual labor and increased productivity. Computerized systems began to appear in the late 20th century, with Point of Sale (POS) and inventory management software enhancing order accuracy, stock control, and financial oversight (Ghosal, 2025).

Entering the 21st century, the proliferation of digital and connected technologies led to the development of smart kitchens. Recent work highlights the pedagogical significance of AI in language learning (Shukla et al., 2025). IoT-enabled appliances and sensors monitor cooking processes and kitchen conditions in real time, optimizing energy use and food safety compliance. More recently, Artificial Intelligence (AI) and robotics have begun to transform kitchens by automating routine culinary tasks and enabling data-driven menu customization (Roy, 2023) by can supporting chefs by handling repetitive or physically demanding tasks, thereby improving consistency and operational efficiency (Bowen & Morosan, 2018). The Hospitality Industry is experiencing a seismic digital shift that is revolutionizing the way businesses operate, engage with customers, and deliver services. With changing customer needs, hospitality companies are moving away from legacy systems, embracing cloud solutions to drive operating efficiency, enable their people, and strengthen B2B relationships. This technology innovation is transforming key business operations and transforming the guest experience (Bhuiyan, 2024).

While recognizing the imperative for digital innovation, many industry executives admit that their organizations are lagging behind in adopting advanced technologies. This generates ongoing challenges, such as having to do software updates regularly, upgrading infrastructure, and training staff. True digital transformation for hospitality is more than just accepting new technologies—it involves a strategic platform that connects front-end customer experiences with optimized back-end operations (Chakraborty, 2025). *Tourism sustainability depends on effective waste disposal systems* (Shukla et al., 2025.Most importantly, the objective is not to eliminate the human aspect of hospitality, but rather to enhance it. When properly implemented, digital technologies can customize services, streamline internal processes, and allow employees to focus more on genuine guest interactions. Sustaining this balance between innovation and personal interaction is critical to the industry's continued growth and relevance in a rapidly evolving digital world (Law, 2024).

CURRENT TRENDS IN SMART KITCHEN AUTOMATION AND AI APPLICATIONS

Artificial intelligence (AI) and smart automation are reshaping the operations of hotel kitchens. The rise of artificial intelligence in hospitality is particularly evident in the kitchen, where AI-driven culinary tools are streamlining operations, improving safety, and enhancing consistency. These innovations are not just gadgets they're becoming essential components of modern hotel kitchens (Güngör & Yücel Güngör, 2024).

Smart Appliances and IoT Integration

These smart devices can adapt, tracking energy usage to enhance power efficiency (Smart fryers can notify chefs when it's time to replace oil, guaranteeing ideal frying conditions; IoT-enabled ovens can preheat to the desired temperature at a set time), avoid spoilage and minimize waste (Smart refrigerators and freezers can modify their internal temperature according to their fullness, IoT-enabled thermometers can check food temperatures in real-time and notify you if something drops outside the safe range), gather data to redesign menus (collect numerous data points, from usage habits to performance statistics*). The integration of AI technologies requires careful adaptation (Yadav et al., 2024).* Moreover, these intelligent devices ensure food safety and quality consistently. In addition, these smart devices stagflations food safety and quality at all times. With this innovative technology, hotels can provide an even higher to while navigational costs (Maddala, 2025) (Nanda, 2025).

AI-Driven Inventory and Waste Management

AI algorithms track ingredient usage in real-time, predict demand, and automate reordering. Systems like Winnow and Kitro use computer vision to monitor food waste and suggest operational changes to reduce costs and environmental impact. AI-driven analytics are assisting restaurants in addressing food waste. "By examining sales data, AI can enhance our production schedules to minimize waste and boost efficiency," states Vongerichten. AI technologies can create more intelligent production plans based on historical sales trends, allowing restaurants to prepare precisely the right amount of food to satisfy demand while reducing surplus. Furthermore, these tools are becoming more advanced. IHG Luxury properties are assessing an

AI system that employs smart cameras at waste bins to identify discarded food. As Gingerich notes, these technologies will offer insights aimed at "reducing waste, thus enhancing profitability, guest satisfaction, and sustainability" (BPM, 2025) (Stephens, 2025).

Automated Cooking Assistants and Voice-Activated Assistants

The Automated Cooking Assistants are becoming a new trend in Smart Hotel Kitchen for preparing meals with minimal human involvement. The companies are developing the automated cooking assistants to decrease reliance on labor with focusing on enhancing hygiene and efficiency (Berezina K. C., 2018). AI assistants embedded in kitchen systems enable chefs to manage appliances, verify recipes, and record data through voice commands, thus promoting hygiene, sanitation and allowing for more effective multitasking. Even, these Machine learning technologies are evaluating guest preferences, seasonal trends, and dietary habits to refine menus, suggest dishes, and customize offerings to individual tastes—thereby improving the guest experience (Kumagai, 2017).

Robotic Sous-Chefs and Automated Stations

AI-powered ovens are revolutionizing the way we cook by providing unparalleled precision, efficiency, and convenience. With sensors, machine learning, and real-time data analysis, these devices can automatically adjust time, temperature humidity to ensure perfectly cooked food every time. They can also recognize different types of food and suggest the most suitable cooking modes, reducing the risk of overcooking or undercooking. Additionally, AI-driven kitchen assistants can recommend recipes based on available ingredients and dietary needs, offer step-by-step instructions, integrate with inventory systems to propose menus based on stock availability, and even provide voice-activated controls for hands-free operation (Kim, 2025). With these innovations, the culinary world is experiencing a transformation, making it easier than ever before to produce delicious and consistent meals (Robosouschef, n.d.) (Learn, 2024).

Sensor-Based Systems for Temperature, Hygiene, and Food Safety

To prioritize safety and cleanliness AI-enabled sensors have been introduced to constantly monitor temperature, hygiene and traceability. The continuous tracking of cold hot holding and cooking temperatures ensures compliance with health regulations. (Jiang W, 2025). For added assurance of cleanliness, sensors detect surface contamination or ensure hand washing compliance, smart the air quality in hotel kitchens to verify their healthful AI trackless from source to plate, flagging any expiration risks along the way. Relying on these sensor-based systems, hotels can rest assured that their culinary operations meet the highest standards of safety and hygiene (Revelou, 2025).

PERSONALIZED GUEST EXPERIENCES

In today's fast-paced world, where customer expectations are always changing, the hospitality and service industries are on a constant quest to find fresh and exciting ways to elevate guest experiences. At the heart of this transformation is the integration of Artificial Intelligence (AI) into personalization strategies. This blend of technology and personalization isn't just a passing trend; it's a game-changing movement that's redefining how businesses connect with their guests.

Creating tailored culinary journeys with AI is revolutionizing the hospitality industry by enabling hotels to provide personalized dining experiences for their guests. By analyzing gustatory needs, AI algorithms can create detailed profiles of each guest, allowing hotels to greet returning guests with dishes or suggest meals aligned with specific dietary preferences, allergies, religious dietary restrictions, and distinct taste profiles (Berezina K. C., 2019). The implementation of artificial intelligence-powered chat bots has revolutionized the dining experience, enabling personalized culinary encounters that cater to individual tastes and preferences from the outset. As these systems evolve and adapt over time, their recommendations are refined with each successive interaction, ensuring a consistently exceptional dining experience. By harnessing the capabilities of AI-powered chatbots, establishments can provide guests with a truly bespoke and memorable culinary experience that meticulously addresses their every need and preference (Farah Zahidi, 2024) (Karagiannis, 2024).

Envision entering a cutting-edge restaurant where the tables transcend their conventional role as mere surfaces, instead serving as dynamic, interactive hubs that significantly enhance the gastronomic experience. With the integration of smart tables equipped with high-resolution touch screens or advanced sensors, patrons can embark on a bespoke culinary journey, replete with personalized touches. They can peruse interactive, digital menus that not only showcase an array of delectable dishes but also provide invaluable nutritional data, including detailed information on macronutrients, micronutrients, and potential allergens, thereby ensuring that every selection is an informed one (Tech, 2024). As they deliberate, they can instantly place their orders, which are then seamlessly transmitted to the kitchen via a wireless network, streamlining the service and minimizing wait times. Moreover, these intelligent tables can synchronize with personal devices, such as smart watches or fitness trackers, analyzing vital health metrics, including calories expended, blood sugar levels, and other pertinent data, and suggesting dishes that are tailored to individual dietary needs and preferences. Additionally, they offer multilingual support, featuring real-time language translation and visual previews of each dish, thereby making them an inclusive and welcoming feature for guests from diverse linguistic and cultural backgrounds, effectively breaking down language barriers and elevating the overall dining experience to new heights (Alexander Seifert, 2021) (Shei RJ, 2022).

With customized culinary experience that is carefully tailored to your individual gastronomic preferences and wellness goals. The AI has developed the advanced virtual tasting menus (Chen, 2023), which continuously adapts preferences and feedback in real-time, undergoing a process of iterative refinement to ensure an exceptional dining experience e.g. Mood-based menu, low-carb Mediterranean" or "North Indian spice trail, Health scoring, Virtual chef recommendations (Hazim Ryad Momani, 2025) (Chairote Yaiprasert, 2024). The creation of these personalized culinary itineraries involves the aggregation of a diverse range of data points, including your past dining behavior, cultural background, and current emotional state, as determined through advanced sentiment analysis techniques. This innovative approach enables you to explore a wide range of dishes virtually, gaining insight into the nuances of ingredients, origin stories, and chef's notes prior to making your selection, thereby enhancing engagement and personalization. By leveraging the power of artificial intelligence-driven insights, you can indulge in a truly immersive and tailored culinary experience that caters to your unique needs and preferences, redefining the boundaries of gastronomic excellence (McKinsey, 2024).

Additionally, AI plays a central role in designing smart menus that adapt to guest behavior, trends, and inventory, such as predicting high-performing dishes based on time of year, and event calendars, optimizing menus for freshness and sustainability, and tracking sales performance to update menus in real-time. With the help of AI, hotels can elevate their culinary offerings beyond generic service and deliver unforgettable experiences for their guests.

SUSTAINABILITY AND EFFICIENCY

As the hospitality and culinary industries expand, so too do the demands on environmental resources. From food production and storage to preparation and disposal, the carbon footprint of modern kitchens is significant. In response, a new generation of kitchens is turning to artificial intelligence (AI) to lead the charge toward sustainability and operational efficiency. The AI is transforming hotel kitchens by boosting efficiency, guaranteeing food safety, enriching customer experiences, and making substantial contributions to sustainability in alignment with the United Nations Sustainable Development Goals (SDGs), particularly Goal 2 (Zero Hunger), Goal 3 (Good Health and Well-being), Goal 9 (Industry, Innovation and Infrastructure), and Goal 12 (Responsible Consumption and Production) (Ekta Kumawat, 2025). AI-driven technologies are reshaping inventory tracking, reducing waste, optimizing energy use, and enabling predictive analytics for smarter demand forecasting and resource allocation. In traditional kitchens, inventory tracking often relies on manual checks or basic digital logs, which are prone to human error and inefficiency. Predictive analytics stands out as one of the most revolutionary applications of artificial intelligence in the realm of culinary sustainability. By analyzing historical service patterns, seasonal trends, customer profiles, and external influences such as weather conditions and local events, AI systems can effectively forecast diner traffic and meal demand (Dhal & Kar, 2025). This capability allows for the optimization of staffing levels and shift schedules, as well as the recommendation of appropriate batch preparation sizes and procurement timelines. Such insights significantly minimize the risks of under-preparation and overproduction, making operations more efficient. (Bennett, 2025), as AI technology becomes increasingly accessible, even smaller kitchens and home stays can take advantage of straightforward plug-and-play solutions that resonate with their operational scale and values. For instance, cloud kitchens are utilizing AI dashboards to oversee multiple brands from a single location, maximizing the use of shared ingredients and space. Similarly, eco-conscious hotels are incorporating AI waste tracking systems that align guest menu selections with portion control, enhancing both sustainability and customer satisfaction. In resource-limited areas, local restaurants are employing AI to fine-tune storage conditions for seasonal produce, effectively reducing spoilage. Thus, AI emerges not merely as a catalyst for innovation but as a vital partner in environmental stewardship. By harnessing intelligent systems to streamline inventory management, minimize waste, and optimize energy use, kitchens can adopt a model that is both economically viable and environmentally responsible (Dhal & Kar, 2024). The growth of casual dining reflects changing urban lifestyles (Shukla & Sharma, 2020). The future of sustainable gastronomy does not hinge on compromising tradition or flavor; rather, it lies in equipping chefs and staff with intuitive, efficient, and scalable tools that promote responsible practices. In this way, the culinary industry can contribute to nourishing not only individuals but also the planet itself (Helen Onyeaka, 2025).

In contemporary culinary environments, the implementation of effective inventory management systems is paramount. The integration of advanced technologies, such as artificial intelligence (AI), plays a pivotal role in optimizing kitchen operations. By leveraging real-time data analytics and predictive modeling, (McKinsey, 2024) AI-enabled systems can accurately track ingredient utilization and forecast depletion rates, thereby facilitating the automated generation of optimized procurement orders. This proactive approach mitigates the risks of overstocking and spoilage, ensuring a more streamlined and efficient supply chain management process.

Furthermore, the strategic deployment of barcode and radio-frequency identification (RFID) technologies, in conjunction with sensor inputs, enables precise monitoring of inventory levels. This, in turn, allows for the implementation of data-driven decision-making processes, ultimately enhancing the overall efficacy of kitchen operations. In addition to optimizing inventory management, AI also contributes significantly to waste reduction initiatives (Kubáňová et al., 2022). By analyzing plate waste, portion sizes, and leftover trends, AI systems can provide actionable insights, enabling kitchens to implement precise portioning strategies and modify menus to minimize food waste. The integration of AI with kitchen scales and cameras facilitates a comprehensive analysis of pre- and post-consumption waste, allowing for the identification of frequently discarded items and the implementation of targeted waste reduction measures (Onyeaka, 2022) (Kanwal Gul, 2025).

Moreover, AI-powered systems can optimize energy consumption patterns in kitchen environments, adjusting energy usage in accordance with operational demands. During periods of reduced activity, AI systems can reduce power consumption, while also ensuring that smart appliances, such as ovens, refrigerators, and dishwashers, are only activated when necessary. The provision of valuable insights via AI dashboards enables kitchens to monitor and mitigate their carbon footprint, as well as energy consumption patterns, ultimately contributing to the creation of a more sustainable and efficient food service operation (Armin Tayefeh, 2024). (Siham Rekiek, 2024), defines guest demand forecasting leverages artificial intelligence to analyze historical data, booking trends, weather conditions, holidays, and local events, enabling accurate predictions of meal demand. This capability assists chefs in determining the appropriate quantity and variety of dishes to prepare, thereby minimizing the risk of overproduction. Additionally, dynamic menus can be tailored on a daily basis to align with anticipated customer traffic and evolving dietary preferences, ensuring that offerings remain relevant and appealing.

Resource management is enhanced through the optimization of staff scheduling based on predicted kitchen workloads, which helps to decrease idle time and mitigate employee burnout. Furthermore, the system forecasts the necessary ingredients and their quantities, promoting a zero-waste procurement strategy. In the event of potential supply chain disruptions or shortages, it also provides suggestions for alternative ingredients, ensuring that operations can continue smoothly without compromising quality (Kanwal Gul, 2025).

The integration of sustainability practices is a key focus, as the system promotes the use of seasonal, locally sourced, and low-carbon-footprint ingredients, aligning with eco-friendly kitchen initiatives. It also generates comprehensive sustainability reports for hotel management, aiding in compliance with green certification standards and reinforcing the establishment's commitment to environmental responsibility.

HUMAN + MACHINE COLLABORATION IN AI-ENHANCED HOTEL KITCHENS

The convergence of artificial intelligence and culinary arts necessitates a nuanced understanding of the complementary role that AI plays in augmenting human creativity, rather than supplanting it. By providing support with tasks such as ingredient pairing, dietary customization, and presentation recommendations, AI enables culinary professionals to devote their attention to innovative and artistic endeavors. While AI can facilitate the documentation of traditional recipes and enhance precision in tasks like timing and measurement, the inherent value of human intuition, gastronomic expertise, and artistic flair remains indispensable, particularly in the realms of fine dining and culturally nuanced cuisines where heritage and tradition are deeply ingrained (Group, 2025).

Furthermore, the automation of repetitive tasks, including food preparation and temperature monitoring, allows chefs to focus on high-value activities such as creative presentation, menu development, and interpersonal engagement with guests. Ultimately, the dining experience extends beyond the culinary realm, encompassing emotional resonance and interpersonal connection. The presence of human staff brings a unique dimension of warmth, narrative, and cultural context to the dining experience, rendering each meal a distinctive and memorable occasion. Seal et al. (2025) argue that food blogs shape tourism and hospitality narratives. By synergistically combining the operational efficiencies afforded by AI with the emotional intelligence, ethical acumen, and creative expertise of human culinary professionals, it is possible to craft a truly exceptional gastronomic experience that not only satisfies the palate but also nourishes the human spirit (ETHospitalityWorld, 2025).

CASE STUDIES OF AI ENHANCED HOTEL KITCHENS

McDonald's has taken significant strides in enhancing its drive-thru experience by leveraging artificial intelligence in its operations and marketing strategies. The pivotal moment came in 2019 when the fast-food giant acquired Dynamic Yield, a company renowned for its expertise in personalization and decision logic technology. This strategic acquisition enabled McDonald's to utilize AI to tailor food recommendations based on various factors, including the time of day, weather conditions, current restaurant traffic, and popular menu items. As a result, the company has seen a notable increase in sales through personalized upselling, while also improving customer satisfaction by providing quicker and more relevant service.

In a similar vein, Domino's Pizza has effectively integrated AI into its marketing and customer interaction processes. A standout feature of this initiative is the AI-driven voice assistant named "Dom," which streamlines the phone ordering process for customers. Furthermore, Domino's employs AI to optimize its marketing campaigns by analyzing customer data to create personalized promotions and offers. This approach has led to an enhanced customer experience characterized by efficient order processing and has significantly boosted conversion rates through targeted marketing efforts.

Starbucks has also embraced AI to enrich the customer experience, particularly through its loyalty program and mobile application. The Starbucks app gathers data on individual customer preferences and purchasing behaviors, which AI algorithms then analyze to provide personalized recommendations and offers. This level of personalization extends to the company's email marketing campaigns, where customers receive tailored promotions that resonate with their interests. Consequently, Starbucks has experienced increased customer engagement and loyalty, alongside a rise in average spending per visit, driven by these customized suggestions.

Chili's Grill & Bar India has embraced artificial intelligence to enhance its menu offerings and pricing strategies by closely analyzing customer preferences. Through the examination of sales data and customer feedback, AI enables Chili's to identify which dishes resonate most with patrons, allowing the restaurant to strategically promote popular items while considering the removal of less favored options. Furthermore, AI plays a crucial role in establishing optimal pricing by taking into account factors such as demand fluctuations and competitive pricing, ultimately leading to a more tailored dining experience for customers. The implementation of these AI-driven strategies has resulted in increased sales due to improved menu selections and enhanced profitability through refined pricing approaches.

The Cheesecake Factory has turned to predictive analytics powered by artificial intelligence to streamline its inventory management processes. By scrutinizing historical sales data and forecasting future demand, the restaurant chain can significantly minimize food waste while ensuring that high-demand items remain readily available for customers. This proactive approach not only enhances operational efficiency but also contributes to a more sustainable business model by reducing excess inventory and associated costs. As a result, The Cheesecake Factory has experienced a notable decrease in food waste and an overall improvement in the effectiveness of its inventory management practices.

CHALLENGES, ETHICAL CONSIDERATIONS, AND THE PATH FORWARD

In the ever-evolving realm of gastronomy, the convergence of innovation, cultural heritage, and cutting-edge automation creates a landscape that is both exhilarating and complex. As culinary establishment's worldwide spanning bustling urban centers to remote cultural havens embrace new technologies, they encounter not only challenges related to efficiency and precision but also profound questions about identity, tradition, and the indispensable human touch in the culinary arts. (Kashyap et al., 2023). Sustainable tourism policies are vital in mountainous regions (Shukla et al., 2025). At the heart of culinary heritage lies a profound appreciation for age-old techniques that have been lovingly handed down through generations. In regions celebrated for their rich food cultures, such as the enchanting valleys of Himachal Pradesh, the charming kitchens of Tuscany, or the meticulous sushi counters of Japan, food becomes more than sustenance; it transforms into a narrative of storytelling, cultural identity, and ritualistic practices (Y, 2025).

The advent of automation manifested through robotic arms, AI-driven recipe generators, and smart ovens provokes a critical inquiry: Can machines genuinely honor and preserve these cherished traditions? This question transcends mere nostalgia; traditional cooking often hinges on instinctive adjustments, cultural significance, and a keen awareness of seasonal and local ingredients. While automation excels in providing consistency and speed, it may struggle to replicate the intricate artistry that human chefs infuse into their culinary creations. Furthermore, one of the most significant hurdles in adopting automation within culinary settings is the financial strain it imposes (Irtiqa Shabir, 2023). The integration of robotic kitchen systems, AI-enhanced appliances, and advanced storage solutions demands substantial capital investment, which can be particularly daunting for small businesses or family-run kitchens that prioritize tradition. Beyond the initial costs, ongoing expenses such as maintenance, energy consumption, and the necessity for customization to accommodate regional cuisines further complicate the landscape, making the transition to automated culinary practices a multifaceted challenge (Nichifor, 2025).

The financial constraints faced by many local food artisans often create a landscape where larger chain restaurants and upscale hospitality venues can more readily implement automation technologies. This trend risks widening the gap between those dedicated to preserving culinary traditions and those driven by commercial interests. Even among establishments that recognize the potential benefits of automation, the challenge of human adaptation remains significant. Chefs and kitchen staff may experience reluctance or resistance to adopting new tools, perceiving them as threats to their roles and the authenticity of their culinary artistry (David Graham, 2020). The apprehension surrounding automation in the culinary sector is rooted in several factors, including fears of job displacement, unfamiliarity with digital tools, and a lack of emotional engagement with automated processes. To address these issues effectively, it is crucial to cultivate a culture of collaboration, framing automation as a complement to human skills rather than a replacement. This strategy requires the development of inclusive training programs that provide sufficient time for staff to adjust, as well as the active involvement of culinary professionals in the design of technological solutions, ensuring that their expertise and experiences inform the tools they will ultimately utilize. Furthermore, the ethical dimensions of incorporating machines into culinary practices are complex and merit thorough examination (Seyitoğlu F., 2025). A primary concern is the potential for cultural dilution, as the automation of recipes could lead to the homogenization of diverse regional cuisines into a single, standardized format. This raises critical questions about whether machine-learning algorithms, which are designed to optimize recipes for taste and marketability, might inadvertently disregard the intricate nuances and cultural significance that characterize traditional dishes. Additionally, the issue of intellectual property becomes pertinent when AI systems draw from a chef's distinctive style or regional culinary heritage, highlighting the need for careful consideration of ownership and recognition in an increasingly automated culinary landscape (Islam, 2024).

The successful replication of a beloved family recipe by a robotic system raises important questions about ownership and attribution, should the credit go to the algorithm that executed the task or to the cultural heritage that inspired it? (Kim, 2025)This dilemma is compounded by the palpable tension between the potential for workforce displacement and the empowerment of culinary professionals. While concerns about robots replacing human chefs are legitimate, the ethical implementation of automation should prioritize the enhancement of human roles rather than their eradication. By delegating repetitive and labor-intensive tasks to machines, chefs can focus their talents on more creative and complex aspects of their work, such as plating, storytelling, engaging with customers, and conducting research and development, ultimately enriching the overall culinary experience (Jozef Youssef, 2023). The case study titled Snapshots: Tradition Meets Tech illustrates the intriguing convergence of culinary tradition and contemporary technology across various regions. In bustling cities like Bengaluru and Mumbai, cloud kitchens are utilizing artificial intelligence to predict demand and prepare meals accordingly, while still employing local chefs to ensure that iconic dishes like biryani and dal makhani maintain their authentic flavors. Likewise, some conveyor-belt sushi restaurants have embraced robotic chefs capable of roll making with impressive accuracy, yet the final presentation is often left to skilled human chefs, creating a hybrid model that honors both innovation and craftsmanship. In the mountainous areas of Himachal, select boutique homestays have incorporated smart stovetops and timers to address the unique challenges of high-altitude cooking, all while preserving ancestral recipes that embody the essence of 'pahadi' cuisine (Kashyap, 2024). As we move forward, it is crucial to navigate the changing landscape of culinary practices with a commitment to principles that promote a harmonious relationship between automation and tradition.

It is essential to prioritize co-creation rather than mere replacement when integrating new culinary technologies, which necessitates the active participation of chefs and cultural guardians in both the design and testing phases of these tools (Seyitoğlu F. F.-F., 2025). This collaborative approach ensures that technological advancements serve to enhance the culinary arts rather than undermine them. Moreover, cultural sensitivity should be a fundamental principle in the development of algorithms, as they must reflect the rich tapestry of regional cuisines and their unique contexts. Transparency is equally important in the implementation of these technologies; fostering open communication with both kitchen staff and patrons regarding the role of machines in food preparation can cultivate trust and a deeper understanding of the culinary process (Vânia Gisele Bessi, 2022). Hospitality providers face ethical dilemmas in birthright tourism (Shukla et al., 2025).

Additionally, equitable access to these innovations must be a priority, with support for smaller kitchens through initiatives such as subsidies, open-source resources, or communal automation facilities. As the boundaries between culinary heritage and technological progress continue to merge, the kitchens of tomorrow are not just theoretical constructs but are manifesting as real changes in restaurants, hotels, and home stays. Innovations like voice-activated cooking stations, AI-assisted wine pairings, and sustainable cooking methods are rapidly transforming the landscape of gastronomy into a domain that is not only digital and intelligent but also deeply interconnected (Mohammad Nurul Alam, 2025). The Voice-command cooking systems are revolutionizing the culinary landscape by introducing hands-free intelligence that enhances kitchen efficiency and hygiene. Picture a chef immersed in their craft, effortlessly instructing their appliances with simple voice commands, such as setting the oven temperature or timing the stirring of a sauce (Buhalis, 2023) (Adrian Iftene, 2020). This innovative technology, driven by advanced natural language processing and smart kitchen integration, allows for seamless operation of cooking devices, real-time monitoring of culinary progress through smart sensors, and dynamic recipe guidance tailored to the chef's needs. By eliminating the necessity to touch knobs or screens, these systems not only streamline multitasking but also maintain a cleaner cooking environment. In high-demand settings like hotel kitchens or culinary schools, these voice-command systems serve as invaluable assistants, always attentive and prepared to support the chef's endeavors (Hazem Golshany, 2025).

The emergence of AI sommeliers is transforming the way diners experience wine and beverage pairing, making this intricate art accessible to a broader audience.

These sophisticated systems leverage extensive databases to analyze a multitude of flavor profiles, regional characteristics, and individual preferences, including dietary restrictions, to recommend ideal pairings. By continuously learning from guest feedback, AI sommeliers refine their suggestions over time, adapting to local availability and even considering the context of the dining experience, such as the mood or accompanying music (Lu Lu, 2024). This technology democratizes fine dining, enabling establishments in mid-tier or remote locations to offer personalized beverage recommendations that rival those of a trained sommelier. The zero-waste systems are becoming essential in modern kitchens, driven by the urgent need for sustainability and linking it to certain sustainable development goals. The future of culinary spaces is increasingly reliant on advanced technologies such as data analytics, artificial intelligence, and smart sensors, all aimed at reducing waste throughout the cooking process. For instance, inventory AI can accurately forecast demand, helping to prevent overstocking and subsequent food waste (Ilhaam & Omar, 2024). Smart composting units are designed to monitor organic waste and offer suggestions for repurposing ingredients, such as creating vegetable stock from peelings. Additionally, portion AI analyzes plate returns to recommend ideal serving sizes, further minimizing excess. The integration of block chain technology enhances transparency by tracking ingredients from their origin to the plate, ensuring ethical sourcing and real-time traceability. In the realms of culinary tourism and hospitality, adopting zero-waste practices is becoming a significant marketing advantage, as consumers increasingly seek not only exceptional dining experiences but also a commitment to responsible food systems (Claudia Cozzio, 2023).

As these technological advancements reshape the culinary landscape, it is imperative to prepare culinary staff for this new era. This preparation involves a comprehensive overhaul of training curricula at culinary schools and hospitality institutes, which must now include essential skills such as digital literacy to navigate and troubleshoot various interfaces, as well as data-driven decision-making to leverage insights effectively. Furthermore, a foundational understanding of basic programming or automation logic will empower future chefs to customize recipes intelligently. Equally important is the emphasis on ethical technology use in food production, ensuring that culinary professionals are not only skilled but also responsible in their practices. In addition to technical knowledge, soft skills will play a crucial role in the collaboration between humans and AI; chefs will need to exercise judgment when AI systems encounter limitations, creatively interpret algorithmic recommendations, and harmonize their intuition with intelligent systems to elevate the culinary arts rather than standardize them. To facilitate this transition, hands-on exposure through simulated smart kitchen labs, augmented reality-assisted cooking modules, and real-time feedback tools should be integral to professional training, while internships in AI-integrated kitchens can provide invaluable experience and demystify the technology for aspiring chefs (Kym Man, 2024) (Jinjin Huang, 2025).

The evolution of smart kitchens is not confined to urban centers or large-scale franchises; rather, it holds significant potential for enhancing culinary practices in rural and traditional settings. By implementing these technological advancements thoughtfully, local food establishments can thrive. For instance, a roadside dhaba could benefit from smart preservation technologies that prolong the freshness of locally sourced ingredients, thereby reducing waste and improving food quality. Considering, a Himachali home-cook could leverage AI-powered dietary applications to create customized meal experiences for visiting tourists, showcasing regional flavors while catering to individual dietary needs. Furthermore, the preservation of local culinary traditions, which may be at risk of fading away, can be supported through the creation of interactive recipe databases that digitally archive these practices and make them accessible to a broader audience. The success of these initiatives hinges on designing solutions that are user-friendly, support local languages, and can be adapted to fit diverse financial and geographical contexts (Dianyuan Wang, 2024; Zhe Wang, 2025).

As we look toward the future of culinary arts, (Öztürk, 2024) it becomes evident that the role of the chef is evolving into a multifaceted identity that blends artistry with technological proficiency. The kitchen of the future will not seek to replace the human element but rather to enhance it, allowing chefs to express their creativity through a combination of traditional techniques and modern tools. This new breed of chef will engage with their craft not only through hands-on cooking but also by utilizing sensors, digital interfaces, and software applications, all while remaining true to the essence of flavor, heritage, and emotional connection that food embodies (Ethan Pancer, 2025). As we navigate this transformative landscape, one fundamental truth persists: the bond between humans and food is timeless. While technology may alter our cooking methods, it cannot change the intrinsic motivations behind why we cook. The challenge lies in ensuring that these smart kitchens foster a warm and inviting atmosphere, where technological intelligence complements the soulful art of cooking rather than overshadowing it (Kim, 2025) (Chundong Zheng, 2024).

CONCLUSION

The incorporation of Artificial Intelligence into hotel kitchens signifies a pivotal shift within the hospitality sector. As discussed in this chapter, AI transcends its role as a mere automation tool; it serves as a transformative force that redefines culinary management, enhances guest satisfaction, and advances sustainability initiatives. The evolution of hotel kitchens from basic mechanization and early digital systems to sophisticated, interconnected environments illustrates this change. Modern AI-powered culinary technologies, including robotic sous-chefs and sensor-driven food safety systems, are setting new standards for precision, consistency, and operational efficiency. Additionally, the utilization of real-time data analytics enables hoteliers to craft highly personalized dining experiences through adaptive menu planning that responds to guests' needs. The impact of AI on sustainability is particularly significant, as intelligent inventory management, waste reduction techniques, and predictive analytics empower hotel kitchens to minimize their environmental impact while simultaneously controlling costs. These innovations are not merely responses to operational hurdles; they represent a commitment to ethical and sustainable business practices. Central to this evolution is the collaboration between humans and machines. Rather than replacing chefs, AI enhances their creative capabilities, allowing culinary professionals to concentrate on innovation, storytelling, and cultural representation. Global case studies demonstrate that with appropriate training and collaborative strategies, AI can serve as a partner in preserving culinary heritage while embracing future advancements. However, challenges remain, including substantial initial investments, the necessity for technical training, ethical considerations regarding data privacy, and concerns about maintaining the personal touch in hospitality. Addressing these issues requires thoughtful leadership that prioritizes not only technological investment but also the development of personnel and effective change management strategies.

The future of smart kitchens is brimming with innovative potential, featuring advancements such as voice-activated cooking, AI-driven wine pairing suggestions, and the development of closed-loop zero-waste systems. As artificial intelligence continues to progress, it is essential for the hospitality sector to equip culinary professionals with the skills necessary to excel in this evolving landscape, where traditional culinary practices intersect with cutting-edge technology and human intuition is complemented by machine intelligence. Ultimately, the concept of the Digital Chef should not be viewed as a substitute for the artistry and expertise of human chefs; rather, it embodies the resilience and adaptability inherent in the hospitality industry. By thoughtfully and ethically integrating AI into their operations, hotel kitchens can not only safeguard their future but also enhance the core values of service, creativity, and sustainability that are the hallmarks of exceptional hospitality.

A significant debate in contemporary kitchens revolves around the impact of automation on culinary authenticity. Contrary to the fear that technology might undermine traditional practices, many chefs are discovering that digital tools can actually enhance and preserve these time-honored techniques. For example, precise digital controls can ensure that sauces achieve the perfect reduction every time, while stable fermentation conditions can be maintained with the help of smart devices. In culinary education, these innovations are being utilized to teach students traditional methods with greater accuracy; sensors in cookware allow aspiring chefs to measure their heat management against that of seasoned professionals. Additionally, advanced ordering systems can tailor menus to individual dietary needs, and AI-driven sommelier applications provide wine pairings based on previous selections. Some hotels are even experimenting with a collaborative approach to meal preparation, where artificial intelligence proposes a foundational concept that chefs then interpret creatively. In this landscape of human-machine collaboration, the core of culinary arts remains intact, emphasizing food as a medium for connection, memory, and significance. Rather than diminishing tradition, technology offers fresh avenues for its preservation, expansion, and reinterpretation. As chefs and machines collaborate, a new culinary philosophy is emerging—one that honors the past, engages with the present, and envisions the future. In this dynamic kitchen environment, the enduring truth is that, regardless of technological advancements, the essence of a dish is ultimately crafted by human hands. While the role of technology continues to grow, chefs remain irreplaceable; machines may analyze flavor profiles, but they lack the ability to taste, and while they can replicate processes, they cannot convey intention. Human chefs infuse their creations with emotional depth, cultural significance, and sensory insight that machines are yet unable to replicate. Increasingly, chefs liken themselves to conductors of an orchestra, where automation sets the tempo, and the chef adds unique improvisational flourishes that elevate the culinary experience.

REFERENCES

Adrian Iftene, D. T. (2020). Eye and Voice Control for an Augmented Reality Cooking Experience. *Procedia Computer Science*, *176*, 1469–1478. DOI: 10.1016/j.procs.2020.09.157

Alexander Seifert, C. V. (2021). The use of wearables and health apps and the willingness to share self-collected data among older adults. *Aging and Health Research*.

Armin Tayefeh, A. A. (2024). Reducing energy consumption in a factory and providing an upgraded energy system to improve energy performance. Cleaner Energy Systems, 100-124.

Bautista, M. G. A. N. J. (2023). Automated Cooking Systems: Benefits, Challenges, and Future Directions. IEEE 15th International Conference on Humanoid, Nanotechnology, Information Technology, Communication and Control, Environment, and Management (pp. 1-6). Coron, Palawan, Philippines: IEEE.

Bennett, L. (2025). How AI Technologies Are Revolutionizing Inventory Management in the Supply Chain. *Preprints*.

Berezina, K. C. (2018). Robots, Artificial Intelligence, and Service Automation in Restaurants. Robots, Artificial Intelligence, and Service Automation in Travel, Tourism and Hospitality, 185-219.

Berezina, K. C. (2019). Robots, Artificial Intelligence, and Service Automation in Restaurants. Robots, Artificial Intelligence, and Service Automation in Travel, Tourism and Hospitality, Emerald Publishing Limited, 185-219.

Bhuiyan, K. A. (2024). Consumer attitude toward using artificial intelligence (AI) devices in hospitality services. Journal of Hospitality and Tourism Insights, 968-985.

Bowen, J., & Morosan, C. (2018). Beware hospitality industry: The robots are coming. *Worldwide Hospitality and Tourism Themes*, *10*(6), 726–733. DOI: 10.1108/WHATT-07-2018-0045

BPM. I. (2025). Using artificial intelligence to tackle food waste. Retrieved 04 04, 2025, from https://www.infosysbpm.com: https://www.infosysbpm.com/blogs/services/using-artificial-intelligence-to-tackle-food-waste.html

Buhalis, D. a. (2023). Voice assistants in hospitality: Using artificial intelligence for customer service. *Journal of Hospitality and Tourism Technology*.

Chairote Yaiprasert, A. N. (2024). AI-powered in the digital age: Ensemble innovation personalizes the food recommendations. *Journal of Open Innovation*.

Chakraborty, P. (2025). *Digital Transformation in Hospitality.* Digital Disruption in Hospitality, AI, and Emerging Technologies. DOI: 10.1108/978-1-83608-796-020251009

Chen, J. W. (2023). A Systematic Literature Review of Virtual Technology in Hospitality and Tourism (2013–2022). SAGE Open .

Chundong Zheng, L. Z. (2024). A showdown in the kitchen: Exploring consumers' preferences for robot-made versus human-made foods at different stages of dietary restraint. *Journal of Retailing and Consumer Services.*

Claudia Cozzio, G. V. (2023). Toward an integration of blockchain technology in the food supply chain. *Journal of Business Research.*

David Graham, A. A. (2020). Open kitchens: Customers' influence on chefs' working practices. *Journal of Hospitality and Tourism Management*, *45*, 27–36. DOI: 10.1016/j.jhtm.2020.07.011

Dhal, S. B., & Kar, D. (2024). Transforming Agricultural Productivity with AI-Driven Forecasting: Innovations in Food Security and Supply Chain Optimization. *Forecasting*, *6*(4), 925–951. DOI: 10.3390/forecast6040046

Dhal, S. K., & Kar, D. (2025). Leveraging artificial intelligence and advanced food processing techniques for enhanced food safety, quality, and security: A comprehensive review. *Discover Applied Sciences*, *7*(1), 7. DOI: 10.1007/s42452-025-06472-w

Dianyuan Wang, M. Z. (2024). Fruits and vegetables preservation based on AI technology: Research progress and application prospects. *Computers and Electronics in Agriculture.*

Dimitrios, B., & Rosanna, L. (2018). Smart hospitality—Interconnectivity and interoperability towards an ecosystem. *International Journal of Hospitality Management*, •••, 41–50.

Ekta Kumawat, A. D. (2025). Artificial intelligence through the lens of hospitality employees: A systematic review. *International Journal of Hospitality Management*, *124*, 103986. DOI: 10.1016/j.ijhm.2024.103986

Ethan Pancer, T. J. (2025). Robots in the kitchen: The automation of food preparation in restaurants and the compounding effects of perceived love and disgust on consumer evaluations. *Appetite.* PMID: 39490695

ETHospitalityWorld. (2025, 04 25). AI vs. human chefs: Can technology replicate the creativity of fine dining maste. Retrieved 5 2, 2025, from https://hospitality.economictimes.indiatimes.com/news/speaking-heads/ai-vs-human-chefs-can-technology-replicate-the-creativity-of-fine-dining-masters/121330773: https://hospitality.economictimes.indiatimes.com/news/speaking-heads/ai-vs-human-chefs-can-technology-replicate-the-creativity-of-fine-dining-masters/121330773

Farah Zahidi, B. B. (2024). Embracing the new era: Artificial intelligence and its multifaceted impact on the hospitality industry. *Journal of Open Innovation.*

Ghosal, S. (2025, 4 25). How AI & Automation Are Transforming Cloud Kitchens. Retrieved 4 25, 2025, from https://www.restaurantindia.in: https://www.restaurantindia.in/article/how-ai-automation-are-transforming-cloud-kitchens.12598

Group, S. E. (2025). Implementing Sustainable Hospitality in Your Business. Retrieved 07 2025, 07, from https://www.shms.com: https://www.shms.com/en/news/sustainability-in-hospitality/

Güngör, O., & Yücel Güngör, M. (2024). Automation in gastronomy: Use of smart cooking systems in industrial kitchens. *Worldwide Hospitality and Tourism Themes, 16*(2), 190–201. DOI: 10.1108/WHATT-03-2024-0067

Hazem Golshany, Y. N. (2025). IoT-enabled smart kitchen technologies and their impact on food storage, preparation, and culinary experiences: A systematic review. *Food Research International, 213,* 116557. DOI: 10.1016/j.foodres.2025.116557 PMID: 40436590

Hazim Ryad Momani, S. B.-R.-Q. (2025). PERCEPTIONS OF ARTIFICIAL INTELLIGENCE IN MENU DEVELOPMENT AND PROFITABILITY IN CULINARY ARTS. In Xi'an Shiyou Daxue Xuebao (Ziran Kexue Ban)/ Journal of Xi'an Shiyou University, Natural Sciences Edition, 61–85 *Zenodo.*

Helen Onyeaka, A. A.-S. (2025). Journal of Agriculture and Food Research. Artificial intelligence in food system: Innovative approach to minimizing food spoilage and food waste .

Ilhaam, A., & Omar, H. R. (2024). Using blockchain technology to achieve sustainability in the hospitality industry by reducing food waste. *Computers & Industrial Engineering.*

Irtiqa Shabir, K. K. (2023). Carbon footprints evaluation for sustainable food processing system development: A comprehensive review. *Future Foods : a Dedicated Journal for Sustainability in Food Science.*

Islam, M. T. (2024). Green Revolution in the Hospitality Industry: A Deep Dive Into Sustainable Food Waste Management. In Sustainable Disposal Methods of Food Wastes in Hospitality Operations (pp. 110-126). IGI Publisher. DOI: 10.4018/979-8-3693-2181-2.ch008

Jiang, W. L. C. (2025). *Advancements in Intelligent Sensing Technologies for Food Safety Detection*. Research A SCIENCE PARTNER JOURNAL. DOI: 10.34133/research.0713

Jinjin Huang, M. Z. (2025). AI-based processing of future prepared foods: Progress and prospects. *Food Research International*. PMID: 39849794

Jobstreibizer, J. B., Beliaeva, T., Ferasso, M., Kraus, S., & Kallmuenzer, A. (2025). The impact of artificial intelligence on business models: A bibliometric-systematic literature review. *Management Decision*, *63*(13), 372–396. DOI: 10.1108/MD-10-2024-2309

Jozef Youssef, C. S. (2023). Gastromotive dining: Using experiential multisensory dining to engage customers. *International Journal of Gastronomy and Food Science*.

Kanwal Gul, V. D. (2025). Analyzing food waste perceptions and its solutions: A pilot study on a non-WEIRD sample. *Food Quality and Preference*.

Karagiannis, M. (2024, 02 27). Food Allergies and AI: Creating Safe Dining Experiences for Guests. Retrieved 7 11, 2025, from https://medium.com/hotel-tech/food-allergies-and-ai-creating-safe-dining-experiences-for-guests-3c6b7b5a70ea: https://medium.com/hotel-tech/food-allergies-and-ai-creating-safe-dining-experiences-for-guests-3c6b7b5a70ea

Kashyap, V. (2024). *The Creative Tourism - A New Approach of Sustainable Practices a Case Study of Kinnaur*. Atlantis press., DOI: 10.52202/074664-0019

Kashyap, V. S., Sharma, A., Katoch, H., & Bisht, R. (2023). The Comprehension of the Nichar Community for the Development and Growth of the Destination as a Creative Tourism Destination? *ASEAN Journal on Hospitality and Tourism*, *21*(1), 109–123. DOI: 10.5614/ajht.2023.21.1.08

Kim, H. C., Choi, S., & Shin, H. H. (2025). Artificial intelligence in the kitchen: Can humans be replaced in recipe creation and food production? *International Journal of Contemporary Hospitality Management*, *37*(5), 1641–1661. DOI: 10.1108/IJCHM-04-2024-0549

Kubáňová, J. K., Kubasáková, I., Čulík, K., & Štítik, L. (2022). Implementation of Barcode Technology to Logistics Processes of a Company. *Sustainability (Basel)*, *14*(2), 790. DOI: 10.3390/su14020790

Kumagai, S. (2017). Cooking assistant service utilizing an interactive robot. *IEEE/SICE International Symposium on System Integration (SII)*. DOI: 10.1109/SII.2017.8279351

Kym Man, J. A. (2024). That Looks Like My Kitchen!" – Personalized context by usage frequency and familiarity influences consumer perception and liking of chicken nuggets in VR. *Food Research International.*

Law, R. L., Lin, K. J., Ye, H., & Fong, D. K. C. (2024). Artificial intelligence research in hospitality: A state-of-the-art review and future directions. *International Journal of Contemporary Hospitality Management, 36*(6), 2049–2068. DOI: 10.1108/IJCHM-02-2023-0189

Learn, S. I. (2024, 6 12). https://somedayillearn.com/kitchen-robot/. Retrieved 6 4, 2025, from The Kitchen Robot Revolutionizing Meal Prep: https://somedayilllearn.com/kitchen-robot/

Lu Lu, M. H. (2024). AI robots over sommeliers? Exploring the service provider effect on diners' wine ordering decisions at restaurants. *International Journal of Hospitality Management, 122*, 103879. DOI: 10.1016/j.ijhm.2024.103879

Maddala, V. S. (2025). Revolutionizing industrial kitchen appliances: How data-driven supply chains. Global Journal of Engineering and Technology Advances, .

McKinsey. (2024). *Beyond the hype:Capturing the potential of AI and gen AI in tech, media, and telecom*. Quatam Black AI by McKinsey.

Mohammad Nurul Alam, M. O. (2025). Understanding consumer behavior in organic food restaurants: Insights from the push-pull-mooring framework. *Environment, Development and Sustainability.*

Nanda, R. (2025). How AI and IoT Are Shaping the Next Generation of Smart Kitchens. Retrieved from https://techstrong.ai/: https://techstrong.ai/articles/how-ai-and-iot-are-shaping-the-next-generation-of-smart-kitchens/

Nichifor, B. Z., Zait, L., & Timiras, L. (2025). Drivers, Barriers, and Innovations in Sustainable Food Consumption: A Systematic Literature Review. *Sustainability (Basel), 17*(5), 2233. DOI: 10.3390/su17052233

Onyeaka, H. T., Tamasiga, P., Nwauzoma, U. M., Miri, T., Juliet, U. C., Nwaiwu, O., & Akinsemolu, A. A. (2022). Using Artificial Intelligence to Tackle Food Waste and Enhance the Circular Economy: Maximising Resource Efficiency and Minimising Environmental Impact: A Review. *Sustainability (Basel), 15*(13), 10482. DOI: 10.3390/su151310482

Öztürk, B. (2024). Art of food: Systematic literature review of culinary creativity. *International Journal of Gastronomy and Food Science*.

Revelou, P.-K. T., Tsakali, E., Batrinou, A., & Strati, I. F. (2025). Applications of Machine Learning in Food Safety and HACCP Monitoring of Animal-Source Foods. *Foods*, *14*(6), 922. DOI: 10.3390/foods14060922 PMID: 40231903

Robosouschef. (n.d.). https://robosouschef.com/pages/about-us. Retrieved 06 26, 2025, from https://robosouschef.com/

Roy, B. K. (2023). *Advancements in arena technology: Enhancing customer experience and employee adaptation in the tourism and hospitality industry*. Smart Tourism.

Seal, P. P., Shukla, A., Bhutia, L. D., & Yadav, N. (2025). The role of food in food blogs: A netnographic analysis of Indian food bloggers' posts on social media. *Asia-Pacific Journal of Innovation in Hospitality and Tourism*, *14*(1), 115–136.

Seyitoğlu, F., Fusté-Forné, F., Yiğit, S., & Engin, S. (2025). The role of technology in the skills and creativity of chefs. *European Journal of Tourism Research*, *39*, 3912. DOI: 10.54055/ejtr.v39i.3840

Seyitoğlu, F. F.-F., Fusté-Forné, F., Yiğit, S., & Engin, S. (2025). Robot chefs: The impacts, compatibility and suitability. *British Food Journal*, *127*(1), 307–323. DOI: 10.1108/BFJ-07-2024-0705

Shei, R. J., Holder, I. G., Oumsang, A. S., Paris, B. A., & Paris, H. L. (2022). Wearable activity trackers-advanced technology or advanced marketing? *European Journal of Applied Physiology*, *122*(9), 1975–1990. DOI: 10.1007/s00421-022-04951-1 PMID: 35445837

Shukla, A., Meeprom, B., Khunasathitchai, K., & Yadav, N. (Eds.). (2025). *AI's Role in Language Learning and Communication*. IGI Global Scientific Publishing., DOI: 10.4018/979-8-3373-5681-5

Shukla, A., & Sharma, N. (2020). An analysis of casual dining outlet: A Himachal-based study. *International Journal of Advanced Science and Technology*, *29*(3 Special Issue), 1546–1557.

Shukla, A., Yadav, N., Berry, K., Sharma, N., & Uniyal, M. C. (2025). Ethical Responsibilities of Hospitality Providers in Prenatal Tourism and Birthright Citizenship. In Chourasia, B. (Ed.), *Exploration of Prenatal Tourism and Birthright Citizenship* (pp. 1–20). IGI Global Scientific Publishing., DOI: 10.4018/979-8-3693-8206-6.ch001

Shukla, A., Yadav, N., Khan, R. H., Kaewdang, P., & Srihirun, J. (2025). Mountains Under Pressure: Smart Solutions for Overtourism. In Valeri, M., Singh, A., Tyagi, P., & Haghi, A. (Eds.), *Balancing Mountain Tourism, Cultural Heritage, and Environmental Stability* (pp. 401–416). IGI Global Scientific Publishing., DOI: 10.4018/979-8-3693-8764-1.ch027

Shukla, A., Yadav, N., Khunasathitchai, K., Bakshi, I., & Sharma, N. (2025). Waste Management Outlook and Future Directions in Rural Touristic Areas. In Albattat, A., Singh, A., Tyagi, P., & Haghi, A. (Eds.), *Solid Waste Management and Disposal Practices in Rural Tourism* (pp. 495–522). IGI Global Scientific Publishing., DOI: 10.4018/979-8-3693-9621-6.ch020

Siham Rekiek, H. J. (2024). Prediction of Booking Trends and Customer Demand in the Tourism and Hospitality Sector Using AI-Based Models. *International Journal of Advanced Computer Science and Applications*.

Sojecka, A. A.-R. (2024). Innovative High Pressure Processing (HPP, HPT) for the 21st-Century Sustainable Society. Foods. Food Preservation in the Industrial Revolution Epoch, 13(19).

Stephens, R. (2025, 02 21). https://www.foodandwine.com/. Retrieved 06 02, 2025, from https://www.foodandwine.com/restaurants-ai-artificial-intelligence-technology-11683792

Tech, K. F. (2024, 08 30). Cloud Kitchen Startup: An Historical Overview and its Impact in India. Retrieved 06 2025, 03, from https://www.kouzinafoodtech.com/blog/which-startup-first-started-cloud-kitchen-concept-india: https://www.kouzinafoodtech.com/blog/which-startup-first-started-cloud-kitchen-concept-india

Tuomi, A. T., Tussyadiah, I. P., & Stienmetz, J. (2020). Applications and Implications of Service Robots in Hospitality. *Cornell Hospitality Quarterly*, *62*(2), 232–247. DOI: 10.1177/1938965520923961

Vânia Gisele Bessi, P. T. (2022). Creativity in gastronomic activities. *International Journal of Gastronomy and Food Science*.

WHO. (n.d.). https://www.who.int/health-topics/food-safety. Retrieved from https://www.who.int/health-topics/food-safety

Y, W. (2025). The Emerging "AI Artists": Breaking the Metacrisis and the Fear of Losing Human Creativity. Nature Anthropology .

Yadav, N., Shukla, A., Khunasathitchai, K., & Kaur, R. (2024). AI Adaptions: Challenges and Strategies for the Hotel and Tourism Industry. In Talukder, M., Kumar, S., & Tyagi, P. (Eds.), *Hotel and Travel Management in the AI Era* (pp. 19–38). IGI Global Scientific Publishing., DOI: 10.4018/979-8-3693-7898-4.ch002

Zhe Wang, M. Z. (2025). AI technology in smart drying of foods: A critical review of research and applications. *Innovative Food Science & Emerging Technologies*.

Chapter 3
Ghost Kitchens and the Transformation of Future Food Services

Mohammad Badruddoza Talukder
https://orcid.org/0009-0008-1662-9221
International University of Business Agriculture and Technology, Bangladesh

Dil Afrin Swarna
https://orcid.org/0009-0000-9176-920X
National University, Bangladesh

Firoj Kabir
https://orcid.org/0009-0001-3014-3163
Daffodil International University, Bangladesh

Shivansu Sachan
https://orcid.org/0000-0002-9893-8018
CSJM University, India

ABSTRACT

The emergence of Ghost kitchens, called cloud kitchens, is causing a radical change in the food service sector. These off-site locations avoid the conventional dine-in restaurant concept by specializing only in meal preparation for delivery. Ghost kitchens provide some advantages, such as increased speed and convenience, cost-effectiveness, and scalability, by using developments in automation, artificial intelligence, and innovative kitchen appliances. They also support sustainability by reducing food waste and streamlining delivery routes. Despite quality control, brand impression, and regulatory compliance obstacles, Ghost kitchens have a bright future. New developments like virtual branding, communal kitchen areas, and

DOI: 10.4018/979-8-3373-4042-5.ch003

connection with smart city infrastructure point to future expansion and creativity. Ghost kitchens have the potential to completely change the food industry by giving customers efficient and customized dining experiences and giving food service operators adaptable and affordable business options.

INTRODUCTION

The food service sector is now undergoing a paradigm change that is being pushed by the progression of technology as well as the changing needs of consumers. Ghost kitchens, often called cloud kitchens, ghost kitchens, or dark kitchens, are at the vanguard of this transition (Das et al., 2024). Ghost kitchens, in contrast to conventional restaurant settings, are off-site facilities solely reserved for cooking meals for delivery. Food service providers can simplify procedures and minimize overhead expenses by utilizing this revolutionary concept, eliminating the requirement for physical eating areas and front-of-house operations (Kumar et al., 2024).

Several factors have contributed to the proliferation of Ghost kitchens, including the growth in popularity of food delivery platforms and the rising appetite for eating alternatives that are quick and available on demand. Using cutting-edge technology, these kitchens can improve efficiency, preserve quality, and cater to the ever-changing requirements of contemporary consumers (Talukder & Kumar, 2023). Ghost kitchens are outfitted with sophisticated technologies that optimize every facet of food preparation and delivery. These tools include automation, robots, and analytics powered by artificial intelligence and electronic kitchen equipment.

In investigating the future of food service, it is essential to have a solid understanding of the development, advantages, difficulties, and new trends related to Ghost kitchens. This chapter investigates the primary elements contributing to the proliferation of Ghost kitchens, their influence on the food service sector, and the potential they offer for innovation and expansion. Through the examination of these components, we can acquire an understanding of how Ghost Kitchens are set to revolutionize the culinary scene. This provides a view into a future in which efficiency, convenience, and sustainability are more important than ever before.

Objectives of the Chapter

1. To provide a comprehensive understanding of the development and technological innovations driving the growth of Ghost kitchens.
2. To explore the advantages of Ghost kitchens, such as cost efficiency and scalability, while also addressing the associated challenges like quality control and regulatory compliance.

3. To highlight emerging trends and opportunities within the Ghost kitchen industry, focusing on virtual brands, collaborative spaces, and integration with smart city infrastructure.

Theoretical Overview

To get a more in-depth comprehension of the emergence of Ghost kitchens and their influence on the food service sector, it is possible to investigate the notion of Ghost kitchens via various theoretical lenses. We can better understand the dynamics at play if we incorporate theories from consumer behavior, company strategy, and the adoption of new technologies.

THE ADOPTION OF NEW TECHNOLOGIES AND THE SPREAD OF INNOVATION

Diffusion of Innovations According to Rogers, Theory plays a significant role in explaining the increase in popularity of Ghost kitchens. According to Rogers, the spread of innovations across a population follows a pattern that may be predicted. Several criteria determine this pattern, including relative advantage, compatibility, complexity, trialability, and observability characteristics. Ghost kitchens offer several advantages over conventional restaurant models (Mohammad et al., 2024). These advantages include lower costs, more efficiency, and greater flexibility. The fact that they are compatible with the growing demand for food delivery among consumers and that they may be tested through pilot projects and operations on a small scale has further aided their acceptance.

THE ADVANTAGES OF UTILIZING BUSINESS STRATEGIES IN THE MARKET

It is possible to analyze the competitive environment in which Ghost Kitchens operates with the assistance of Porter's Five Forces Framework. The threat of new entrants, the bargaining power of suppliers and consumers, the danger of replacements, and the competition within the sector are all factors that this model investigates (Kumar et al., 2024). Using Ghost kitchens helps to lessen the barriers to entry for new food service companies since they eliminate the requirement for premium real space and a large number of employees. Existing firms can also use them to diversify their operations and generate new income streams without making significant extra investments.

The company's Resource-Based View (RBV) emphasizes the strategic significance of using one's resources and talents to obtain a favorable position in the market. Using technical resources like automation, artificial intelligence, and data analytics by Ghost Kitchens allows for optimizing operations and providing higher service (Talukder et al., 2024a). These technical skills act as crucial differentiators in a highly competitive sector.

SHIFTS IN CONSUMER BEHAVIOR AND DEMAND

Maslow's Hierarchy of Needs and Consumer Behavior Theories are two more lenses through which one may investigate the proliferation of Ghost kitchens. Convenience, variety, and prompt service are becoming increasingly important to customers as they progress up the hierarchy of their purchasing power. Providing a wide array of cuisines brought to customers' doorsteps promptly is one of the ways that Ghost kitchens respond to higher-order demands.

According to the Theory of Planned Behavior (TPB), customers' attitudes, subjective standards, and perceived behavioral control all play a role in determining whether or not they intend to utilize meal delivery services. A combination of factors, including the growing acceptance of food delivery services in society and the positive attitudes towards convenience and diversity, has increased demand for Ghost kitchens (Talukder, 2020).

INTEGRATIVE PERSPECTIVE

Through the integration of different theoretical frameworks, we can acquire a thorough understanding of the factors that are driving the rapid growth of Ghost kitchens. Theories of technology adoption explain the fast dissemination of technologies, corporate strategy models emphasize the strategic advantages of technologies, and consumer behavior theories demonstrate how demand dynamics are constantly altering (Talukder, 2024). When taken as a whole, these viewpoints offer a solid theoretical basis for understanding Ghost Kitchens' current and future trajectory in the food service business.

LITERATURE REVIEW

Evolution of Kitchens Located in Ghosts

Centralized food production units developed for large-scale events and institutional meals gave rise to Ghost Kitchens. Cousins, Foskett, and Pennington (2002) detailed that these kitchens' initial focus was maximizing productivity and carrying out large-scale manufacturing. The development of delivery-only kitchens that cater to the growing need for quick, on-demand food services has been made possible by the emergence of digital platforms, resulting in a dramatic transformation of this paradigm (Pariza, 2020).

What is the Impact of Digital Platforms for Food Delivery?

Digital meal delivery companies such as Uber Eats and DoorDash have significantly aided the proliferation of Ghost kitchens. Because of these platforms, a solid market for food delivery-only services has been developed, which has led to an increased need for kitchen models that are both efficient and scalable (Pariza, 2020). The food service business has been revolutionized due to integrating these platforms with Ghost Kitchens, which has aligned it with the inclinations of modern consumers toward convenience and speed (Talukder et al., 2024c).

Provision of Automation in Ghost Kitchens

Ghost kitchens have become increasingly dependent on automation technology, improving operational efficiency and uniformity. In their article, Berman and Macdonald (2019) explore the introduction of robotic chefs and automated culinary equipment, eliminating the need for human labor and lessening the likelihood of human mistakes. This technological integration makes it possible to exercise exact control over the processes involved in food preparation, guaranteeing that very high quality and efficiency criteria are met (Talukder et al., 2023).

In the Field of Data Analytics and Artificial Intelligence

Optimizing many parts of Ghost Kitchen operations relies heavily on artificial intelligence and data analytics. Kumar and Sinha (2021) investigated how artificial intelligence algorithms can forecast inventory requirements, simplify menu options using real-time data, and improve delivery logistics. These innovations make it possible for Ghost kitchens to function more effectively, lowering the amount of waste produced and enhancing overall efficiency.

Internet of Things-Enabled Home Appliances

The Internet of Things (IoT) enables innovative kitchen technology, such as appliances, to provide real-time monitoring and control, ensuring food quality and safety. Using smart ovens and refrigerators that can monitor cooking processes and storage conditions, Wang et al. (2022) emphasize the benefits of these appliances. These appliances may warn staff members of possible problems before they have an impact on the quality of the food. This technology is necessary to keep the standards in Ghost kitchens high.

Effectiveness of Ghost Kitchens in Terms of Cost

A key benefit of using Ghost kitchens is their cost-effectiveness. In their article, Yeo and Wahid (2020) highlight that modern kitchens reduce the need for expensive real estate and substantial front-of-house labor, which reduces this financial model. Foodservice operators can increase their investments in technology and quality, which ultimately results in an improvement in their total service offering.

Ability to Scale and Expansion of the Market

The scalability of Ghost Kitchens is unique, making it possible to expand the market while minimizing the associated financial risk rapidly. According to Anderson and Kumar (2021), these kitchens allow food service providers to enter new markets quickly, allowing them to adapt to the demands of the local community without the cost of building full-scale restaurants. To succeed in the highly competitive foodservice market, scalability is necessary.

Both Convenience and Rapidity

Ghost kitchens can meet the needs of modern consumers who are looking for meal delivery services that are both dependable and timely because of the ease and speed they provide. According to Smith et al. (2019), these kitchens are built to maximize efficiency, guaranteeing that orders are made and delivered promptly. The success of Ghost kitchens may be attributed, in large part, to the fact that they prioritize speed and convenience.

Advantages to the Environment

Ghost kitchens have the potential to contribute substantially to the environment's preservation. Through centralized production, Garvey and Lee (2020) highlight how these kitchens maximize resource use and reduce wasted food. This concept lessens the total impact of food service operations on the environment, making it appealing to consumers and regulators concerned about the environment.

Challenges Present in Quality Control

Ghost kitchens have a significant obstacle in the form of the problem of preserving the same level of food quality across several sites. Jones and Lee (2018) emphasize the need to standardize the recipes and procedures to guarantee that everything is consistent. Maintaining quality consistency remains challenging with all the technical breakthroughs, calling for constant attention and creativity.

What Customers Think of Your Brand and Their Loyalty

The lack of a physical eating location can affect how customers perceive the business and how loyal they are to it. According to Green and Brown (2020), it is necessary to have robust online branding and marketing tactics to construct and keep a loyal consumer base from the beginning. To achieve success, they recommend that digital involvement and open communication on the quality and safety of food are essential.

Administration of Regulations

Ghost kitchens face a difficult challenge when navigating the many health and safety regulations across different locations. Martin and Liu (2021) stress the need to comply with local norms to guarantee that activities will go smoothly. Compliance with these requirements is necessary to preserve one's credibility and stay out of legal trouble.

The Concept of Virtual Brands

One of the most critical trends in the food service sector is the proliferation of virtual brands and businesses solely focused on delivery. According to Roberts and Scott (2020), virtual brands make it possible to experiment with different culinary techniques without being constrained by the limitations of an actual shopfront. By allowing operators to experiment with new ideas and cuisines, this strategy encourages innovation and variety in the restaurant industry.

Cooperative Cooking Areas in the Kitchen

The concept of collaborative cooking spaces, in which many food service providers share facilities, is gaining popularity to lower costs and encourage innovation. Davis and Taylor (2021) investigate the advantages of these shared spaces, which include reducing costs and enhancing chances for participation in collaborative actions. This concept assists smaller operators and startups, fostering a dynamic and diversified ecosystem in terms of food service (Talukder, 2024b).

Integration with the Infrastructure of Smart Neighborhoods

The incorporation of Ghost kitchens into the infrastructure of smart cities gives prospects for optimizing delivery routes and reducing transportation congestion. It is highlighted by Garcia and Chen (2022) that smart city technologies have the potential to improve the efficiency of food delivery systems, hence enhancing the overall quality of service and ensuring its continuing sustainability.

Individualization and Adjustment of Specifics

With the advancements in artificial intelligence and machine learning, fully personalized eating experiences are now possible. In their article, Smith et al. (2021) explore how these technologies might be used to personalize menu items to each consumer's tastes, increasing customer satisfaction. Personalization is becoming an increasingly significant difference in the highly competitive food service industry.

Sustainability, as well as the Optimization of Resources

Johnson and Miller (2022) investigate how Ghost kitchens contribute to sustainability by maximizing resource utilization and minimizing wasted food. It is possible to improve inventory management and use resources more effectively through centralized manufacturing, which helps promote environmental goals and appeals to environmentally sensitive consumers.

Dynamics of the Workforce

The influence that automation will have on employment opportunities in the food service industry is an essential research topic. A study by Harris and Gomez (2021) investigates the impact of greater automation in Ghost kitchens on labor dynamics, including the displacement of jobs and the requirement for reskilling programs.

To ensure the long-term viability of Ghost kitchens, it is vital to strike a balance between the growth of technology and the sustainability of labor.

Delivery Drivers and the Service-Based Economy

There is a strong connection between the advent of the gig economy and the proliferation of Ghost kitchens, particularly the reliance on delivery drivers. In their study, Lee and Hwang (2022) explore the significance of fair labor practices and the requirement for rules to protect gig workers. Maintaining the viability of the gig economy concept is necessary to ensure the continuous expansion of Ghost kitchens.

Attitudes of Customers and Their Trust

Regarding the success of Ghost Kitchens, consumers' perception and acceptability are essential. The findings of Kim and Park (2021) indicate that the most important characteristics that influence customer trust are quality assurance and openness in business operations. Increasing customer confidence and loyalty may be accomplished via effective communication on the quality and safety of food.

Branding and Marketing Strategies and Methods

When it comes to establishing consumer trust and loyalty, effective branding and marketing tactics are essential. In their study on promoting Ghost kitchens, Singh and Jain (2022) highlight the significance of digital engagement and transparent communication tools. Strong branding can differentiate operators in a very competitive market.

The Education of Consumers

Educating customers about the advantages and precautions that Ghost Kitchens takes can increase their acceptance and pleasure with the product. It has been suggested by Poon and Lam (2021) that consumer education campaigns have the potential to resolve concerns over the quality and safety of food, hence increasing general trust in the model.

Effect on the Economy

There is a considerable economic influence that Ghost kitchens have on the sector of food service establishments. It is revealed by Zhang and Li (2023) that these kitchens help the industry's growth by enabling operators to enter new geographic regions with a cheaper initial expenditure. Innovation and development are both supported by this strategy, which in turn drives economic progress.

An Examination of the Market and Experimentation

Because of its adaptability, Ghost kitchens can conduct market research and experiment with creative new ideas. Operators may use Ghost kitchens to test new menu items and culinary trends, obtaining insights into consumer tastes without the danger of typical restaurant setups. According to Li and Zhao (2023), Ghost kitchens are a technique that operators can use.

Concerns Relating to Law and Regulation

Much complexity is involved in navigating Ghost Kitchens' legal and regulatory framework. In their study, Smith and Thompson (2023) describe the difficulties of attempting to comply with various health and safety standards, zoning restrictions, and food safety requirements. Compliance with regulatory requirements is essential to prevent legal complications and to keep operational integrity intact.

Constantly Changing Regulatory Structure

The regulatory framework for Ghost kitchens is evolving in response to the sector's expansion. In their study, Johnson and Green (2023) investigate how new rules are formulated to address rising concerns, such as food safety and labor practices. For operators to continue to be compliant and competitive, they must stay ahead of changes in regulatory requirements.

Effects on Conventional Dining Establishments

The proliferation of Ghost kitchens does have repercussions for conventional dining establishments. Williams and Anderson (2023) investigate how these kitchens offer an alternate model that may coexist with conventional eating venues. These kitchens complement services like delivery and catering without needing extra physical locations.

The Landscape of Competition

Ghost kitchens are reshaping the competitive environment of the food service business, which is now experiencing growth. Thompson and Miller (2023) examine how these kitchens make it possible for smaller companies and startups to compete with major brands by lowering the barriers to entry and encouraging innovation.

Expectations of the Consumers

The proliferation of Ghost kitchens has resulted in a shift in customers' expectations regarding food service. The authors Jones and Taylor (2023) underline the fact that customers today place a higher priority on convenience, speed, and quality, and they anticipate delivery services that are dependable and smooth. That Ghost Kitchens can meet these expectations is necessary for their success.

In the Future, Research will Focus on

Whether Ghost kitchens can be scaled up and maintained over the long term should be the subject of future study. According to Kim and Lee (2023), further research is required to investigate the influence that recent technical developments, alterations in regulatory policies, and shifting customer preferences have had on the development of Ghost kitchens. To handle new difficulties and make the most of new possibilities in the food service business, it will be necessary to continue conducting research.

Table 1.

Author(s) & Year	Title	Objective / Focus	Key Findings	Relevance to Ghost Kitchens
Zhang et al. (2022)	*The Rise of Ghost Kitchens: Business Model Innovation*	Explore the ghost kitchen model and its scalability	Ghost kitchens improve operational efficiency and reduce overhead costs	Shows how the model is reshaping traditional food service
Kim & Park (2021)	*Digital Transformation in the Food Service Industry*	Analyze how tech impacts food service operations	Tech tools like AI, apps, and IoT drive ghost kitchen efficiency	Supports integration of tech in ghost kitchen operations
Sharma et al. (2023)	*Consumer Perception Toward Virtual Restaurants*	Investigate consumer trust and experience in delivery-only kitchens	Consumers demand quality, safety, and transparency in ghost kitchen services	Highlights consumer trust as a moderator
Garcia & Lee (2020)	*Sustainable Urban Food Systems*	Discuss urban food delivery trends and sustainability	Ghost kitchens contribute to reduced urban food footprints	Ties ghost kitchens to urban sustainability goals

continued on following page

Table 1. Continued

Author(s) & Year	Title	Objective / Focus	Key Findings	Relevance to Ghost Kitchens
Jones & Miller (2021)	*Operational Challenges in Cloud Kitchen Models*	Identify challenges in logistics and staffing	Managing multiple brands and tech platforms poses coordination challenges	Emphasizes limitations and improvement areas
Oliveira et al. (2022)	*Customer Satisfaction in Delivery-Only Food Models*	Examine service quality in ghost kitchen models	Timely delivery and digital interaction are key to satisfaction	Reinforces operational priorities in ghost kitchens
Tan & Wilson (2023)	*Regulatory Gaps in Online Food Services*	Analyze legal and safety regulations for ghost kitchens	Lack of clear policies creates risk for both consumers and operators	Points to note for regulatory frameworks

Source: Author's compilation

FINDINGS

A full grasp of Ghost Kitchens' history, technical achievements, benefits, problems, and future trends may be gained from the conclusions of the literature review on Ghost Kitchens. The most important findings are summarized as follows:

Evolution of Kitchens Located in Ghosts

Since their inception as conventional centralized food production facilities for large-scale events and institutions, Ghost kitchens have seen considerable transformation, transitioning into delivery-only kitchens that respond to the expectations of modern consumers for convenience and acceleration. A significant contribution to this shift has been made by incorporating digital food delivery systems, allowing widespread acceptance and market development (Kumar et al., 2024).

Effects of Digital Platforms on Society

An extensive market for Ghost kitchens has been formed due to the proliferation of digital meal delivery companies such as Uber Eats and DoorDash. The customer preferences for on-demand food delivery services are aligned with the capabilities of these platforms, which allow for seamless integration and operational efficiency.

Progress in Technological Innovation

Automation, artificial intelligence, data analytics, and appliances enabled by the Internet of Things have all contributed to Ghost kitchens' transformation. These technologies allow for increased operational efficiency, consistency, and food safety, guaranteeing high standards and maximizing the utilization of available resources.

Efficiency in Terms of Cost and Scalability

It is possible to achieve excellent cost-effectiveness with Ghost kitchens since they eliminate the requirement for costly real estate and many front-of-house employees. Their concept is scalable, enabling quick market development with minimum financial risk, supporting growth and innovation in the food service business.

Both Convenience and Rapidity

Convenience and speed are given priority in the design of Ghost kitchens, which allows them to satisfy the needs of modern consumers for meal delivery services that are both rapid and dependable—this emphasis on efficiency guarantees that customers will have timely and satisfying experiences.

Maintaining a Sustainable Environment

Ghost kitchens contribute to the preservation of the environment by maximizing the utilization of resources and minimizing the amount of wasted food. Through their centralized manufacturing strategy, they can improve inventory management and reduce energy use, which makes them appealing to customers and regulators concerned about the environment.

Challenges Present in Quality Control

One of the most critical challenges that Ghost kitchens continue to face is ensuring that the quality of their cuisine is consistent across numerous sites. Standardized recipes and methods are necessary to guarantee uniformity and fulfill customers' expectations.

What Customers Think of Your Brand and Their Loyalty

The lack of a physical eating location can affect how customers perceive the business and how loyal they are to it. It is necessary to have strong internet branding and marketing tactics and honest information regarding the quality and safety of food to develop and keep consumers' confidence.

Administration of Regulations

Ghost kitchens face a difficult challenge when navigating the many health and safety regulations across different locations. Consistent adherence to local rules is necessary to preserve the integrity of operations and stay out of legal trouble.

Virtual Organizations and Spaces for Collaborative Work

It is a crucial trend that virtual brands and communal culinary spaces are becoming increasingly popular. Virtual brands allow experimenting with different culinary techniques without physical limitations. At the same time, collaborative spaces provide potential for innovation and cost savings through shared facilities.

Integration with the Infrastructure of Smart Neighborhoods

The incorporation of Ghost kitchens into the infrastructure of smart cities has the potential to optimize delivery routes and minimize traffic congestion, improving the overall efficiency and sustainability of service delivery.

Individualization and Adjustment of Specifics

As artificial intelligence and machine learning continue to advance, they have made it possible to create highly personalized eating experiences that cater to individual tastes and increase user pleasure. Personalization is becoming an increasingly significant difference in the highly competitive food service industry.

The Gig Economy and the Dynamics of the Workforce

Automation will substantially influence employment in the food service business, raising worries about job displacement and the necessity of educational programs that will help people retrain. Because of the importance of the gig economy, particularly the reliance on delivery drivers, the expansion of Ghost kitchens is essential, which in turn requires the implementation of fair labor policies.

Attitudes of Customers and Their Trust

Several essential aspects influence consumer acceptability and confidence in Ghost kitchens. These factors include transparency in operations, quality assurance, and practical information sharing. The happiness and loyalty of customers can be increased by providing them with information about the advantages and safety precautions.

Impact on the Economy and the Market

The ability of food service providers to enter new geographic areas with a cheaper initial expenditure is one of the ways that Ghost kitchens contribute to the expansion of the industry. This approach encourages innovation, expansion, and economic development in the food service business.

Legal and Regulatory Obstacles to Overcome

The ever-evolving regulatory environment for Ghost kitchens necessitates that operators be current with changes and ensure that they are under health and safety rules, zoning laws, and food safety requirements.

Effects on Conventional Dining Establishments

Ghost kitchens provide an alternate model that may coexist with conventional eating places. These kitchens offer ancillary services like delivery and catering without needing extra physical locations.

The Landscape of Competition

Ghost kitchens are altering the competitive landscape of the food service sector. These kitchens are allowing smaller operators and startups to compete with big brands by lowering the hurdles to entry and encouraging innovation.

Expectations of Consumers

For Ghost kitchens to be successful, they must meet the ever-changing expectations of customers about the convenience, timeliness, and quality of their services. The fact that consumers today place a higher priority on delivery services that are dependable and smooth makes it very necessary for operators to concentrate on efficiency and customer satisfaction.

In the Future, Research will Focus on

Further study is necessary to investigate the long-term viability, scalability, and influence of technical improvements, legislative shifts, and consumer trends on the development of Ghost kitchens. For the future food service business to thrive, addressing new difficulties and maximizing the emerging possibilities will be essential.

DISCUSSIONS

Changes in Ghost Kitchens and Their Growing Popularity

A significant paradigm change is taking place in the food service business due to the fast development of Ghost kitchens. The industry's response to shifting customer needs and technical improvements is seen in the change from traditional centralized kitchens to delivery service models. The provision of a solid infrastructure for order administration and delivery logistics has been made possible by digital food delivery platforms, which have allowed for the acceleration of this shift. Convenience, speed, and digital integration are becoming increasingly popular today, and their progression mirrors those tendencies.

The Integration of Technology and the Efficiency of Daily Operations

Incorporating cutting-edge technology like automation, artificial intelligence, and the Internet of Things into Ghost kitchens has completely transformed preparing and delivering meals. While artificial intelligence and data analytics optimize inventory management, menu offers, and delivery logistics, automation has the effect of lowering labor costs and minimizing the impact of human error. Internet of Things-enabled equipment improves real-time monitoring and control, which in turn ensures the quality and safety of food. These technological improvements are essential to preserving a high level of operational efficiency and satisfying the growing demand for food delivery services that are both dependable and speedy.

The Advantages and Difficulties of Utilizing Ghost Kitchens

Ghost kitchens provide many advantages, including cost-effectiveness, scalability, and environmental sustainability. They can lower both their operating expenses and their financial risk by doing away with the requirement for costly real estate and many front-of-house employees. By maximizing the utilization of resources and reducing the amount of wasted food, the centralized production model helps to preserve the environment. Nevertheless, ensuring that the food quality is constant throughout several different sites continues to be a considerable difficulty. Standardized recipes and procedures are necessary to maintain uniformity and satisfy consumers' expectations. It is essential to have excellent online branding and marketing tactics since the lack of a physical eating area influences the business's perception and the customer's loyalty.

Considerations Regarding Legality and Compliance with Regulations

Ghost kitchens face a difficult challenge when navigating the many health and safety regulations across different locations. Consistent adherence to local rules is necessary to preserve the integrity of operations and stay out of legal trouble. Operators must be at the forefront of changes in the regulatory framework and ensure that they conform with health and safety standards, zoning rules, and food safety requirements as the system advances. The current regulatory environment presents Ghost Kitchens with several obstacles as well as possibilities since there is the potential for compliance to increase credibility and consumer confidence.

Dynamics of the Workforce and the Gig Economy Workforce

As a result of the influence that automation has had on employment in the food service industry, there are worries over job displacement and the requirement for programs that provide skill retraining. Even though automation lowers labor costs and boosts efficiency, it still requires a focus on workforce sustainability. It is essential for the expansion of Ghost kitchens that the gig economy, and more specifically, the reliance on delivery drivers, take center stage. To protect gig workers and ensure the continued viability of this employment model, rules and laws about fair labor practices are required.

The Perceptions of Consumers and Various Marketing Strategies

Regarding the success of Ghost Kitchens, the perception and acceptability of consumers are pretty important. Several elements, including transparency in business operations, quality assurance, and effective communication, may significantly impact a consumer's trust. The happiness and loyalty of customers can be increased by providing them with information about the advantages and precautions that Ghost Kitchens offer. To construct and sustain a loyal consumer base for your brand, it is vital to have effective branding and marketing tactics highlighting these features. Marketing strategies must include digital engagement and transparent communication to succeed in the highly competitive food service business.

The Opportunities and Trends of the Future

The Ghost kitchen industry has the potential for expansion and innovation, as seen by new trends such as virtual branding, collaborative cooking spaces, and integration with smart city infrastructure. Virtual brands allow experimenting with different culinary techniques without physical limitations. At the same time, collaborative spaces provide potential for innovation and cost savings through shared facilities. It is possible to optimize delivery routes and minimize traffic congestion by integrating with smart city infrastructure, which will ultimately improve service delivery's overall efficiency and sustainability. In the highly competitive market for food service, personalization and customization, made possible by artificial intelligence and machine learning, are becoming crucial differentiators. These technologies offer highly individualized eating experiences based on the tastes of individual customers.

Impact on the Environment and Long-Term Sustainability

Ghost kitchens contribute to the preservation of the environment by maximizing the utilization of resources and minimizing the amount of wasted food. Because of their centralized manufacturing strategy, they can improve inventory management and reduce energy use, which helps them achieve their environmental goals and appeal to environmentally sensitive consumers. Ghost kitchens are in an excellent position to fulfill the demands of customers and authorities increasingly concerned about their environmental impact. This is because of Ghost Kitchens' inherent efficiency and resource optimization.

Effects on Conventional Dining Establishments and the Dynamics of the Market

Ghost kitchens are becoming increasingly popular, which has repercussions for conventional dining facilities. These kitchens provide an alternate model that can coexist with traditional dining venues. Without needing extra physical sites, Ghost kitchens offer supplementary services such as delivery and catering to customers. This strategy encourages innovation and expansion, which drives economic development in the food service industry. By lowering the obstacles to entry and encouraging innovation, Ghost kitchens are reshaping the competitive environment. This makes it possible for smaller operators and startups to compete with major brands.

Continuing Research and Environmental Stewardship

Whether Ghost kitchens can be scaled up and maintained over the long term should be the subject of future study. Investigating the influence of technical breakthroughs, regulatory policy changes, and customer behavior trends will be essential to confront new difficulties and make the most of new possibilities. To fully realize the potential of Ghost Kitchens and guarantee its long-term success in the global foodservice scene, it will be vital to continue research and innovation. The equilibrium between technical progress, regulatory compliance, worker sustainability, and customer happiness will heavily influence the direction Ghost Kitchens will take in the future.

Ghost kitchens are a disruptive movement in the food service business. This transition is driven by technical innovation and shifting consumer needs. Even though they have several significant benefits, they confront obstacles in quality control, brand perception, regulatory compliance, and workforce dynamics. To solve these problems and fully realize the promise of Ghost Kitchens, it will be necessary to do ongoing research, innovate, and manage strategically. Operators must maintain their flexibility and forward-thinking mindset while the sector changes. They must use technical improvements and customer insights to propel growth and ensure sustainability.

RECOMMENDATIONS

1. The operators of Ghost Kitchens should continue investing in cutting-edge technology such as automation, artificial intelligence (AI), and the Internet of Things (IoT). Implementing these technologies increases operational efficiency, enhances food safety, and results in AI and data analytics, which can optimize inventory management and forecast client preferences. At the same time, automation can simplify food preparation and minimize labor costs. Both benefits can be achieved through automation. Consumer electronics that are enabled by the Internet of Things (IoT) offer real-time monitoring and control, which further ensures the dependability and safety of food items (Berman & Macdonald, 2019; Wang et al., 2022).

2. Quality control measures should be improved. Ghost kitchens should apply robust quality control procedures to meet the difficulty of maintaining consistent food quality across numerous sites because of the widespread availability of food. The standardization of cooking procedures and recipes is necessary to achieve uni-

formity. To assist in maintaining high standards, it is helpful to have regular training programs for kitchen personnel as well as frequent quality audits. It is also possible to guarantee that the food satisfies the needed standards and the expectations of the customers by utilizing technology for real-time quality monitoring and feedback (Yeo & Wahid, 2020; Green & Brown, 2020).

3. Ghost restaurants must concentrate on developing and sustaining client loyalty by implementing powerful online branding and marketing techniques. This is because they do not have an actual eating place. It is possible to improve consumers' perceptions of a brand by effectively using social media, producing engaging content, and communicating openly about the quality and safety of food. It can also increase awareness and client confidence through collaborations with well-known food delivery services and influential individuals. According to Kim and Park (2021) and Singh and Jain (2022), to attract and keep clients, it is beneficial to highlight the advantages of Ghost kitchens, including the convenience, speed, and sustainability of the kitchens.

4. The ability to successfully navigate the intricate regulatory landscape is essential for the success of Ghost Kitchens. Operators must maintain awareness of local health and safety rules, zoning laws, and food safety quality requirements. Establishing a compliance team solely committed to compliance or working with legal professionals can help ensure that all applicable requirements are followed. Keeping operators up to speed on regulatory changes and best practices may also be accomplished by proactively contacting regulatory agencies and participating in industry forums (Martin & Liu, 2021; Smith & Thompson, 2023).

5. The importance of focusing on workforce sustainability and fair labor standards cannot be overstated because automation and the gig economy are becoming increasingly integrated into Ghost kitchen operations. Providing displaced workers with the opportunity to enhance their careers and participate in retraining programs can assist them in making the move to new positions. To keep a dependable workforce, it is essential to ensure that gig workers, particularly delivery drivers, are paid fairly and provided with favorable working conditions. According to Harris and Gomez (2021) and Lee and Hwang (2022), implementing policies that encourage work-life balance and employee well-being can also boost worker satisfaction and retention.

Figure 1. Conceptual framework (Author compilation)

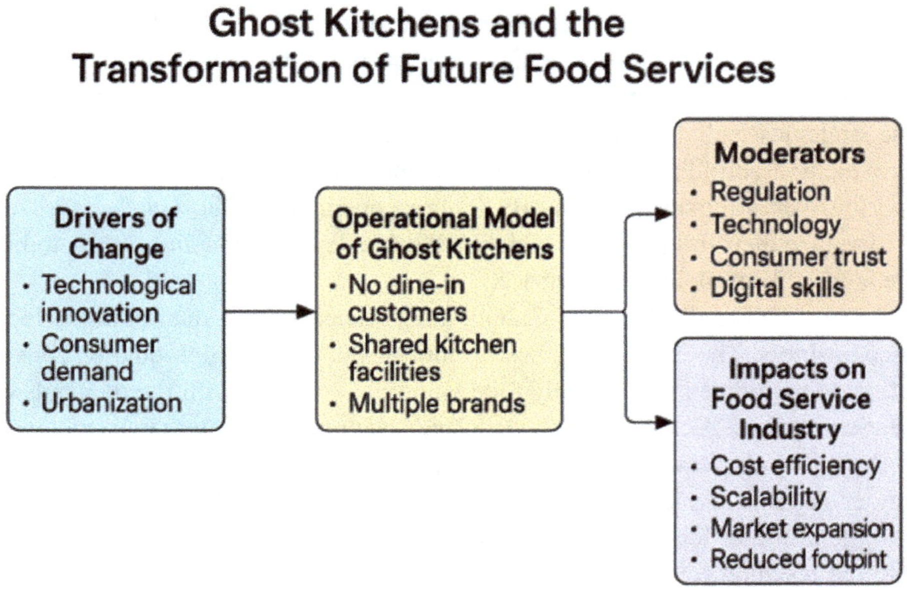

CONCLUSION

Technological advancements and the ever-changing requirements of customers are driving this revolutionary change in the food service business. Ghost kitchens reflect this transition. Using cutting-edge technology such as automation, artificial intelligence, and the Internet of Things, these kitchens provide several essential benefits, including cost-effectiveness, scalability, and environmental sustainability. The proliferation of digital platforms for food delivery has further hastened their acceptance, making it possible for operators to satisfy the modern consumer's need for convenience and speed.

However, several constraints must be solved to bring about the effective deployment of Ghost kitchens. The preservation of constant food quality over several different sites, the guarantee of regulatory compliance, and the cultivation of a staff that can be maintained over time are all essential to their long-term survival. Without a physical eating place, it is equally necessary to have strong online branding and marketing techniques to establish and maintain client loyalty. Considering the findings of the literature study, it is clear that there is a pressing requirement for continuous research and innovation to investigate the long-term viability and scalability of Ghost kitchens. The influence of technical improvements, changes in regulatory policies, and trends in consumer behavior should be the primary focus of research in the future. This research should also address growing difficulties and capitalize on new possibilities. Operators may be assisted in navigating the intricacies of this new paradigm by putting into action the tactics that have been advised. These strategies include investing in sophisticated technology, improving quality control methods, building robust branding, maintaining regulatory compliance, and supporting workforce sustainability. By doing so, operators of Ghost kitchens will be able to realize their full potential, which will drive development and sustainability in the landscape of global food service.

Ghost kitchens will likely bring about a significant transformation in the food service business. The capacity of these companies to provide high-quality, convenient, and environmentally friendly food solutions is in line with the demands of current consumers. As a result of the industry's evolution, operators must maintain their flexibility and forward-thinking and capitalize on technical improvements and customer insights to succeed. Ghost kitchens have the potential to play a crucial role in the future of food service, providing a viable and sustainable alternative to traditional restaurant models. This potential may be realized via innovative management and continual innovation.

REFERENCES

Anderson, R., & Kumar, A. (2021). Scalability in the food service industry: The rise of Ghost kitchens. *International Journal of Hospitality & Tourism Administration*, *22*(1), 45–62. DOI: 10.1080/15256480.2020.1831485

Berman, M., & Macdonald, A. (2019). Integrating AI and Automation in Food Service Operations. *Journal of Culinary Science & Technology*, *17*(1), 30–45. DOI: 10.1080/15428052.2018.1554200

Cousins, J., Foskett, D., & Pennington, A. (2002). *Food and Beverage Service*. Routledge., DOI: 10.1201/9780203393582

. Das, I. R., Talukder, M. B., & Kumar, S. (2024). Implication of Artificial Intelligence in Hospitality Marketing. *Utilizing Smart Technology and AI in Hybrid Tourism and Hospitality*. IGI Global, USA. DOI: 10.4018/979-8-3693-1978-9.ch014

Davis, E., & Taylor, G. (2021). Collaborative kitchen spaces: Reducing costs and fostering innovation. *International Journal of Gastronomy and Food Science*, *24*, 100340. DOI: 10.1016/j.ijgfs.2021.100340

Garcia, L., & Chen, Y. (2022). Integration with Smart City Infrastructure: Opportunities for Ghost Kitchens. *Smart Cities*, *5*(2), 320–332. DOI: 10.3390/smartcities5020019

Garcia, P., & Lee, D. (2020). Sustainable food systems in urban economies: The rise of delivery-only kitchens. *Sustainability*, *12*(19), 7889. DOI: 10.3390/su12197889

Garvey, M., & Lee, C. (2020). Environmental sustainability in the cloud kitchen model. *Sustainability*, *12*(8), 3234. DOI: 10.3390/su12083234

Green, P., & Brown, R. (2020). Innovations in Food Service Technology: Impact on Ghost Kitchens. *Journal of Foodservice Business Research*, *23*(1), 5–18. DOI: 10.1080/15378020.2020.1712582

Harris, M., & Gomez, R. (2021). Workforce Dynamics in the Gig Economy: Implications for Ghost Kitchens. *Human Resource Management Review*, *31*(2), 100743. DOI: 10.1016/j.hrmr.2020.100743

Johnson, E., & Miller, T. (2022). Environmental Impact and Sustainability in Food Service Operations. *Environmental Science & Policy*, *135*, 127–139. DOI: 10.1016/j.envsci.2022.06.006

Johnson, R., & Green, M. (2023). The evolving regulatory framework for Ghost kitchens. *Regulation & Governance*, *17*(3), 287–304. DOI: 10.1111/rego.12358

Jones, K., & Taylor, M. (2023). Meeting evolving consumer expectations with Ghost kitchens. *Journal of Service Management, 34*(2), 115–130. DOI: 10.1108/JOSM-11-2022-0445

Jones, M., & Miller, B. (2021). Operational challenges in cloud kitchen models: Staffing, logistics, and brand management. *Journal of Operations in Hospitality, 5*(2), 92–105.

Jones, P., & Lee, R. (2018). Branding and Customer Loyalty in Food Delivery Services. *International Journal of Hospitality Management, 75*, 98–107. DOI: 10.1016/j.ijhm.2018.03.009

Kim, J., & Park, S. (2021). Building consumer trust in cloud kitchens through transparency. *Journal of Consumer Behaviour, 20*(5), 978–990. DOI: 10.1002/cb.1910

Kim, S., & Lee, D. (2023). Long-Term Sustainability and Scalability of Ghost Kitchens. *Food Policy, 112*, 102468. DOI: 10.1016/j.foodpol.2023.102468

Kim, S., & Park, M. (2021). Digital transformation in the food service industry: The role of automation and delivery platforms. *International Journal of Contemporary Hospitality Management, 33*(6), 1803–1821. DOI: 10.1108/IJCHM-10-2020-1133

Kumar, P., & Sinha, M. (2021). IoT and Food Safety: The Future of Ghost Kitchens. *International Journal of Food Science & Technology, 56*(4), 1022–1030. DOI: 10.1111/ijfs.14925

Kumar, S., Talukder, M. B., Kabir, F., & Kaiser, F. (2024). Challenges and Sustainability of Green Finance in the Tourism Industry: Evidence from Bangladesh. In Taneja, S., Kumar, P., Grima, S., Ozen, E., & Sood, K. (Eds.), (pp. 97–111). Advances in Finance, Accounting, and Economics. IGI Global., DOI: 10.4018/979-8-3693-1388-6.ch006

Kumar, S., Talukder, M. B., & Pego, A. (Eds.). (2024). *Utilizing Smart Technology and AI in Hybrid Tourism and Hospitality*. IGI Global., DOI: 10.4018/979-8-3693-1978-9

Kumar, V., & Sinha, S. (2021). AI in food service: Enhancing operational efficiency through data-driven insights. *International Journal of Hospitality Management, 92*, 102772. DOI: 10.1016/j.ijhm.2020.102772

Lee, J., & Hwang, S. (2022). Fair Labor Practices in Food Delivery Services. *Labor Studies Journal, 47*(1), 22–36. DOI: 10.1177/0160449X211056865

Li, Y., & Zhao, Q. (2023). Market testing and experimentation in Ghost kitchens. *Journal of Business Research, 148*, 347–359. DOI: 10.1016/j.jbusres.2022.07.016

Martin, D., & Liu, Y. (2021). Regulatory challenges for Ghost kitchens: Navigating local health standards. *Food Policy*, *99*, 101936. DOI: 10.1016/j.foodpol.2021.101936

Oliveira, C., Santos, J., & Freeman, K. (2022). Customer satisfaction and service quality in virtual food service businesses. *International Journal of Hospitality Management*, *102*, 103146. DOI: 10.1016/j.ijhm.2022.103146

Pariza, M. W. (2020). The Evolution and Impact of Ghost Kitchens. *Food Service Technology*, *20*(3), 150–160. DOI: 10.1002/fst2.245

Pariza, T. (2020). The rise of cloud kitchens: How digital platforms are changing the food delivery landscape. *Journal of Culinary Science & Technology*, *18*(3), 278–295. DOI: 10.1080/15428052.2020.1710767

Poon, A., & Lam, T. (2021). Educating consumers about Ghost kitchens: Enhancing acceptance and satisfaction. *Food Quality and Preference*, *92*, 104207. DOI: 10.1016/j.foodqual.2021.104207

Roberts, A., & Scott, R. (2020). Virtual brands and the future of food delivery. *Journal of Foodservice Business Research*, *23*(1), 45–59. DOI: 10.1080/15378020.2020.1713764

Sharma, L., Banerjee, R., & Gupta, A. (2023). Consumer perception and satisfaction in ghost kitchen experiences: A delivery-focused study. *Journal of Consumer Behaviour*, *22*(1), 112–127. DOI: 10.1002/cb.2001

Singh, R., & Jain, P. (2022). Effective branding strategies for cloud kitchens. *Marketing Intelligence & Planning*, *40*(2), 167–182. DOI: 10.1108/MIP-05-2021-0263

Smith, A., & Thompson, J. (2023). Legal and regulatory challenges in operating Ghost kitchens. *Food Law Review*, *14*(2), 183–200. DOI: 10.2139/ssrn.3467354

Smith, J., & Thompson, K. (2023). Navigating Local Food Safety Regulations: A Guide for Ghost Kitchens. *Journal of Environmental Health*, *85*(2), 45–55. DOI: 10.1080/00207233.2022.2120456

Smith, L., Robinson, K., & Lee, J. (2019). Enhancing convenience and speed in food delivery services. *Journal of Service Research*, *21*(4), 345–359. DOI: 10.1177/1094670519845789

Talukder, M. B. (2020). The Future of Culinary Tourism: An Emerging Dimension for the Tourism Industry of Bangladesh. I-Manager's. *Journal of Management*, *15*(1), 27. DOI: 10.26634/jmgt.15.1.17181

Talukder, M. B. (2024). Implementing Artificial Intelligence and Virtual Experiences in Hospitality. In *Innovative Technologies for Increasing Service Productivity* (pp. 145–160). IGI Global., DOI: 10.4018/979-8-3693-2019-8.ch009

Talukder, M. B., & Kumar, S. (2023). The Effect of Food Service Quality on Customer Satisfaction in the Hotel Industry: A Conceptual Paper. *ResearchGate,* Available: https://www.researchgate.net/publication/371503829_The_Effect_of_Food_Service_Quality_on_Customer_Satisfaction_in_the_Hotel_Industry_A_Conceptual_Paper

. Talukder, M.B., & Kumar, S. (2024c). Revisiting intention in food service outlet of five-star hotels:A quantitative approach based on food service quality. Sport i Turystyka. Œrodkowoeuropejskie Czasopismo Naukowe, 7(1), 137–156. DOI: 10.16926/sit.2024.01.08

Talukder, M. B., Kumar, S., & Das, I. R. (2024a). *Food Wastage on the Economic Outcome: Evidence From the Hotel Industry. Sustainable Disposal Methods of Food Wastes in Hospitality Operations.* IGI Global., DOI: 10.4018/979-8-3693-2181-2.ch005

Talukder, M. B., Kumar, S., & Das, I. R. (2024b). Mindful Consumers and New Marketing Strategies for the Restaurant Business: Evidence of Bangladesh. In Ramos, C., Costa, T., Severino, F., & Calisto, M. (Eds.), *Social Media Strategies for Tourism Interactivity* (pp. 240–260). IGI Global., DOI: 10.4018/979-8-3693-0960-5.ch010

Talukder, M. B., Kumar, S., & Das, I. R. (2024). Perspectives of digital marketing for the restaurant industry. In *Advancements in socialized and digital media communications* (pp. 118–134). IGI Global Scientific Publishing.

Talukder, M. B., Kumar, S., Sood, K., & Grima, S. (2023). Information Technology, Food Service Quality and Restaurant Revisit Intention. *International Journal of Sustainable Development and Planning*, *18*(1), 295–303. DOI: 10.18280/ijsdp.180131

Tan, E., & Wilson, G. (2023). Regulatory gaps and health compliance in the ghost kitchen industry: A global perspective. *Journal of Foodservice Business Research*, *26*(1), 14–30. DOI: 10.1080/15378020.2022.2109876

Thompson, J., & Miller, A. (2023). Innovation and Growth in the Food Service Industry: The Role of Ghost Kitchens. *Journal of Foodservice Management*, *25*(2), 75–89. DOI: 10.1080/23356460.2023.2161231

Wang, H., Li, J., & Zhang, Y. (2022). IoT-enabled appliances in modern kitchens: Enhancing food safety and quality. *Journal of Food Engineering*, *293*, 110338. DOI: 10.1016/j.jfoodeng.2020.110338

Williams, L., & Anderson, C. (2023). Ghost kitchens and their impact on traditional restaurant models. *International Journal of Hospitality Management*, *107*, 103275. DOI: 10.1016/j.ijhm.2023.103275

Yeo, M., & Wahid, N. (2020). Cost efficiency in cloud kitchens: An economic analysis. *Foodservice Business Review*, *15*(2), 125–136. DOI: 10.1177/2158244019834237

Zhang, H., Kumar, R., & Thomas, A. (2022). The rise of ghost kitchens: Business model innovation and disruption in the food service sector. *Journal of Hospitality and Tourism Technology*, *13*(4), 523–538. DOI: 10.1108/JHTT-03-2022-0045

Zhang, X., & Li, Z. (2023). Economic impacts of Satelite kitchens: A case study. *Economics & Business Review*, *13*(1), 68–82. DOI: 10.18559/econbusrev.2023.1.5

Chapter 4 Automation in Ghost Kitchens: Exploring the Role of Robotics in Revolutionizing Food Service Operations

Gurucharan Singh
https://orcid.org/0009-0006-3432-6270
Maharishi Markandeshwar University, Mullana, India

Rakesh Sharma
https://orcid.org/0000-0002-3762-1500
Maharishi Markandeshwar University, Mullana, India

ABSTRACT

Ghost kitchens, sometimes called cloud or dark kitchens, are a fast growing subset of the food service industry that focusses solely on delivery and does not have any dining areas. These facilities offer a cost-effective and scalable strategy to meet the growing demand for online meal delivery. However, they face substantial challenges, such as the need for stringent quality control and sanitary measures, intense operating pressure, and a lack of available workers. In order to tackle these difficulties, automation—especially through robotics—is becoming an essential tool for creativity and efficiency. In this chapter, we look at how robotics is changing the face of ghost kitchen operations by automating mundane but necessary tasks like chopping, cooking, plating, packaging, and cleaning. Autonomous cooking systems, robotic arms, and AI-driven inventory and order management technologies are all investigated. This chapter delves into how modern technologies are changing the nature of the workforce, looking at the pros and cons of job loss and new skill acquisition.

DOI: 10.4018/979-8-3373-4042-5.ch004

INTRODUCTION

The hospitality and food service sector in the world is facing a paradigm shift, most of it being as a result of technological advancement and evolving tastes and preferences of consumers. One of the most important innovations is the emergence of ghost kitchens or cloud kitchens, virtual kitchens, or dark kitchens. These restaurants have no physical restaurant where people can dine but solely prepare food to be delivered using digital applications (Sigala, 2018). The model has seen a quick spread, especially in urban markets, where the increased demand of online delivery food services are matched with limitations of high real-estate prices and the changing life style to convenience-based consumption (OECD, 2021). This trend has increased with the growing involvement of robotics in ghost kitchens. Robotics is being implemented in a variety of processes, including food preparation, packaging, visual inspection and even in the last-mile delivery process. Robots can be used to repeat and complete tasks precisely and quickly, which means that reliance on human labor is lessened, that fewer errors occur, and that food is hygienic (Ivanov and Webster, 2020). As an example, burgers, fry foods, or beverage dispensing are assembled by the help of robotic arms, customer experience is being improved by autonomous delivery robots and drones, which are being tested in metropolitan areas (Cohen, 2021). Not only is the dynamism of the operations being transformed by these advancements, but they are also transforming the expectation of customers of quality of service in a highly competitive food delivering market.

The implementation of robotics in ghost kitchens was accelerated by the COVID-19 pandemic even more. Ghost kitchens were a savior to restaurants and business owners amid the limitation of dine-in services, providing hygienic and contactless services (Zhang et al., 2023). Robotics was critical in combating the wave of delivery demands as well as in maintaining compliance to protective measures and labor shortages due to lockdowns. Automation has become central to the post-pandemic recovery as companies aim to find a strong and scalable model to meet the changing conditions of the market (Kraus et al., 2022). Economically, robotics can save on labour costs and maximize on resources. Robot-powered ghost kitchens also tend to employ fewer workers than a conventional restaurant, thus being economical to investors and operators (Choudhury et al., 2021). Additionally, robotic technologies enable operators to ensure the quality of the products remains consistent, and this aspect is becoming more significant when it comes to sustaining brand loyalty in the digital market. Meanwhile, implementation of robotics also adds some problems concerning costs, recalibration of workers, standardization of technologies, and perceptions of robot-ready food in the consumer market (Gursoy et al., 2019).

Ghost kitchens have added to the greater discourse of smart hospitality and service automation on an academic level, including the role of robotics. Researchers have stressed that robotics do not only drive efficiency but change the very notion of value creation by creating new means of interaction between technology, service providers, and consumers (Buhalis and Amaranggana, 2015; Gretzel et al., 2015). The main part of a digitally controlled ecosystem of the specific example of ghost kitchen is robotics, supported through the assistance of artificial intelligence (AI), the Internet of Things (IoT), and cloud-based management tools. In total, all of these inventions will facilitate smarter inventory control, predictive analytics demand forecasting, and monitoring the real-time kitchen processes (Celtek, 2023). This chapter seeks to analyze critically the role of robotics in ghost kitchens, their opportunities and issues. It scouts the technological landscape, the financial factors, the consumer behaviour and the prospect of robotics-based food service ideas. When referring to ghost kitchens in the backdrop of the overall digital transformation of the hospitality industry, the chapter can be positioned in the literature that is increasingly interrelated in terms of robotics, smart technologies, and the changing food service industry. The necessity to adopt sustainable practices, develop workforce, and policy interventions to make sure that the adoption of robotics will benefit every stakeholder is also discussed.

Figure 1. Cloud kitchen

The above image is the simplified structure of a cloud kitchen broken down into six interdependent process which are kitchen, online order, quality control, reduce cost, customer data, and delivery. This systematic stream is one of the reasons why ghost kitchens exist in a high-tech interconnected food service environment with rapid turnover. Especially with the help of automation and robotics, each stage contributes to the efficiency in operations, customer satisfaction, and scalability of businesses (Goh & Lee, 2020).

Kitchen

The rear-of- house cooking kitchen of any cloud kitchen is its keystone. In contrast to the traditional restaurants, this would simply be a food preparation and packaging facility which lacks the elements that are related to customers such as the dining rooms, waiting areas, and reception desks. The kitchen is made as small as possible, modular and shared with other various virtual brands, also known as multi-brand kitchens (Bechar & Vigneault, 2017). And in this specific case robotics will play a crucial role; robotic cooking machines such as robotic fryers, robotic grills and robotic stirrers will allow to standardise complex meal preparation, maintain food hygiene and reduce cook time by a factor of hundreds.

Online Order

Cloud kitchens are deeply integrated in the operational ecosystems of digital ordering systems. Customers place their orders solely through mobile applications, websites, or third-party aggregators such as Zomato, Swiggy, Uber Eats, or DoorDash. Companies are able to gather and manage large volumes of orders concurrently thanks to its digital-first interface, which allows them to do so even across multiple brands within a single kitchen. The use of artificial intelligence in order management systems allows for the prediction of preparation times, the assignment of orders to the appropriate kitchen stations, and the notification of delivery teams. Due to the absence of conventional point-of-sale systems, cloud kitchens are able to freely experiment with various marketing methods, loyalty programs, and real-time customer interaction, all of which are dependent on available web platforms (Chen & Wu, 2023).

Quality Control

Maintaining a consistent level of quality, taste, and hygiene is of the utmost importance, particularly in situations where orders are fulfilled without direct touch with the customer. Automation helps to increase quality control by doing tasks like as precise cooking, controlling portion sizes, and monitoring temperature through the use of embedded sensors and smart devices. Using vision systems that are powered by artificial intelligence, it is possible to detect defects in plating, packaging, or portioning, thereby ensuring that the food satisfies the recognised brand specifications. One of the advantages of robotics is that it eliminates the possibility of human error. This is especially beneficial during times of high activity, when hand fatigue could potentially affect quality (Wang & Zhao, 2022).

Reduce Cost

The cloud kitchen concept offers a number of advantages, one of the most significant being cost effectiveness. Through the elimination of client service areas, the reduction of labour requirements, and the improvement of inventory control, these kitchens significantly reduce operating expenses. Automation not only helps to reduce the amount of labour that is required, but it also helps to reduce operating expenses that are associated with salaries, training, and human error. In addition, cloud kitchens have the advantage of centralised purchasing and production, which, when paired with data analytics, helps to reduce the amount of food that is wasted and guides decisions about the supply chain (Chen & Wu, 2023).

Customer Data

Favourite cuisines and frequency of ordering to geographic demand and preferences when it comes to delivery, digital ordering enables cloud kitchens to gather immense consumer data. Artificial intelligence and machine learning tools allow one to analyse this data to guide hyper personalised marketing campaigns and pricing and product development. Dynamic menu, cross selling opportunities and real time feedback loops can also be used with the knowledge of consumer behaviour. This information can be used to adjust the amount of time, amount of spice, or amount of food that the robotics systems prepare to a specific client profile and hence provide a more unique dining experience.

Delivery

Final cloud kitchen value chain is last-mile delivery, and this step determines how happy the clients will be significantly. Alterations in the redefinition of food delivery to consumers are automated packing lines, smart food lockers, and integration of drone or delivery robots. Error-free and swift hand-off of orders ensured through automation lowers the number of handling mistakes and the time taken to wait. Advanced configurations of robotic systems coordinate delivery lines, sort batches of orders, and control dispatch times to maximise fleet use and improve delivery efficiency.

One of the key causes of early acceptance of automation in food service activities was the beginning of the COVID-19 outbreak. With hygiene concerns first and social distances restrictions guiding, contactless service methods became top priority. Ghost kitchens' delivery-centric approach helped them to meet this demand. The addition of robotics enhanced their ability to run safely and efficiently in a small area (Goh & Lee, 2020). Emerging as a practical solution to ensure both production and public health are automated kitchens, where machines handle most food preparation and handling tasks. Apart from aspects linked to pandemics, robotics investment has more technological and financial justification (Chen & Wu, 2023). First of all, a scarcity of skilled chefs, rising pay demands, and high turnover rates define the worldwide food service labour market as it is under severe duress. Robotics offers a way to lessen these pressures by helping to consistently and without exhaustion do labour-intensive tasks. Second, by running nonstop without stops, robots reduce downtime and boost throughput. These characteristics can significantly raise the quality of services in environments like high-volume delivery operations where speed and precision are quite crucial (Choudhury, 2019).

Besides, robotics and artificial intelligence introduce a different perspective on data-driven decision-making. Smart kitchens are able to aggregate and analyze large volumes of data about consumer preferences, consumption of food items, equipment usage, and data on deliveries. Based on this, it is possible to predict the demand, optimize the distribution of resources, and tailor offers to specific consumers (Patel and Kumar, 2024). Besides, artificial intelligence and data analytics are the sources of the problem of ethical use of customer data, algorithmic bias, and privacy of the data. Proper governance frameworks should be developed to facilitate the preservation of consumer trust and patronage of digital responsibility (Wang and Zhao, 2022). One more important factor is the legal dimension governing the use of robots in commercial kitchens. The rules of responsible and safe use of automation technologies should be developed by food safety authorities, labour boards, technology regulators, and other parties collaborating (Chen and Wu, 2023). All policies on quality control, fail-safe conditions, hygiene of the machine and liability must be considered in the event of a problem. Clarity of regulation will be totally essential to foster innovation and self-confidence in the stakeholders as ghost kitchens are increasingly dependent on autonomous technology.

The chapter attempts to explore the increasing intersection of ghost kitchens and robotic automation in detail. It starts with highlighting the reasons to create ghost kitchen and proceeds with delineating their basic ideas and working principles. It then considers the various types of automation currently in operation, ranging all the way to robotic kitchen appliances to artificial intelligence driven inventory management systems. The chapter illustrates how important operational issues are being resolved using technology through the use of sweeping case studies and real-life examples (Davies and Williams, 2022). The following chapter focuses on the undoubted advantages of automation increased efficiency, higher food safety, and increased scalability. It also addresses the practical obstacles to adoption such as technological challenge, resistance to change and high start-up costs. The importance of workforce change is especially drawn to in analysing how human jobs are changing in reaction to mechanisation and what skills will be required in the automated kitchen of the future. Going forward, the chapter provides a visionary outlook of the coming generation of ghost kitchens facilities which are fully automated, with integrated AI, as well as with the ability of operation on a 24/7 basis with minimum human input. It analyses modern concepts such as virtual and augmented reality-enabled immersive ordering experiences, predictive maintenance, and cobots that is, modern concepts. Blending current trends and emerging opportunities will assist the chapter to paint a complete picture of what the future of food service may be. Finally, the automation of ghost kitchens marks the beginning of a new era in the development of the hotel and restaurant industries (Pires and Silva, 2021). It captures a more of a societal drift to technical integration, performance in the work place and comfort in the digital world. Since we are on the verge of the Fourth Industrial Revolution, it is not only vital but necessary to comprehend the role that robots play in the process of revolutionising the cooking methods. This chapter provides some of the simplest research on that transition, as well as recommendations and resources to allow players to use automation to its fullest potential to create a more resilient, efficient and creative food environment.

KEY TECHNOLOGICAL INTERVENTIONS IN GHOST KITCHENS

Ghost kitchens (also known as cloud kitchens or dark kitchens) have been disruptive to the global food service industry by taking advantage of technology to streamline production, distribution, and experiences of customers. Ghost kitchens have no physical restaurants and are extremely dependent on online platforms and logistics of delivery, unlike traditional restaurants. The main driver of these operations has been robotics and automation which is redefining how food is prepared, packaged and delivered. There are a number of technological interventions that characterize the success of ghost kitchen such as robotic cooking arms or artificial intelligence (AI) and Internet of Things (IoT).

Robotic Cooking Arms and Automated Food Preparation

The most natural application of robotics in ghost kitchens is the use of robotic cooking arms capable of cooking with the minimal human intervention. Companies such as Miso Robotics in its Flippy robot have already altered the restaurant format of quick-service by mechanizing processes such as frying, grilling, and assembling food (Liu et al., 2024). These robot systems enhance efficiency of operations by being consistent in tastes, portion control and level of hygiene. They also reduce the use of human labor, which has turned out to be less and more costly particularly following the pandemic (Celtek, 2023). These kinds of robotic systems are paired with digital menus in ghost kitchens to dynamically optimize the utilization of ingredients and cooking time depending on real-time flows of orders. In addition to efficiency, the robotic arms in cooking minimize wastage of food by measuring and monitoring food systems which adjust to inventory. They also promote safety as they reduce the risks of burns and accidents inherent in the traditional kitchen work and are therefore particularly handy in large volume kitchens. In addition, the other systems also possess long services, which are not tired and thus 24-hours services in delivery-based models. With the ongoing evolution of AI and machine learning, the new generation of robotic cooking arms is not only capable of repeating specific actions but should also adapt recipes, personalize dishes, and adjust them to diet requirements. Robotics adoption in ghost kitchens will also grow in future as the demands of consumers become faster, more hygienic and personalized, which changes the future of the food service.

Artificial Intelligence and Machine Learning

Ghost kitchens make use of AI and machine learning and this is the core of predictive analytics. Using the analysis of consumer behavior and previous sales and the impact of external factors, e.g., the weather or local events, AI algorithms can deduce the trends and improve the management of the inventory (Buhalis, 2020). This minimizes the level of food wasted and ensures that when their kitchens become busy, they will have sufficient to serve their clients. One can also engage AI-based recommendation systems to tailor the experience of customers on delivery apps such as Zomato, Swiggy, or Uber Eats by offering meal suggestions based on their diet, previous orders, or health (Gretzel et al., 2015). The operational side of any ghost restaurant can be streamlined with AI technology to reduce robotic workflow scheduling, minimise downtime, and improve throughput. In addition to that, there is an increasing use of AI in dynamic pricing systems, as a result of which restaurants will have the opportunity to set menu prices dynamically based on demand, competition, and ingredient availability. Natural language processing (NLP) and chatbot are also used to help in customer interaction and this makes their queries, ordering changes and complaints more efficient and hence customers are more satisfied. Besides, machine learning is applied to detect fraudulent transactions and safe online payment, which is a priority in the online delivery ecosystem. With a more developed type of a system, in the long-term, it will be more than optimizing, and will be able to act more as an autonomous decision-maker, such that ghost kitchens will be able to self regulate, predict new food trends, and introduce new menu items with minimum human intervention. Not only are these innovations more profitable but ghost kitchens are also the first to be more data-driven as well as customer-focused.

Internet of Things (IoT) and Smart Sensors

The IoT technologies can be used to integrate ghost kitchens to operate as ecosystems. Smart refrigerators, ovens, and storage units have sensors that monitor the temperature and humidity and inventory in real-time (OECD, 2021). Through these sensors, maintaining food safety and regulatory compliance is easily achieved whereby staff are alerted about anomalies in time before it escalates. One more is energy management system that works based on IoT and reduces operating expenses due to the effective use of electricity and water. IoT will also create seamless communication between all the devices present in robotized kitchens, causing them to work together, and hence the cooking robots, packaging system, and the delivery drones will not experience any inconveniences (Neuhofer et al., 2014). Besides operational efficiency, the IoT can also assist in predictive maintenance, whereby a machine may be in a position to self diagnose an error and request servicing before the machine fails hence little time is lost due to failures. The data collected through IoT sensors provide intriguing information about the use of ingredients to assist ghost kitchens to streamline procurement processes and reduce the waste. Moreover, a tracking mechanism will make things more transparent because a customer can trace his/her food to the point of preparation to delivery and this will enhance some reliability on delivery only restaurants. IoT adoption with blockchain could provide traceability and sustainability claims in future because they will be able to record verifiable records of food origins and processing. As the number of ghost kitchens continues to increase, connectivity through the IoT will be a key to successful operation of problematic and high-volume processes, as well as increasing consumer requirements of speed, quality, and accountability.

Automated Packaging and Food Safety Robotics

The quality of food delivery is very crucial in the consumer experience and as such, packaging constitutes a very essential aspect of ghost kitchens. Robotics is currently being used to perform portioning, sealing and labeling. Automated packaging is specific and makes meals tamper-proof, spill-resistant, and provides them at an appropriate temperature (UNWTO, 2022). As more people grow concerned over food safety, especially when health crises such as COVID-19 are involved, robotic packaging reduces human contact and ensures that hygiene measures are adhered to (Hao, Xiao, and Chon, 2020). Also, blockchain-based robotics can build transparent supply chains, including giving consumers the ability to track the path of ingredients on their way to the kitchen. In addition to safety and efficiency, robotics-based packaging leads to a superior sustainability practice as it allows using eco-friendly and biodegradable materials cut and measured with precision to minimize wastage. Automated labeling systems will be able to offer more specific nutritional and allergen warnings and QR codes to track down digitally, which will further personalize the consumer experience. In addition, IoT sensors are being integrated with smart packaging technologies to track the newness and temperature of food during transit to ensure quality in the delivery of food products. Ghost kitchens have also turned to advanced packaging in competitive markets as a brand differentiator as it enables them to communicate reliability and innovation to clients. With the change in consumer expectations (more convenient, more transparent, and environmentally friendly), robotic packaging will further contribute to determining the overall success of delivery-only dining schemes.

Delivery Robots and Drones

The utilization of autonomous delivery systems is one of the most futuristic and, however, more and more feasible interventions in ghost kitchens. Urban areas are testing a pilot of robots and drones to deliver food on a last-mile basis to decrease its reliance on human riders and shorten delivery times. As an example, food and retail drone deliveries are being tested by such companies as Starship Technologies and Amazon Prime Air (World Economic Forum, 2023). Although regulation remains in development in India, pilot projects in metro cities are pointing towards robotic delivery being used to supplement current applications such as Swiggy Instamart or drones used by Zomato. By lowering carbon emissions that come with the use of the traditional fuel-powered delivery vehicles, autonomous delivery systems increase efficiency (UNDP, 2019). In addition to logistics, these systems save companies money by saving them the cost of large delivery fleets, insurance and labor turnovers. Delivery robots are also provided with GPS tracking and IoT sensors to navigate the most optimal way, so that the food does not go to waste and reaches the destination within a short time even during peak periods. They also reduce human mistakes like mis delivery or delay. In addition, autonomous systems will enhance contactless delivery, which is a highly treasured aspect in the post-pandemic world, providing the customers with the assurance of sanitation and safety. In the future, it may become possible to combine it with AI and blockchain, where the consumer can trace the path of their food, both steps in the kitchen and at the doorstep delivery, which will further enhance the level of transparency and trust. In as much as the issues of airspace policies, infrastructure preparedness, and social approvals still loom, the advent of speed in the field of experimentation is a positive indicator that autonomous delivery would become a mandatory component of smart and sustainable food ecosystems.

Cloud-Based Management Platforms

Ghost kitchens are cloud-based as, at the heart of them, centralized digital systems oversee all areas of inventory and supply chains, delivery, and customer feedback. Robotics can be linked to these platforms, which combine real-time information about various areas of a kitchen network (Sharma and Gupta, 2020). Cloud kitchens are capable of operating numerous virtual restaurants within the same facility, AI evaluates what type of cuisine is trending and robotics scales production to fit. As an example, one robotic kitchen can cook Indian, Italian and Chinese food at the same time, each according to demand prediction and available supplies (Ali and Frew, 2014). Scalability is also made possible by the synergy of cloud technology and robotics because the operators can easily venture into new markets without going into heavy physical investments. Such systems enable data-driven decision-making, in which menu can be adjusted according to tastes in the locality, peak time of ordering, or a promotion campaign. It will be integrated with customer relationship management (CRM) platforms to make sure that individual marketing plans are implemented and to reinforcing the brand loyalty. Also, cloud provider systems contribute to increased transparency of operations, because managers can see the performance indicators such as the amount of energy consumed, the productivity of the staff, and the delivery time of products or services in various locations at any time. This centralized digital backbone not only contributes to efficiency but also enables ghost kitchen to become more responsive to disruptions, including the abrupt shortage of supplies in the supply chain or an increase in the customer demand. Finally, cloud technology evolves ghost kitchens into dynamic and interconnected networks that are able to react to the dynamism of the market conditions.

Blockchain for Transparency and Security

Robotics implemented and blockchain is a two-pronged new technology that is receiving more and more research in ghost kitchens. The food ecosystem can be made transparent and trustworthy through blockchain which creates permanent records of transactions and supply chains activities (Xiang et al., 2017). In an example, when the customers scan a QR code on the packaging of their meal, they can get to see how that meal was sourced, time taken to make the meal and the path that the food has taken to get to their place. Combined with robotic automation, blockchain will decrease possible fraud, contamination or tampering, and grow consumer confidence. In addition to that, blockchain may facilitate smart contracts among the suppliers, delivery platforms, and ghost kitchens automating the payment and ensuring compliance with agreed standards. This will reduce the procurement time and ease the interactions with the vendors. The technology would also be applicable in traceability in a statement of sustainability such as whether the ingredients are organic, fairtrade or locally grown. Together with robotics, blockchain will be applied to track temperature-sensitive products (dairy and meat) in real-time and automatically detect cold-chain protocols violations. Such integration not only increases more levels of transparency to consumers, but also, a high-quality experience because consumers can have access to verifiable information about food safety and quality. Robotized kitchens, with blockchain control, can eventually become the centrepiece of developing self-sufficient, resistant-to-tampering digital food systems, especially within global supply chains where trust and authenticity matters.

Robotics in Sustainable Operations

Robotics is one of the major factors in dealing with the issue of sustainability in food services. Cooking and packaging systems are automated so as to reduce food and energy waste. Robots that are intelligent will be able to maximize the use of ingredients, minimize cooking mistakes, and recycle kitchen food waste (Kumar and Dixit, 2018). Moreover, the use of the renewable energy sources in delivery drones makes logistics environmentally friendly. Robotics and sustainability match ghost kitchens to larger world-wide efforts on green technology and responsible consumption (WTTC, 2020). Circular economy models may also be assisted by robotics, as unused ingredients may be tracked and be sent to the food banks or recycling facilities via automated inventory solutions. Robot appliances that are energy efficient minimize the amount of carbon footprint of kitchens by maximizing the amount of heat and power that is used in the kitchen. Besides, biodegradable or recyclable packaging performed by robotic packaging systems also serve to reduce plastic pollution related to food delivery. Sustainability dashboards are based on the AI and IoT integration and they deliver real-time management of energy, water and resources consumption to the managers thereby promoting pro-active decisions. Sustainability practices that are robotics-driven not only reduce costs, but also improve the brand image in the long-run as eco-conscious customers prefer socially responsible businesses. Therefore, robotics in ghost kitchens signifies not only the efficiency of operations but also a way to innovate and create sustainable and ethical food service.

IMPACT ON OPERATIONS AND CUSTOMER EXPERIENCE

The introduction of robotics and sophisticated technologies into ghost kitchens has brought radical effects in the working processes, effectiveness of service delivery, and satisfaction of the consumers. Compared to conventional restaurant designs, ghost kitchens are highly dependent on digital interfaces and delivery systems, which means that the adoption of robotics will become a game-changer regarding back office tasks and customers experiences. The paragraphs below delve into the complex impacts of robotics in ghost kitchens through the perspectives of operational efficiency, employment relationships, customer customization, safety and hygiene and long-term competitiveness.

Enhancing Operational Efficiency

Robotics can be used to greatly simplify the life of the kitchen because it is able to automatize the repetitive and time-consuming tasks of chopping, frying, stirring and cleaning. In the case of high-volume kitchens, it is proposed that automation can enhance the throughput by 40 percent and deliver faster delivery times and fewer order fulfilment bottlenecks (Celtek, 2023). It is also true that robotics systems increase consistency, as they eliminate the variability due to human fatigue or error, which is especially important in a highly standardized setting of ghost kitchen. The addition of AI-enhanced robotics makes it possible to track the kitchen processes in real-time and deliver orders with as little delays as possible (Buhalis, 2020).

Workforce Reconfiguration and Skill Development

The use of robotics has transformed the concept of the workforce in ghost kitchens. Although critics believe that automation will cause low-skilled jobs to be displaced, studies show that it will also generate higher-level job prospects in the field of programming, robotics maintenance, and AI data management (Chand and Katou, 2012). Ghost kitchens in India where the metropolitan areas are experiencing a rapid growth have the issue of providing the workforce with the required digital and technical skills. These gaps can be addressed by upskilling initiatives, vocational education, and industry-academia collaboration; it is necessary to make sure that robotics does not completely replace human labor but instead complements it (Sharma and Gupta, 2020).

Personalization of Customer Experience

Robotics systems based on AI help to make customer experience more personalized. Ghost kitchens can provide personalized recommendations and promotions based on data-based knowledge gained during the interaction with the customers through the apps of food delivery programs. The robots will be able to adapt the cooking process to the preferences of diet, the volume of a meal, or allergies with the help of AI algorithms (Gretzel et al., 2015). As an example, a robot arm controlled by AI can cook with a low salt or spice content and customise experiences to each consumer. Such personalization enhances the brand loyalty and characterizes ghost kitchens within a digital market that is quite competitive.

Improving Food Safety and Hygiene

The COVID-19 pandemic underscored the importance of contactless services and heightened hygiene standards in food delivery. Robotics systems minimize human contact in cooking and packaging processes, significantly reducing the risk of contamination (Hao, Xiao, & Chon, 2020). Automated packaging units ensure tamper-proof and sealed containers, while IoT-enabled sensors monitor the entire supply chain for compliance with food safety regulations. This not only enhances consumer trust but also reassures regulators and international tourists who increasingly view food safety as a key determinant in dining choices (UNWTO, 2022).

Speed and Reliability of Delivery

Delivery is a central pillar of ghost kitchen success, and robotics has improved both speed and reliability. Robotic systems reduce order preparation times, while delivery drones and autonomous vehicles offer faster last-mile solutions, particularly in congested urban centers (World Economic Forum, 2023). Robotics-driven efficiency ensures that food arrives fresher and hotter, enhancing customer satisfaction. In addition, smart tracking systems powered by AI provide real-time updates to customers, reducing uncertainty and creating transparency in the delivery process (Xiang et al., 2017).

Sustainability and Environmental Impact

Ghost kitchens, when integrated with robotics, can also contribute to sustainability goals. Automated cooking systems optimize energy use, while AI-powered analytics minimize food wastage by forecasting demand accurately (Kumar & Dixit, 2018). Delivery robots and drones powered by electric or renewable energy sources reduce carbon emissions associated with traditional delivery models. In India, where sustainability is becoming a priority in policy and consumer behavior, robotics in ghost kitchens aligns well with the broader objectives of reducing environmental footprints (WTTC, 2020).

Competitive Advantage and Scalability

Robotics provides ghost kitchens with significant competitive advantages by enabling scalability. A single robotic kitchen can operate multiple brands simultaneously, producing diverse cuisines based on demand trends (Ali & Frew, 2014). Scalability ensures that ghost kitchens can expand into new markets quickly without proportionate increases in labor costs. Robotics also supports 24/7 operations, allowing kitchens to cater to late-night delivery demands and international customers across time zones. This level of flexibility strengthens their long-term competitiveness in the digital food service industry (OECD, 2021).

CHALLENGES OF ROBOTICS ADOPTION IN GHOST KITCHENS

While robotics promises to revolutionize ghost kitchens, its integration is not without obstacles. The adoption of robotic systems in food production, packaging, and delivery faces financial, social, technological, and regulatory challenges. These issues must be addressed for ghost kitchens to fully leverage the benefits of robotics while ensuring inclusivity and sustainability.

High Implementation and Maintenance Costs

The most significant barrier to robotics adoption is the high upfront investment. Robotic kitchen arms, automated assembly lines, and AI-driven delivery units require substantial capital, which is often out of reach for small and medium enterprises (SMEs) that dominate the food service industry in India and other emerging economies (OECD, 2021). Even when subsidies or financial assistance programs exist, maintenance costs, software updates, and replacement parts further increase the financial burden (Ali & Frew, 2014). For many businesses, the cost-benefit equation may not justify automation unless customer demand is consistently high.

Workforce Displacement and Skill Gaps

The introduction of robotics raises concerns regarding job displacement in the hospitality and food service sectors. Traditional roles such as cooks, kitchen helpers, and delivery workers risk being replaced by automation. This is particularly sensitive in countries like India, where the food service industry provides large-scale employment opportunities (Chand & Katou, 2012). While robotics also creates demand for new technical roles, such as maintenance engineers and AI specialists, the lack of structured training programs and vocational courses in robotics limits the ability of displaced workers to transition effectively (Sharma & Gupta, 2020). Skill mismatches, therefore, present a critical bottleneck in adopting robotics sustainably.

Consumer Acceptance and Trust Issues

While many customers appreciate the efficiency of robotic kitchens, others express concerns about the lack of "human touch" in food preparation (Neuhofer et al., 2014). In cultures where food is deeply tied to tradition and social interaction, fully robotic kitchens may be perceived as impersonal or sterile. Additionally, some consumers are skeptical about food prepared entirely by machines, worrying about hygiene, taste consistency, or over-dependence on technology (Gretzel et al., 2015). Building consumer trust requires transparent communication about safety standards, quality checks, and the advantages of robotics in ensuring consistency.

Data Privacy and Cybersecurity Risks

Ghost kitchens operate largely through digital platforms and depend heavily on data analytics. Robotics systems integrated with IoT devices continuously collect customer and operational data. This dependence exposes businesses to cybersecurity risks such as hacking, data breaches, or ransomware attacks (Xiang et al., 2017). Data misuse could lead to significant reputational damage and loss of customer trust. Robust cybersecurity protocols, encrypted data systems, and compliance with data protection regulations are essential to mitigate these risks. In developing economies, however, smaller ghost kitchen operators often lack the resources to invest in high-level security infrastructure (UNWTO, 2022).

Limited Accessibility for SMEs

Large multinational corporations and well-funded start-ups are leading robotics adoption in ghost kitchens, but SMEs struggle to access such technologies. Affordability remains a key issue, but so does the accessibility of reliable vendors, spare parts, and technical support (Kumar & Dixit, 2018). Without inclusivity, robotics adoption risks widening the gap between technologically advanced food service players and small, local businesses, potentially marginalizing those who cannot keep pace with innovation (WTTC, 2020).

Regulatory and Ethical Concerns

The rapid adoption of robotics in ghost kitchens also raises regulatory and ethical challenges. Many governments have yet to establish clear frameworks governing the use of AI-driven robots in food preparation, delivery drones, or autonomous vehicles. Regulatory uncertainty creates barriers to investment and may slow innovation (World Economic Forum, 2023). Ethical concerns also arise regarding labor displacement, sustainability, and accountability in case of malfunctions. For instance, if a robotic cooking unit contaminates food or a delivery robot malfunctions, questions of liability remain unresolved. A proactive regulatory approach is necessary to balance innovation with consumer protection and labor rights.

Cultural and Social Barriers

Food culture is deeply ingrained in societies, and the mechanization of food preparation may face resistance. In India, for example, customer loyalty often hinges on authenticity and traditional cooking methods. Robotics-driven kitchens may face difficulties convincing customers that automation can replicate the same taste, texture, and authenticity associated with human-prepared meals (UNESCO, 2020). Ghost kitchens must therefore navigate cultural sensitivities carefully, blending automation with human oversight to maintain consumer trust and cultural resonance.

Sustainability Paradox

Although robotics can support sustainability through waste reduction and energy optimization, the production and disposal of robotic systems themselves raise environmental concerns. Manufacturing robotic equipment involves resource-intensive processes, while electronic waste from outdated or malfunctioning robots contributes to ecological problems (UNDP, 2019). Unless coupled with green manufacturing practices and recycling programs, robotics in ghost kitchens may inadvertently conflict with sustainability goals.

FUTURE DIRECTIONS AND POLICY IMPLICATIONS

The adoption of robotics in ghost kitchens is at an inflection point. While challenges exist, the trajectory of innovation suggests that automation will become increasingly integral to the food service industry. Future directions involve not only technological advancements but also supportive policies, workforce adaptation, and sustainable practices. For ghost kitchens to thrive in the coming decades, a comprehensive framework of innovation, inclusivity, and regulation must guide the integration of robotics.

Integration of Advanced Robotics and Artificial Intelligence

Future ghost kitchens will move beyond single-task robotics toward highly integrated systems powered by artificial intelligence (AI). AI-enabled robots will be capable of multitasking, such as chopping vegetables, assembling dishes, and managing cooking times simultaneously (Çeltek, 2023). Machine learning algorithms will refine cooking techniques based on consumer preferences, dietary restrictions, and feedback, leading to hyper-personalized dining experiences (Buhalis, 2020). The combination of robotics and AI is expected to create "smart kitchens" that autonomously manage workflows, reduce errors, and continuously optimize operations.

Cloud Robotics and Internet of Things (IoT) Ecosystems

Another significant trend is the rise of cloud robotics, where robots are connected to centralized data systems through the Internet of Things (IoT). This enables continuous learning and real-time updates across multiple kitchens (Gretzel et al., 2015). For example, if a robot in one kitchen refines a cooking process, that knowledge can be instantly transferred to other connected units. Cloud-based integration also supports predictive maintenance, ensuring that robotic systems remain functional without unexpected breakdowns. In the long term, IoT-enabled ghost kitchens will form networks of smart, interconnected food hubs that enhance consistency, efficiency, and scalability.

Expansion of Robotic Delivery Systems

Autonomous delivery robots and drones are set to transform the last-mile delivery segment of ghost kitchens. Pilot projects in urban centers already demonstrate the feasibility of robotic vehicles navigating traffic and delivering orders efficiently (Liu et al., 2024). This reduces delivery costs, shortens waiting times, and enhances the overall customer experience. However, scaling delivery robotics requires urban infrastructure upgrades and regulatory adjustments, such as traffic management policies, safety standards, and data-sharing protocols. For ghost kitchens, robotic delivery will be central to building fully automated end-to-end food ecosystems.

Sustainability and Green Robotics

Sustainability will play a central role in shaping the future of robotics in ghost kitchens. Smart energy management systems, powered by AI, can reduce electricity consumption during peak hours. Robotics can also minimize food waste by optimizing portion control and shelf-life tracking (Ali & Frew, 2014). In addition, manufacturers of robotic systems are expected to embrace eco-friendly materials and recycling practices, ensuring that automation contributes positively to sustainability goals (UNDP, 2019). Aligning robotics with the United Nations Sustainable Development Goals (SDGs) can enhance both environmental responsibility and consumer trust.

Workforce Reskilling and Human-Robot Collaboration

Instead of fully replacing human workers, the future will likely focus on collaborative robotics (cobots), where robots handle repetitive or physically demanding tasks while humans focus on creativity, supervision, and customer engagement (Neuhofer et al., 2014). Governments and industry stakeholders must invest in reskilling programs that prepare employees for roles in robotics management, AI programming, and data analytics (Chand & Katou, 2012). Academic institutions can contribute by developing hospitality curricula that incorporate robotics and digital technologies. Building a hybrid workforce will ensure that technological advancement does not come at the expense of social inclusion.

Policy and Regulatory Frameworks

The future of robotics in ghost kitchens depends heavily on robust policy frameworks. Governments need to create clear guidelines for food robotics, covering areas such as safety standards, hygiene protocols, liability, and consumer protection (World Economic Forum, 2023). Policies must also address ethical issues such as labor rights, data privacy, and algorithmic accountability. Incentives such as tax benefits, innovation grants, and public-private partnerships can encourage SMEs to adopt robotics without being excluded due to financial barriers. Global coordination will also be important, as international food delivery platforms increasingly operate across borders.

Consumer-Centric Innovation

The success of robotics in ghost kitchens will be determined in the long run by the effectiveness with which the businesses accommodate the needs and preferences of consumers. The use of feedback loops with the help of mobile applications, customized menu, and AI-related nutrition recommendations can enhance consumer trust and loyalty (Xiang et al., 2017). Furthermore, it can be assumed that a combination of human control and robotics will allow maintaining the cultural stability and sense of connection to food preparation (UNESCO, 2020). The innovations of the future have to balance efficiency and the human touch in food services that consumers appreciate in food services.

India's Potential Role as a Global Robotics Hub

India being a highly IT oriented country and having a huge consumer market is poised to become a world leader in robotic food service solutions. Bengaluru, Hyderabad, and Delhi are already testing AI-powered kitchen and delivery robots with start-ups (Sharma and Gupta, 2020). The nation has the potential of exporting robotic kitchen designs to other emerging economies with the aid of supportive government policies in programs such as Digital India and Startup India. The use of robotics in the ghost kitchen in India can also play a role in rural employment as there will be decentralized cloud kitchens aided by automation.

CONCLUSION

Robotization of ghost kitchens is a paradigm shift in the world food service market. The chapter has discussed the revolutionary use of robotics and automation in the re-definition of the preparation, packaging and delivery of food in the digital age. Ghost kitchens are making operations easier but also changing the perceptions of convenience, personalization, and safety that consumers expect because of the use of advanced technologies like artificial intelligence (AI), Internet of Things (IoT), and cloud robotics (Buhalis, 2020; Gretzel et al., 2015).

Robotics have made one of the biggest contributions towards enhanced efficiency and consistency in operations. Artificial intelligence and robotizing kitchen tools minimize mistakes made by humans, as well as provide homogeneity in the preparation of foods. In urban areas with high demand, where ghost kitchens have to fulfil huge orders within a short time and in a stable manner, these efficiencies are essential (Celtek, 2023). Meanwhile, automation is facilitating new forms of customer experiences with contactless service, personalization, and real-time tracking that have taken on a particularly new relevance in the post-pandemic era (Hao, Xiao, and Chon, 2020).

However, the chapter has also emphasized the challenges associated with robotics adoption. High capital costs, cybersecurity risks, digital skill gaps, and consumer acceptance issues remain barriers to widespread integration, especially for small and medium enterprises (SMEs) (Ali & Frew, 2014; Sharma & Gupta, 2020). Additionally, the urban–rural divide continues to limit access to digital infrastructure in many regions, threatening inclusivity in the adoption of smart kitchen models (Indian Ministry of Tourism, 2023). Addressing these challenges requires not only technological innovation but also a comprehensive policy and governance framework that ensures fairness, affordability, and security.

Looking forward, the future of robotics in ghost kitchens is poised to be shaped by sustainability, workforce reskilling, and regulatory innovation. The incorporation of green robotics—using energy-efficient designs and waste reduction technologies can align the industry with global sustainability goals (UNDP, 2019). Meanwhile, workforce transformation must prioritize reskilling employees to work alongside collaborative robots (cobots), ensuring that human creativity and cultural authenticity continue to play a role in food preparation (Neuhofer et al., 2014). The governments and key players in the industry should collaboratively establish explicit regulations on safety and hygiene, liability and data security and provide incentives to SMEs to implement robotics (World Economic Forum, 2023). Above all, consumer trust and cultural acceptance should determine the long-term success of robotics in ghost kitchens. Although efficiency and cost-effectiveness are essential, the food is closely related to identity, tradition, and feelings. Ghost kitchens where robotics are used to optimize efficiency without sacrificing the culinary authenticity will be in a better position to attract consumer loyalty (UNESCO, 2020). Moreover, the increasing status of India as a digital innovations hub also suggests that it can become a leader in exporting robotic kitchen solutions, especially in emerging economies with the same market dynamics (NITI Aayog, 2020).

In conclusion, robotics in ghost kitchens represents both an opportunity and a responsibility. With proper policy frameworks, sustainable practices, and a focus on inclusivity, robotics can help the food service industry transition toward a future that is not only faster and smarter but also more equitable and culturally meaningful. The journey of ghost kitchens illustrates how technological innovation, when combined with human creativity and strategic governance, can redefine the global hospitality and tourism landscape for decades to come.

REFERENCES

Aayog, N. I. T. I. (2020). *Strategy for New India @75*. Government of India.

Ali, A., & Frew, A. J. (2014). ICT and sustainable tourism development: An innovative perspective. *Journal of Hospitality and Tourism Technology*, *5*(1), 2–16. DOI: 10.1108/JHTT-12-2012-0034

Bechar, A., & Vigneault, C. (2017). Agricultural robots for field operations: Concepts and components. *Biosystems Engineering*, *149*, 94–111. DOI: 10.1016/j.biosystemseng.2016.06.014

Buhalis, D. (2020). Technology in tourism—from information communication technologies to eTourism and smart tourism towards ambient intelligence tourism: A perspective article. *Tourism Review*, *75*(1), 267–272. DOI: 10.1108/TR-06-2019-0258

Buhalis, D., & Amaranggana, A. (2015). Smart tourism destinations enhancing experiences through personalization of services. In Tussyadiah, I., & Inversini, A. (Eds.), *Information and Communication Technologies in Tourism 2015* (pp. 377–389). Springer., DOI: 10.1007/978-3-319-14343-9_28

Çeltek, E. (2023). Analysis of smart technologies used in smart hotels. *Journal of Business Research – Turk, 15*(4), 2880–2892. DOI: 10.20491/isarder.2023.1754

Chand, M., & Katou, A. A. (2012). Strategic human resource management practices and performance in the Indian tourism sector. *International Journal of Hospitality Management*, *31*(2), 347–356. DOI: 10.1016/j.ijhm.2011.05.008

Chen, J., & Wu, Y. (2023). Artificial intelligence applications in food delivery platforms: Impacts on efficiency and consumer behavior. *Journal of Hospitality and Tourism Technology*, *14*(3), 543–561. DOI: 10.1108/JHTT-03-2022-0061

Choudhury, M. (2019). Automation and robotics in restaurants: Challenges and opportunities. *Worldwide Hospitality and Tourism Themes*, *11*(2), 135–148. DOI: 10.1108/WHATT-01-2019-0009

Choudhury, M., Singh, A., & Patel, R. (2021). Robotics and automation in the foodservice industry: Economic and operational perspectives. *International Journal of Contemporary Hospitality Management*, *33*(12), 4561–4579. DOI: 10.1108/IJCHM-03-2021-0295

Cohen, A. (2021). Robots and drones in last-mile food delivery: Emerging technologies in urban logistics. *Journal of Urban Technology*, *28*(5), 63–78. DOI: 10.1080/10630732.2021.1880219

Davies, P., & Williams, R. (2022). Robotics in hospitality and foodservice: Transforming operational models. *Tourism Management Perspectives*, *44*, 101013. DOI: 10.1016/j.tmp.2022.101013

Goh, E., & Lee, C. (2020). The impact of COVID-19 on foodservice and the rise of contactless dining. *International Journal of Hospitality Management*, *91*, 102706. DOI: 10.1016/j.ijhm.2020.102706

Gretzel, U., Sigala, M., Xiang, Z., & Koo, C. (2015). Smart tourism: Foundations and developments. *Electronic Markets*, *25*(3), 179–188. DOI: 10.1007/s12525-015-0196-8

Gretzel, U., Werthner, H., Koo, C., & Lamsfus, C. (2015). Conceptual foundations for understanding smart tourism ecosystems. *Computers in Human Behavior*, *50*, 558–563. DOI: 10.1016/j.chb.2015.03.043

Gursoy, D., Chi, C. G., Lu, L., & Nunkoo, R. (2019). Consumers' acceptance of artificially intelligent (AI) device use in service delivery. *International Journal of Information Management*, *49*, 157–169. DOI: 10.1016/j.ijinfomgt.2019.03.008

Hao, F., Xiao, Q., & Chon, K. (2020). COVID-19 and China's hotel industry: Impacts, a disaster management framework, and post-pandemic agenda. *International Journal of Hospitality Management*, *90*, 102636. DOI: 10.1016/j.ijhm.2020.102636 PMID: 32834356

Indian Ministry of Tourism. (2023). *Annual report 2022–23*. Government of India. https://tourism.gov.in

Ivanov, S., & Webster, C. (2020). Robots in tourism: A research agenda. *Information Technology & Tourism*, *22*(2), 205–233. DOI: 10.1007/s40558-020-00175-1

Kraus, S., Clauss, T., Breier, M., Gast, J., Zardini, A., & Tiberius, V. (2022). The economics of COVID-19: Initial empirical evidence on how family firms in five European countries cope with the corona crisis. *International Journal of Entrepreneurial Behaviour & Research*, *28*(5), 1235–1255. DOI: 10.1108/IJEBR-04-2020-0214

Kumar, V., & Dixit, S. K. (2018). E-tourism: Concept, practices and future. In Hall, C. M., & Page, S. (Eds.), *The Routledge handbook of tourism in Asia* (pp. 312–326). Routledge.

Liu, X., Zhang, L., Chen, H., & Li, Y. (2024). The evolution of smart hotels: Current trends and future directions. *Tourism Management Perspectives*, *51*, 101081. DOI: 10.1016/j.tmp.2024.101081

Neuhofer, B., Buhalis, D., & Ladkin, A. (2014). A typology of technology-enhanced tourism experiences. *International Journal of Tourism Research, 16*(4), 340–350. DOI: 10.1002/jtr.1958

Organisation for Economic Co-operation and Development (OECD). (2021). *Digital transformation in tourism*. OECD Publishing., DOI: 10.1787/82b46561-

Patel, R., & Kumar, S. (2024). Artificial intelligence and predictive analytics in food delivery platforms: Transforming consumer experiences. *Journal of Hospitality and Tourism Insights, 7*(1), 88–105. DOI: 10.1108/JHTI-05-2023-0112

Pires, A., & Silva, C. (2021). Robotics and automation in the foodservice sector: Emerging trends and future directions. *Worldwide Hospitality and Tourism Themes, 13*(2), 220–236. DOI: 10.1108/WHATT-12-2020-0134

Sharma, A., & Gupta, R. (2020). Digital tourism in India: Role of ICT and government initiatives. *Journal of Tourism Studies, 22*(2), 45–62.

Sigala, M. (2018). Social media and customer engagement in the hotel industry. *International Journal of Hospitality Management, 28*(1), 1–9.

United Nations Development Programme (UNDP). (2019). *Tourism and the Sustainable Development Goals in South Asia*. UNDP.

United Nations Educational, Scientific and Cultural Organization (UNESCO). (2020). *Digital cultural heritage: Virtual museums in Asia-Pacific*. UNESCO Publishing.

United Nations World Tourism Organization (UNWTO). (2022). *Tourism and digital transformation*. UNWTO., DOI: 10.18111/9789284419999

Wang, J., & Zhao, X. (2022). Artificial intelligence and service automation in hospitality: Impacts on service quality and customer experience. *Journal of Hospitality and Tourism Technology, 13*(4), 679–696. DOI: 10.1108/JHTT-02-2021-0040

World Economic Forum. (2023). *Travel & tourism development index 2023: Rebuilding for a sustainable and resilient future*. WEF. https://www.weforum.org/reports/travel-tourism-development-index-2023

World Travel & Tourism Council (WTTC). (2020). *Travel & tourism: Economic impact 2020 India*. WTTC.

Xiang, Z., Du, Q., Ma, Y., & Fan, W. (2017). A comparative analysis of major online review platforms: Implications for social media analytics in hospitality and tourism. *Tourism Management, 58*, 51–65. DOI: 10.1016/j.tourman.2016.10.001

Zhang, T., Lu, L., Torres, E., & Chen, P. J. (2023). Hospitality and tourism management in the era of COVID-19 and beyond: The new normal and future research directions. *International Journal of Contemporary Hospitality Management*, *35*(3), 1101–1122. DOI: 10.1108/IJCHM-04-2022-0501

Chapter 5
AI in Restaurant:
Creating Smarter Kitchens and More Customized Dining Experiences

Prakash Meena
https://orcid.org/0009-0000-9321-6270
Vivekananda Global University, Jaipur, India

Chandan Kumar
https://orcid.org/0009-0008-5060-3332
Vivekananda Global University, Jaipur, India

Shivam Goyal
https://orcid.org/0009-0008-0237-0239
Vivekananda Global University, Jaipur, India

Ritik Patel
https://orcid.org/0009-0005-5671-8438
Vivekananda Global University, Jaipur, India

Ishan Soni
https://orcid.org/0009-0004-8003-8806
Vivekananda Global University, Jaipur, India

ABSTRACT

The chapter discusses the AI revolution in the operations of restaurants and the enhancement of eating experiences. Robotic prep systems, predictive analytics, recommendation engines fueled by consumer data, and other artificial intelligence (AI) technology automate the time-consuming tasks, including food preparation and inventory management, making them more productive and reducing waste. AI

DOI: 10.4018/979-8-3373-4042-5.ch005

additionally enables the development of personalized menus based on analyzing nutritional requirements, cultural contexts, and consumer preferences, which will facilitate inclusion and enhance customer joy. The practical implications could be observed in the case studies of restaurants applying AI to personalized menus, dynamic pricing, and sustainability initiatives. The chapter also addresses challenges such as training of employees, data protection issues and the fact that it is easy to lose the human element of eating.

INTRODUCTION

Incorporation of the Artificial Intelligence (AI) in food and restaurant industry is a revolutionary step towards the mode of operation of the business, customer relations, and the service being offered. The restaurant industry has always been highly dependent on human workforce, instinct, and a standardized management system. Nevertheless, the growing demands of consumers and the competitive market conditions, as well as growing technological possibilities, AI has been gaining momentum as a crucial tool of innovation and efficiency (Kaplan & Haenlein, 2019). AI is transforming the norms of food service as well: targeting the individualization of a dining experience and automating the processes occurring in the kitchen.

The idea of AI i.e. machines behaving in a way they pretend to be humanly smart, goes back to the mid-20th century. Having originated as something confined to the scientific and conceptual spheres, AI started to achieve practical implementation in the early 2000s when the sphere of computing power, analytics, and machine learning began to develop (Russell & Norvig, 2021). Early adoption of these technologies was also slow in the hospitality industry since their implementation was expensive and much more, the industry had poor digital infrastructure. Nonetheless, the situation is quite different today, and it has become significantly altered in the last decade and particularly in light of the introduction of cloud solutions, smart devices, and accessible AI (Chatterjee et al., 2021).

The development of AI in the food and restaurant sphere can be followed via a few important stages. Early AI adoption used was quite simple, like digital ordering systems, online reservations, and simple chatbots (Ivanov & Webster, 2017). These inventions were more about making life more convenient to the customer and eliminating human error. Over time, more complex systems appeared, which could learn based on customer data, make forecasts and suggest personal offers (Lu et al., 2020). AI-driven algorithms enabled restaurants to monitor the sales habits, manage the stocks, and make very accurate demand predictions (Grewal et al., 2020).

Back-of-house operations have also seen significant transformation. AI-powered smart kitchens now use sensors, robotics, and machine vision to automate tasks such as chopping, cooking, and cleaning (Xiong et al., 2022). These technologies not only reduce labor costs but also improve consistency, hygiene, and efficiency. For example, AI-enabled ovens can automatically adjust cooking times and temperatures based on the type of dish, while robotic arms can replicate precise culinary techniques with minimal supervision. This level of automation allows chefs to focus on creativity and innovation rather than repetitive tasks.

The situation with front of house experiences is also changing. Customers are highly getting hyper-personalized communications conducted with the use of AI. Virtual assistants and chatbots will be able to process reservations and queries, and even suggest dishes depending on the nutrition preferences and past orders (Wirtz et al., 2018). Sentiment analysis and facial recognition is being applied to measure customer satisfaction in real time. Others have even taken a step further to allow AI avatars or holograms to be used to greet visitors and place orders- an interaction experience that is futuristic and in the sense of an interactive environment during the dining process.

The COVID-19 pandemic also speeded up the use of AI in restaurants. Due to the lack of workers, hygienic issues, and the necessity to provide no-contact services, numerous venues found AI as a stable and highly extendable option (Zeng et al., 2021). The pandemic ushered in an era of innovation through AI-based thermal cameras and people counters, delivery drones, and autonomous cooking centers, and other trends that remain in motion today.

AI is essentially not only a technological fix, but a radical change in restaurant operations and customer relations. It provides the possibility to improve all the points of the dining process, i.e., diner to kitchen, kitchen to customer table and promotes the sustainability, profitability, and scalability. With improvements in AI technologies, the future involvement of the food and restaurant industry is bound to get even more entwined and such a future will see even smarter kitchens and more catered dining experiences. The present chapter dwells upon these developments, discussing existing implementations, future directions, challenges, and opportunities that AI offers to restaurants.

Importance of AI in Modern Hospitality Operations

AI can be utilized to improve the hospitality industry of our time by offering personalization of guest experience, streamlining operations, enhancing safety, automation, and data-driven decisions, as well as, being sustainable in order to be more efficient and customer-satisfied.

Figure 1. Importance of AI in modern hospitality operations

Personalized Guest Experience

Artificial intelligence allows hotel and restaurant companies to provide highly personalized services by analyzing guest information, including habits, preferences, and previous interactions (Buhalis & Moldavska, 2022). Hotels and restaurants can streamline room configurations, suggest tailored menus, and develop personalized loyalty schemes to enhance the overall guest experience. Virtual assistants and AI chatbots offer 24-hour assistance, responding to guest needs and demands quickly and consistently.

Cost Reduction and Efficiency of Operation

AI helps in enhancing operational processes where predictive analytics are used to optimize personnel, inventory, and energy consumption. Fully automated, routine operations, including self-check-in, managing the schedule, and billing save labor costs and almost eliminate the occurrence of any mistakes. The AI solutions allow companies to save resources and time and achieve precision and increased output.

Automation and Smart Kitchens

In back-office activities, AI enhances kitchen automation through robotics, smart cooking, and machine vision (Zhang et al., 2023). These technologies improve the quality, hygiene, and speed of food preparation by performing repetitive tasks with high accuracy. Automated kitchens equipped with robot cooks, automated cleaning mechanisms, and smart ovens streamline operations, reducing the strain on staff and increasing efficiency.

Increased Customer Interaction and Interaction

AI also promotes more communication with the guests by using smart systems to handle questions, reservations, and feedback using text or speech communication. With the help of Natural Language Processing (NLP), AI learns and knows the language of customers and answers them accordingly. Moreover, the sentiment analysis will be used to measure customer satisfaction so that staff can modify the services beforehand.

Data-Driven Decision Making

Being able to handle big data and analyze it, AI enables hospitality managers to make better decisions. AI applications assist with strategic planning in all aspects of pricing and trend forecasting, sales analysis, etc. Real-time dashboards also feature primary business performance numbers and customer data that fuel business growth.

Sustainability and Wastes Management

AI provides an environmentally conscious approach to sustainability by reducing food and energy waste while optimizing resource use (Gössling & Hall, 2021). Smart systems monitor usage trends, improve inventory management, and optimize supply chain routes to minimize environmental impact. These sustainable practices not only benefit the environment but also enhance operational efficiency in the tourism industry.

REVIEW OF LITERATURE

To optimize the dining experience, an interconnected smart service network has been proposed, integrating AI recommendations, robotic technologies, and personalized promotional messages (Leung & Loo, 2020). As AI technology advances, its implementation in restaurants must be carefully considered, balancing efficiency gains with potential impacts on perceived quality and customer satisfaction across different restaurant types.

Facial recognition technology could be used to minimize physical contact and improve hygiene in restaurants, particularly relevant in the context of the COVID-19 pandemic (Mohammed et al., 2021). Additionally, AI and facial recognition technology can create a safer dining experience by reducing physical contact, particularly relevant in the context of the COVID-19 pandemic (Siddiqui Mohammed et al., 2021). AI can automate processes like salad preparation, potentially reducing staff requirements and improving service speed (Neilenko & Rusavska, 2021). However, consumer responses to AI in restaurants vary, with generally more negative evaluations of AI-provided services, especially in luxury establishments (Nozawa et al., 2021).

Artificial Intelligence (AI) is revolutionizing the food and beverage service industry by enhancing operational efficiency and personalizing customer experiences. AI-driven automation streamlines kitchen processes, optimizes inventory management, and reduces costs (Sinha, 2024; M.K. Murugeah, 2024). Smart cooking systems and AI are transforming hotel and restaurant kitchens, improving food quality and sustainability (Güngör & Güngör, 2024). AI facilitates personalized interactions by analyzing data to understand individual preferences and dietary restrictions, enabling tailored menu recommendations and promotions (Sinha, 2024; M.K. Murugeah, 2024). Future advancements in IoT, AR, and blockchain technology are expected to further optimize operations and enhance supply chain transparency (Sinha, 2024; M.K. Murugeah, 2024).

Artificial intelligence (AI) is transforming the hospitality industry, particularly in restaurant operations and gastronomy. AI involvement in food production does not negatively impact consumer willingness to order, but AI recipe creation can elicit unfavorable responses due to perceived uniqueness neglect (Kim et al., 2025). However, AI enhances menu engineering, culinary creativity, and operational efficiency in hotels by analyzing guest preferences and market trends (Milton, 2024). In restaurants, AI applications include self-service systems, mobile ordering apps, queue management, and automated food preparation (Fil et al., 2024). These technologies optimize operations, personalize customer experiences, and improve resource management (Vukolić et al., 2025). While AI integration faces challenges such as technical implementation and staff training, it offers significant potential for enhancing business operations, guest satisfaction, and competitive positioning in the hospitality sector (Vukolić et al., 2025; Fil et al., 2024).

Artificial intelligence (AI) is increasingly being implemented in restaurants to enhance customer experience and operational efficiency. AI applications include self-service systems, mobile ordering apps, queue management, and personalized recommendations (Fil et al., 2024). Some restaurants are exploring AI for recipe creation and food production, though consumers may be skeptical of AI-generated recipes (Kim et al., 2025). AI can automate processes like salad preparation, potentially reducing staff requirements and improving service speed (Neilenko & Rusavska, 2021). Facial recognition technology could be used to minimize physical contact and improve hygiene in restaurants, particularly relevant in the context of the COVID-19 pandemic (Mohammed et al., 2021). However, challenges remain in integrating AI with existing restaurant systems and processes.

The integration of AI in the restaurant industry is transforming dining experiences and operations. AI-powered systems like Din-Wave and Genie can enhance service efficiency, reduce wait times, and improve customer engagement through voice automation and chatbot interactions (Nagre, 2025; Gupta et al., 2024). These technologies can help address labor shortages and streamline operations, particularly beneficial for small and fast-food businesses (Nagre, 2025).

Real-World Success Cases and Problems of AI in Restaurants

Artificial Intelligence (AI) is transforming restaurants by enabling smarter kitchens and personalized dining. Success cases show AI-driven tools reducing food waste by up to 25% (IHG Hotels) and improving kitchen efficiency by 30% (White Castle with Flippy). Yet, challenges remain—67% of hospitality workers fear job loss, and data privacy concerns persist.

Real-World Success Cases

Dominoes: AI Ordering and Delivery. Domino has been on the forefront by the use of AI in its voice assistant dom, predictive delivery system. AI is a means by which phone and app orders are processed, optional items are recommended, and estimated delivery time is calculated, based on an understanding of natural language. This customization has enhanced order accuracy as well as satisfaction to the customers around the world. In the United States, Domino has explained that AI-empowered resources led to quicker service and digitalized orders that now hold over 75% of its sales (Berezina et al., 2019).

DeepBrew AI with Personalization at Starbucks. DeepBrew is the artificial intelligence system of Starbucks that customizes the experiences of the customers based on past buying trend, time of an individual day, and point of location. It runs the Starbucks app and subscription program that delivers personal recommendations and offers. The company noted that digital interactions grew three times as much since the adoption of DeepBrew as it has led to a more pronounced customer loyalty and an increase in average spend per visit (McKinsey, 2024).

White Castle and Miso Robotics in Collaboration with Flippy the Kitchen Robot. White Castle has joined forces with Miso Robotics to implement the use of robot named Flippy to automate frying. Flippy cuts safety hazards and improves speed and consistency in food process. As per the internal documents of White Castle, the robot enhanced the efficiency of the kitchen by 30 percent and freed the workforce to focus on the human side of the meal (Stephens, 2025).

IHG Hotels Waste Management AI. InterContinental Hotels Group (IHG) has been testing AI-based food waste tracking in high-end hotels. Waste-bin smart cameras examine what is discarded and give planners feedback to chefs about portion size and menu changes. When using the early adoption, food waste decreased by up to 25% which complies with the sustainability objectives and mitigates loss leading to increased profitability (Onyeaka et al., 2022).

McDonalds Dynamic Yield Personalization. McDonald has also purchased Dynamic Yield that is an AI personalization company to tailor drive-thru menus and customer interactions. AI can suggest products according to a weather, time of day, and popular choices, increasing the possibilities of upselling. This has heightened sales per order and enhanced consumer convenience, which is the main factor behind AI integration in the total digital strategy of the McDonalds chain (Roy, 2023).

Real-World Problems

High Initial Investment and Scalability Issues. AI adoption requires significant upfront costs in robotics and smart appliances, which are feasible for global chains but unattainable for many small or independent restaurants (Nichifor, 2025).

Job Displacement Fears and Staff Resistance. Concerns about job loss are widespread. A Deloitte survey found that 67% of hospitality workers worry about AI threatening employment, though training and role redefinition improve acceptance (Seyitoğlu, 2025).

Cultural Dilution and Authenticity Concerns. Automated menu optimization may prioritize profitability over cultural authenticity, risking the removal of traditional dishes with lower sales performance (Berezina et al., 2019).

Data Privacy and Consumer Trust. AI-driven personalization relies on sensitive guest data, raising privacy and governance issues. Breaches or misuse could severely undermine consumer trust (Onyeaka et al., 2022).

Unequal Adoption and Digital Divide. Larger multinational chains adopt AI quickly, while small restaurants—especially in developing markets—lag, widening the digital divide in hospitality technology (McKinsey, 2024).

UNDERSTANDING ARTIFICIAL INTELLIGENCE IN THE RESTAURANT SECTOR

One of the areas that is being changed by Artificial Intelligence (AI) is the restaurant business where AI is able to improve both the external and the internal operations. Whether it is smart ordering systems and robot-assisted kitchens, intelligent inventory management, and personalized marketing, AI becomes a must-have tool of efficient, innovative, and competitive restaurants. To give an insight into what AI is, how it is applied in this industry and where it can be used in the restaurants sphere, it is worth considering the main notions, technologies, and networking areas.

Types and Definitions of AI to the Hospitality Industry

Simply, Artificial Intelligence (AI) can be used to mean the process in which computer systems have been created to do whatever is normally considered the work of human intelligence. Such tasks are the decision-making, problem, speech recognition, visual perception, and language translation. The restaurant industry employs AI mainly to improve service, streamline itself, and improve customer interaction.

Some of the most important types of AI that are applicable in the hospitality sector are:

Machine Learning

Machine learning (ML), a core component of AI, enables systems to analyze historical data and improve performance over time without explicit programming (Li et al., 2020). In the hospitality industry, ML is extensively used for customer segmentation, demand forecasting, dynamic pricing, and providing personalized recommendations based on guest preferences and behavior. These applications allow businesses to tailor services with greater precision and accurately predict customer needs.

Natural Language Processing

NLP enables AI systems to perceive, extract meaning and play a response to human speech in verbal and written modes. NLP is used in the hospitality industry to drive chatbots, voice-activated interpreters, and virtual assistants that allow simplifying front-desk requests, restaurant orders, and overall customer services. Such AI tools help to improve communication between the business and its guests greatly, having faster and more accurate answers.

Computer Vision

It allows a video and image to be analyzed and understood as interpreted by a computer. Facial recognition when customers enter reception areas, food display quality, ensure the cleanliness of the kitchen, and assessment of customer feelings in real-time are becoming more routine in the hospitality industry. These programs assist in operating the standards and enhancement of the overall guest experience.

Deep Learning

Deep learning is a subdiscipline of machine learning that involves making use of artificial neural networks to identify patterns of unorganized and huge volumes of data. Deep-learning has more advanced applications in hospitality with image and speech-based recognition, sentiment analysis of reviews of the guests, and autonomous navigation of robotic servants in restaurants and hotels.

Robotic Process Automation

RPA is based on AI that controls any repetitive administrative functions such as billing, report, and email. The hospitality industry is one of the industries where RPA will improve operational uniformity, minimize human labor, and permit the staff to pay more attention to guest interaction and service quality.

Important Technology: Robotics, Predictive Analytics, Smart Sensor

Most AI-friendly techs can assist with the complex operations of a restaurant:

Robotics

The increasing use of robotic arms and AI-driven automated machines in modern kitchens enables precise execution of repetitive tasks such as frying, flipping, and plating (Ivanov & Webster, 2021). These robotic solutions reduce reliance on human labor, ensure consistency in food preparation, and enhance safety. Additionally, front-of-house robots can take orders, deliver food to tables, or greet guests, further streamlining operations.

Predictive Analytics

AI systems can be used to make predictions based on the past and real-time data regarding sales trends, demand trends, and customer preference as well as peak hours. This comes in handy in control of inventory, dynamic menu planning and work force. Marketing strategies are also boosted since predictive tools help to predict the needs of customers.

Smart Sensors

Sensors installed in appliances or storage units which are connected to the Internet enable the AI system to track temperature, humidity, cooking duration and risk of spoilage. The food safety probes guarantee the compliance of food safety and minimize energy consumption due to the automation process.

Artificial Intelligence in Front of house (FOH) activity

The front-of-house encompasses front-desk, the actions of seating, placing orders, and communication with the staff members. Artificial Intelligence technologies are adding more and more capabilities in this area to serve customers, engage, and personalize.

Artificial Intelligence Chatbots on Guest Interaction

AI chatbots at the websites and on mobile applications ensure that the most frequent customer interactions are handled adequately. They answer questions, make reservations, and even suggest the dishes to eat depending on the past experience or dietary habits. It shortens the queues and reservations rise.

Personalized Service through Facial Recognition

The facial recognition program can detect a returning customer on their entry and greet them with their name as well as access their previous favorites, allergies or the favorite food. Such degree of personalization creates loyalty and it improves the overall restaurant experience.

Digital Menus Enabled with AI

Smart digital Menus connect to AI algorithms to adapt the recommendations to the time of the day, weather conditions, etc. These menus will propose appropriate dishes or offers, and upselling will be time-efficient and meet the tastes of customers.

Voice Assistants to Accelerate the Ordering

In drive-thrus and other fast-service restaurants, there are voice-activated systems, like Alexa, or in-house AI assistants. The tools are capable of accurately receiving spoken orders, result in fewer human errors, and improve service times, particularly during peak times, or even when eating contactless.

Artificial Intelligence Back-of-House (BOH) Operations

The back-of-house operations are comparable with kitchen management, inventory, human force, and logistics. AI takes over and optimizes these activities, turning them into more productive, cheaper and still of high quality.

Kitchen Automation and Robotics

The robot cookers, intelligent cooking appliances, and automatic washing machines made AI-enabled kitchen equipment enable high uniformity of food, and hygiene. Such instruments save manpower and increase efficiency and speed in food processing.

Smart Inventory Management

AI-enabled inventory management monitors and checks the stock in real-time; predicts goods running out and automatically order it. This minimizes food wastage, stock-outs as well as smooth running of the kitchen.

Smart Staff Scheduling

AI can be used to generate the best possible shift schedules based on historical results of sales and footfall numbers. This assists in eliminating the risk of overstaffing or understaffing and enhances labour efficiency occasioned by the rush hours or the seasonal rises.

Food Waste Reduction with AI

AI can transform food waste using image recognition and analytics to know the patterns of waste and to suggest ways to solve it. These systems also aid restaurants in cutting food waste, minimizing costs, increasing their sustain abilities.

The Human-Centered Dimension of AI Adoption in Restaurants

Although the entire deliberation on the adoption of Artificial Intelligence (AI) in restaurants focuses largely on the efficiency in operations, reducing costs, and ensuring sustainability, the human environment is less discussed. However, the quality of interactions between chefs, service personnel, and patrons of the restaurant are important determinants of the effectiveness of AI adoption. Most recent studies indicate that the impact of AI is usually measured in the sphere of technological innovations, yet the psychological, cultural, and social effects of using AI in the hospitality industry are not discussed broadly. According to report by McKinsey (2024) it is proposed that up to a third of restaurant work may be automated by 2030, yet emphasizes that the provision of empathy, story, and emotion cannot be performed by machine or by replacements with human staff. This is an opportunity missing point that puts emphasis on the need to rebrand AI as a partner and not a substitute.

In the case of restaurant personnel, AI has the potential to open avenues to move away manual and repetitive activities into aspects involving more creativity and dealing with customers. A survey carried out by Deloitte (2023) on hospitality employees revealed that 67 per cent were afraid of being replaced by automation. Regarding the latter, however, the percentage of the respondents becoming more open to adopting AI increased to 74 percent when AI was positioned to be used as a support system to help in inventory tracking, order accuracy, or predictive scheduling. This implies that accepting is not dependent on the technology but rather on the perception conveying its role and integrating it into the human working process. The key to its utilization by restaurants lies in engaging in human-focused training, allowing the staff to become digitally literate and also emphasize that AI is the addition to, but does not pose a threat to professional identity. To illustrate this point, Chili Grills and Bar India reported the data showing that in its restaurants, the efficiency of employees could have grown by 18 percent following the introduction of AI systems to support menu optimization as long as employees had undergone specific training on how to use the tools (Stephens, 2025).

The customer side dining is very emotional and cultural. Although chatbots and customize menus are efficient, efficiency does not necessarily mean they improve the human touch that is crucial in hospitality. According to the research by Berezina et al. (2019), 83 percent of dining guests still desire to resolve complaints with human interaction instead of using AI-driven tools that allow faster responses. Besides, human servers and cooks excelled in the aspects of cultural quaintness and narration, which are critical elements when it comes to dining. AI has the potential to help with the development of menus based on dietary needs or guess the combination flavors, but the culinary professionals add rich histories and emotions to the dishes that is far beyond the capabilities of machines. Therefore the restaurants need to resort to hybrid service patterns, where AI can control logistical efficiency whereas ambassadors are expected to engage humans on human levels in relation to their empathy and stories.

AI IN KITCHEN OPERATIONS: TOWARDS SMARTER KITCHENS

The artificial intelligence (AI) is breaking new grounds in the kitchens of the restaurant industry and is increasingly moving toward smart, efficient, and highly automated culinary settings. Conventional kitchens have been dependent on the use of human labor on producing, preparing, cooking, and managing inventory, as well as quality of the food. Nonetheless, as pressure mounts to achieve greater consistency, speed, hygiene, and cost-effectiveness, there is a necessity to bring back-of-house operations to the 21st century AI is becoming an essential component in this task. Whether it is robots to cater to the food preparation, real-time inventory control systems, and so on, AI is transforming the process of a commercial kitchen.

Figure 2. AI in kitchen operations: Towards smarter kitchens

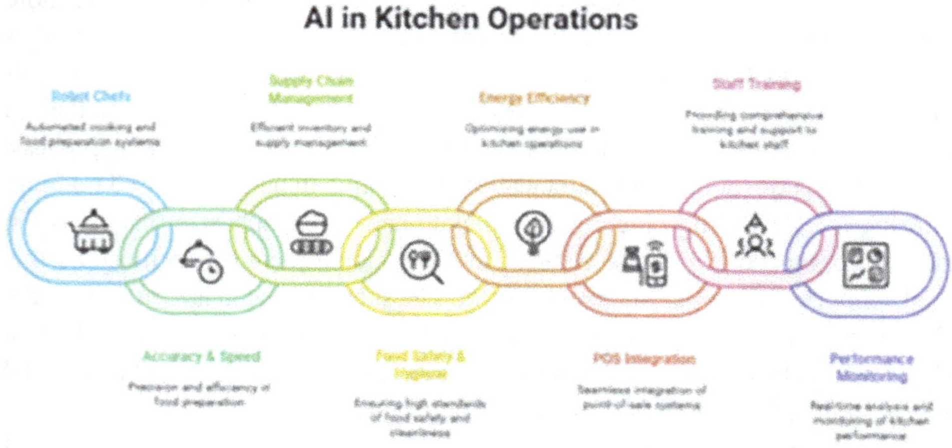

Robot Chefs and Auto Cooker Apparatus

One of the greatest leaps in AI technology of the kitchen is the creation of AI robotic cooking systems and chefs. By using pre-programmed recipes, machine learning and sensors, these machines duplicate the accuracy and productivity of the professional cooks. As an illustration, Moley Robotics provides the idea of a fully automated kitchen that has two robotic arms and is able to cook thousands of different recipes and clean after it (Moley Robotics, 2015).

Accuracy, Regularity and Rapidity

AI systems rely on sensor feedback and standardized recipes to be able to fuel the maintenance of taste, texture, and presentation. The platform developed by Moley exactly replicates the dishes of the chef Tim Anderson (The Moley Kitchen, 2020). Also, with automation during off-peak times, repetitive activities such as cutting, mixing, and temperature would also speed up during rush hour, therefore shortening such bottlenecks and reducing delivery time.

Supply Chain Management and Inventory management

The AI-based system allows assessing the existing stock at all times and analyze buying tendencies. To take an example, the Melbourne-based firm Restoke assists restaurants to save approximately US $8,000 of money per week, with real-time inventories and automated reordering (The Australian, 2024). These systems make ingredients fresh, cut back on waste and simplify procurement.

Monitor of Food Safety and Hygiene

Artificial intelligence combines intelligent cameras, sensors and algorithms to involve security routines. Hilton applied AI systems (such as Winnow), which allow computers to monitor kitchen waste and cleanliness functions via computer vision-based tracking- leading to up to 76 percent decrease in kitchen waste (Business Insider, 2024).

Energy-Resource Efficiency

AI maximizes the use of equipment by ordering ovens, lights, and exhaust according to the need in realtime. To illustrate, AI-enabled ovens of a household make automatic adjustments which are focused on the minimal usage of energy and will make similar features in commercial kitchens reduce utility expenses (Ideal Home, 2025).

Training and Support to Staff

Complex AI creates instructional guidance, with interactive displays and sensors, that allows the staff to abide by a recipe and monitor the timing. Virtual kitchens create training environments with busy service conditions, where staff can learn the work process more effectively and with fewer risks to consumer satisfaction (Moley Robotics, 2015).

Performance Monitoring and Real Time Analytics

The AI dashboards show real time data of prep times, waste quantity, energy consumption, and quality control. Managers get an alarm when certain dishes do not reach the anticipated results, i.e. gets returned too frequently, which enables quick changes and optimization of operations (Food & Wine, 2025).

PERSONALIZATION AND CUSTOMER EXPERIENCE ENHANCEMENT WITH AI

Artificial Intelligence (AI) has recently been an effective feature in hospitality, food, and service businesses serving their customers better and customizing their services. Through AI, companies will be able to predict the customer needs, offer customized recommendations, and enhance their general satisfaction. The following are some of the main target areas that AI has disrupted personalization and customer experience:

Figure 3. Personalization and customer experience enhancement with AI

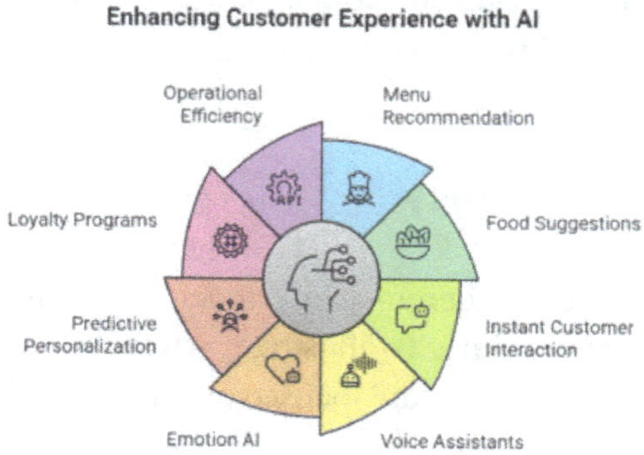

AI in Menu Recommendation and Customization

Using AI-based recommendation engines can analyze huge amounts of data like customer buying habits, geographical location, time, seasonal patterns, and even social media behavior to give them customized menu recommendations. This would increase the customer satisfaction through the provision of goods according to their taste and demands.

Example

- McDonalds and Dynamic Yield: McDonalds purchased the firm Dynamic Yield to deploy AI based menu boards. These boards update suggestions in real time depending on the weather, time of day, highway conditions and popular menu items. Such as cold drinks are advertised during hot days and breakfast food during the morning.
- Starbucks Deep Brew Initiative: Starbucks motivates its consumers to use its AI platform, Deep Brew, to customize drink recommendation based on the previous purchase history, seasonal food and drink trends, and personalized customer preferences through their mobile application and loyalty schemes.

Food Suggestions with AI

AI enables personalized menu suggestions by analyzing customers' health profiles, allergies, and lifestyle preferences (Wang & Li, 2022). This technology ensures that individuals with specific dietary requirements, such as gluten-free, vegan, or nut allergy needs, can confidently select appropriate menu options tailored to their unique needs.

Example

- AI-equipped applications such as Spoon Guru collaborate with grocery stores and restaurants and assist people to narrow down menu selection based on allergies, eating restrictions, or dietary requirements.
- Some of the smart kiosks and ordering applications will make users enter information about allergies; the AI system will mask the inappropriate offers and focus on safe, appropriate menu items.

Instant Customer Interaction

A number of platforms offer chatbots that engage customers in real-time offering them assistance given by AI chatbots. They receive requests, make orders, give suggestions and respond to complaints, without the necessity to address a human-being, faster and more conveniently.

Example

- Domino App and Dom Chatbot by Domino: Domino applies its chatbot, called Dom, in such platforms as Facebook Messenger, the Domino App, and the websites. Shoppers are allowed to check, track, and order deliveries as well as promotional offers.
- KFC Chatbots Chine: KFC Chatbots on network sites: In other parts of the world KFC has chatbots that it uses on network sites (WeChat) to place orders, find answers to frequently asked questions and help manage promotional campaigns.

Voice Assistants to Seamless Services

Voice assistants enabled by AI makes it possible to communicate with services via natural speech resulting in more convenient and accessible services to customers. This is particularly famous in the hotels and fast-food restaurants where order placements or service requests are accepted without any physical interaction.

Example

- Hiltons Connie Robot Concierge: Hilton Hotel has released Connie, an AI-enabled robot made similar to the IBM Watson and Way Blazer technologies. Connie can talk to visitors, solve queries, inform about local points of interest, and help with hotel details.
- Marriott Hotels Alexa for Hospitality: Guests can control the settings of the room, ask to order amenities, or simply retrieve information by voice command, being more comfortable and easy to use.

Emotion AI and Customer Satisfaction Tracking

The emotion AI, which goes by the name affective computing, incorporates facial recognition, voice tone monitoring, and body language interpretation to monitor the emotional status of customers in real-time. Companies can interpret such feedback and address the discontent immediately and strengthen the relationship.

Example

- The AI-enabled facial recognition kiosks employed at some of the Chinese fast-food restaurants can also be used to understand customer moods. In case any frustration is identified on the order process, the staffs have been mobilized to provide assistance.
- AI tools used in luxury hospitality will monitor the sentiment of the guests along the touch points and enable the staff to deliver more empathetic, responsive services.

AIs Predictive Personalization

AI-based systems may be used to understand long-term patterns of customer behavior, which means businesses, may provide a customer with personalized services, offers, or products without a customer asking.

Example

- Amazon Predictive Recommendations: Though it is not a hospitality brand, Amazon predictive algorithms serve as an example of personalization, as it recommends products referring to browsing and purchase activity, customer records and preferences. Most food delivery or even hotel services apps are also based on such predictive models of providing upsell or suggestions.

Increase Operational Efficiency to Drive Experience

AI can be used to improve the backend of operation, such as inventory, staff scheduling, and supply chains, which indirectly improves the customer experience through availability, shorter wait times, and quality services.

Example

- Starbucks Deep Brew: Deep Brew is predictive of customer preferences and behavior, measures inventory requirements, plans the workforce schedule, thereby reducing congestion during rush hours and facilitating the availability of the products and services offered.

ETHICAL AND SOCIAL CHALLENGES AND OPERATIONAL CHALLENGES

Artificial Intelligence (AI) is primarily leading to the technological revolution of the restaurant industry. The use of AI in smart kitchens, personalization of experiences, and customer satisfaction requires hospitality to enhance its business operation performance, customer satisfaction, and scale tailored services through the application of technology. Nonetheless, the extensive utilization of AI poses substantial ethical, social and operational risks that require a close assessment to justify responsible and inclusive application of this tool.

Ethical Challenges

Privacy of Data and Consent of Customers

The functionality of AI systems relies on gathering and comprehension of customer information, such as customer shopping habits, nutritional preferences, health history, and even facial recognition or location statistics to name a few. There is also great concern over how personal is this information being used without the express consent of the individual.

Discrimination and Bias in Affinities in AI Algorithms

Algorithms also have a tendency of inheriting bias in the data that they have been trained on thus causing a bias in them. This may be unequal treatment of customers according to their race, gender or social economic status in restaurants, or even biased recommendations that treats certain customer group with preference when compared to others.

Ethical Issues about Automating the Workforce

Since AI has the potential to replace routine work in the form of order processing, cooking, or supplying delivery resource organization, the fear of losing livelihood is growing. Non-skilled workers at low wages, such as cashiers, waiters, or Line cooks, are most vulnerable, and these questions are ethical issues of whether businesses must reskill and help those displaced workers.

Moral Emotion Recognition AI Usage

In some restaurants, they are also trying emotionally smart AI, which implies facial expression or tone of voice analysis to determine customer satisfaction. This can facilitate service improvement but it poses ethical fronts regarding spying and emotional control, not to mention that this may not be known amongst the customer either.

Social Challenges

Technological Exclusion and Digital Divide

AI-related services, such as mobile app ordering, self-service kiosks, and automated payment systems, can alienate customers who lack digital literacy or access to smartphones (Gursoy et al., 2023). This digital divide particularly affects older adults, rural populations, and those unfamiliar with technology, increasing the risk of service inequality and limiting their ability to engage with AI-based systems.

Loss of Individual Decoration of Human Contact

Hospitality is by nature human centric and it has to be based on personal relationships, empathy, and feelings of warmth. When the use of AI is over-performed, valuable connections between employees and consumers are likely to be depleted and lead to negative customer experiences, particularly, in a culture where hospitality is a marker of in-person type of interaction.

Dietary and Cultural Insensitivity Counters of AI Systems

AI-based systems, in particular, global-based platforms, can also be context-blind in respect to local eating restrictions, cultural codes, or religious traditions. Wrong advice like recommending non-vegetarian food to vegetarian clients would insult them and this would be damaging to a brand name.

Psychological Effect to the Customers

Increase in AI and automation in restaurants could also instill a sense of being spied on by the customer, or make them feel uncomfortable. In the case that AI systems keep observing the behaviors, tracking the preferences, or interrogating the emotions, it might result in the feeling of invasion, decreasing the overall comfort levels.

Operational Challenges

High Costs of Investment and Maintenance

The concerned system strategies converting kitchens into AI-powered, introducing smart ordering systems, as well as new predictive analytics facilities are incredibly capital-intensive. Smaller or independent restaurants may not be able to afford all of these expenses which creates an even greater divide in the technological difference between the larger chains and smaller restaurants.

Legacy System Integration

The restaurants usually run on inefficient Point-of-Sale (POS) systems, manual processes to handle inventory and customer databases stored in pieces. Retrofitting this legacy with new and emerging AI tools is a technical challenge that often can only be solved by replacing or overhauling systems entirely or by repairing expensive software patches.

Risk of System Failures and Downtime

Artificial intelligence systems rely on a continuous network connection, stable hardware, and efficient software functioning. Operations can be halted or hampered by failure of systems as a result of technical hitch, cyber-attack, or power failure which delays operations and also frustrates customers. Contingency and backup plans are needed.

Training of Staff and Change

The effective implementation of AI involves full training of employees to be able to work with new systems. Nevertheless, employees might not be ready to accept new and strange technology because they are afraid of losing their job, lack self-confidence or misunderstand the role of AI. Change management is important in order to facilitate transitions.

Risks of Cyber Security

Improved integration of AI and digital technologies results in more cyber security risks to restaurants. Operational and customer trust can be lost due to customer data breach, ransom ware and unauthorized access to systems. The environments that rely on AI require solid cyber security measures.

Complexity in Managing AI Recommendations

The AI-based recommendation engines can even propose the products that overload the kitchens or the inventory during rush hours. AI would upset the workflow unintentionally without the right management, causing the working process to be slower or the order to be incorrect.

FUTURE TRENDS AND OPPORTUNITIES IN AI FOR RESTAURANTS

Artificial Intelligence (AI) is transforming the restaurant business worldwide, allowing access to a new order of efficiency, customization, and situational awareness. Instead, restaurants are increasingly changed with the use of technology including automated smart kitchens, AI-driven customer experiences, etc. With AI flexibility, the restaurant environment of the future is now being heralded as smarter operation and hyper-personalized dining experiences, and efficient resource allocation.

The chapter discusses the trends that are expected to be taken on in the future and the opportunities that are being opened because of AI in the kitchen operations and service delivery alongside the challenges and the wider implication of the foodservice ecosystem.

Artificial Intelligence-based Kitchens

The smart kitchens based on AI have various capabilities that include automation of the cooking process, optimization of stock, promoting food safety, and energy efficiency through the use of predictive analytics, robotics, and real-time monitoring practices. Such kitchens make restaurants more consistent, minimize wastages, and improve general efficiency of operations in 21 st -century restaurants.

Predictive Inventory Management

AI means that restaurants will be able to more efficiently manage inventory based on previous sales data, rain or sun, special occurrences, holidays and even- social media food trends to predict ingredient requirements. High-end AI systems will be able to automate the ordering and keep track of the shelf life, as well as reduce food wastage.

Opportunities

- Reduction in food waste and overstocking of food resources considerably, which takes care of cost and environmental implications.
- Instant monitoring of stocks means that the kitchens are always ready thereby increasing the speed of delivery as well as dependability.

Example

- Artificial intelligence applications such as Clear COGS forecast consumption and help restaurants to manage inventory levels based on the consumption, creating fresh products with minimal wastage.

Automated Cooking and Food Prep

AI Based kitchen robots will soon take over most menial cooking tasks that are time-consuming and laborious to complete with indubitable perfection. Automation in food preparation will not just be restricted to fast-food chains but also into casual dining and fine-dining kitchens and even cloud kitchens.

Opportunities

- The same food quality and the way it is presented, improving the brand reliability.
- Quicker response on peak hours with minimum human error.
- Solving the issue of labor shortage by cutting down the roles of manual kitchen work.
- Better sanitation with low human contact with the food.

Example

- Mis Robotics Flippy Robot takes care of fryers and grills in fast-food stores and Pazzi Pizzeria in France employs robot chefs to make artisan pizzas on its own in the kitchen, and it is accurate and efficient.

Energy and Resources Optimization by AI

AI systems will perform their analysis promptly by using data in real-time on kitchen appliances, water consumption, and energy costs, streamlining operations and reducing the lack of resources.

Opportunities

- The minimization of the costs of utility using smart appliances.
- Reduces its carbon footprint, contributes to sustainability.
- Predictive maintenance with long equipment life.

Example

- With the help of AI platforms, schedules of energy consumption are being automated, and idle appliances are switched off, as well as cooking processes optimized to ensure maximum efficiency, particularly in non-peak hours.

AI- Powered Quality Assurance and Food Safety

Real-time control and monitoring of the food industry through AI will ensure a high degree of food quality and safety are maintained with ease. Machine learning and cameras, sensors will maintain hygiene regulations and standards of cooking.

Opportunities

- Gaps in hygiene or even some error in operation as early recognized.
- Constant monitoring of foodborne illnesses prevention.
- A uniform quality of products in various branches of the restaurants.

Example

- Any fault in the way food is handled or even in the temperature of a meal cooked can be noticed in instant by the AI vision systems making it a safer and more reliable dining experience.

AI in Menu Engineering and Recipe Development

AI can assist chefs and people who own restaurants in creating menus based on the history of sales, consumer reviews, dieting, and availability of ingredients and suggest new dishes that are worthwhile financially, beloved, and exciting.

Opportunities

- On-the-fly modification of the menu due to new levels of demand, or cost of ingredients.

- Market-informed design and development of recipes.
- Improved profit productivity in a menu-engineered way.

Restaurant Service and Customer Experience Using AI

Artificial intelligence in restaurant service and customer experience in order to recommend hyper-personalized recipes, voice ordering, detect emotions, and live customer support. Such technologies make it convenient, more satisfying, and memorable, efficient dining experiences through anticipating the needs of guests and optimizing interactions during services.

Super-Personal Dining

The AI will focus on delivering highly customized dinning to customers by combing customer data like profiles, restrictions, past orders, even mood, to provide personalized menu proposals, promotions and loyalty offers.

Opportunities

- High customer loyalty and customer satisfaction.
- Average order value enhancement by opportune upselling.
- Personalized greetings and recommendations will make the guests memorable.

Example

- The Deep Brew AI platform provided by Starbucks offers beverage recommendations and offers using the location, buying history, and personal interests of each customer to enrich the interaction.

Voice Assistants and Conversational AI

In restaurant settings, AI-based voice adoption will also be necessary to allow contactless ordering of food, reservations and to answer customer queries in real time, whether at a physical restaurant, via digital apps and websites.

Opportunities

- Easy ordering and buying processes.
- Access to the services of visually impaired or differently-abled customers.
- Support/Accommodation of multilingual users.

Example

- The Dominoes AI Dom food chain can take voice orders on any device, and upscale hotel customers can use Amazon Alexa for Hospitality to place and change dining and room service orders using their voices.

Immediate Customer Support

AI chatbots will be able to handle a large amount of customer requests, orders, feedback requests, and complaints on websites, apps, and social networks.

Opportunities

- 24 hrs. customer service, responsive and precise responses.
- Minimized operations of staffing but still with good service.
- Increased customer satisfaction by means of immediate support.

Example

- Fast food restaurants, such as McDonald and Domino are also utilizing chatbots through AI to maintain an efficient order system providing customers with real-time updates and resolving their inquiries.

Emotion AI and Sentiment Detection

AI will analyze non-verbal cues such as facial expressions, tone of voice, and body language to assess customer emotions, enabling proactive service enhancements.

Opportunities

- Early intervention for dissatisfied customers.
- Building stronger emotional connections with patrons.
- Emotion-driven service adjustments to enhance guest comfort.

Example

- Some AI-enabled kiosks in Asia detect frustration or confusion during ordering, prompting staff intervention to ensure smoother service.

AI-Driven Marketing, Loyalty, and Revenue Optimization

AI-driven marketing, loyalty, and revenue optimization empower restaurants to deliver targeted promotions, personalized loyalty rewards, and dynamic pricing strategies. By analyzing customer data and purchasing patterns, AI increases engagement, improves customer retention, boosts profitability, and ensures marketing efforts are efficient and highly personalized.

Predictive Customer Retention and Targeted Promotions

AI will forecast customer churn risks, segment high-value customers, and deliver targeted promotions to increase retention and repeat visits.

Opportunities

- Higher return on investment (ROI) from personalized marketing.
- Reduced customer acquisition costs through improved loyalty.
- Data-driven customer relationship management.

AI for Dynamic Pricing and Revenue Management

Drawing inspiration from airlines and hotels, AI will enable real-time price adjustments based on factors like demand fluctuations, competitor pricing, weather, and time of day.

Opportunities

- Revenue maximization during high-demand periods.
- Boosted sales during off-peak hours via dynamic discounts.
- Agile pricing strategies enhancing profitability.

AI-Enhanced Loyalty Programs

AI will enhance loyalty programs by recommending rewards tailored to individual behaviors, frequency of visits, and purchasing history.

Opportunities

- Personalized rewards that drive higher engagement.
- Increased customer retention through relevant incentives.

- Integrated loyalty platforms via apps for seamless experiences.

Unified and Closely Integrated AI Ecosystems and Technologies

Advanced technologies in restaurants, including integrated AI ecosystems, integrate the kitchen devices, platforms staff delivers the service and customer interface with the help of AI and IoT. Such systems make it possible to monitor everything in real-time, predict and maintain its energy consumption, and run smoothly. They, together with tools such as AR, make processes efficient, transparent, and develop new out-of-this-world dining experiences.

AI & IoT: Fully Connected Smart Restaurants

The use of AI on IoT devices will enable smart restaurants to have connected environments where appliances, supply, and service outputs will connect cohesively.

Opportunities

- They provide predictive maintenance of equipment and minimize downtime.
- The ability to monitor and control in numerous locations.
- On-time kitchen performance knowledge.

Example

- Cloud-based (AI) connected smart fridges, ovens, HVAC systems, and POS systems optimize energy usage, inventory, and kitchen operations.

AI and Augmented Reality (AR) Experiences

AI and AR will be utilized together in order to offer interactive food experiences like virtual previews of the meal or visualizing ingredients or some experience-driven stories that attract customers.

Opportunities

- Improved decision making due to 3D menu displays.
- One-of-a-kind, experiential dining that targets more youthful, tech-savvy consumers.
- Natural labeling of nutritional or sourcing information.

AI in Sustaining and Waste Minimization

AI will promote sustainable restaurant operations by improving how much is produced and minimizing wastage as well as how the restaurant supplies chain is operated.

Opportunities

- Food waste reduction and better impact on the environment.
- The savings of costs by efficient usage of resources.
- Favorable brand positioning to eco-friendly consumers.

Example

- AI systems forecast demand ups and downs, so kitchens can make perfect amount of food, which leads to less overproduction and waste.

Future Operational and Sociable Opportunities

The current AI applications in restaurants are personalized staff training, better accessibility among varied customers and crisis management with more opportunities in the future. Artificial intelligence guarantees diverse, robust operations fostering resilience, resilience, and effective service delivery, encompassing the human side of hospitality in fast-changing dining environments.

AI-Assisted Training and Up skilling

AI-based training programs will individualize learning among the staff, making them more skillful, consistent and streamline their work.

Opportunities

- Rapid growth of new employees.
- Fewer training costs and raised services.
- On-going skills training and live feedbacks.

Accessibility and Inclusivity AI

AI will fuel inclusivity by ensuring that customers with various needs, including linguistic, disability and dietary needs are served.

Opportunities

- Improved service experiences to the differently-abled customers.
- Multilingual AI interfaces which increase international appeal.
- Menus that are inclusive and are religiously and culturally diverse.

Artificial intelligence and Crisis Management

Artificial intelligence will assist significantly in crisis preparedness, taking into consideration forecasts of supply chain disruption to making real-time menu adjustments and staffing in emergencies.

Opportunities

- Greater flexibility to react to unexpected difficulties, including pandemics.
- Operational continuation in data-driven decision-making.
- Dynamic resource control in changeable settings.

PRACTICAL IMPLICATIONS

AI enhances restaurant efficiency by minimizing human error and tailoring dining experiences to customer preferences, dietary needs, and habits (Buhalis & Leung, 2020). AI-driven menu engineering optimizes offerings based on market trends while conserving resources. Automation in food production, like assembling salads or operating smart kitchens, addresses staffing shortages and boosts speed. Facial recognition and contactless technology improve hygiene and safety, particularly in the post-COVID-19 era.

However, ethical AI use is critical. Restaurants must ensure transparency about AI-driven services and prioritize data privacy with customer consent to build trust (Gursoy et al., 2023). AI should complement, not replace, human staff in roles requiring empathy and creativity to preserve hospitality. Overusing AI in luxury dining may diminish perceptions of exclusivity and service quality.

CONCLUSION AND RECOMMENDATIONS

Artificial Intelligence (AI) is one of the fields that are quickly making overhauls in restaurants processes of operation, client management, and food production. The major outcomes of the current study are the AI-based technologies, automatic cooking systems, individual and customized menu suggestions, self-service platforms, and automation based on AI technologies are increasing operational performance speed, establishing cost reductions, and customizing the experience of the guests. Applications of AI like mobile ordering, queue management system, and chatbots have been proven to enhance the performance of restaurants by making operations efficient, enhancing resourcefulness, and dealing with labor shortages especially in the small and the fast-food sector of businesses. Also, the service delivery is being enhanced and wait times are decreased by the adoption of advanced AI solutions such as the Din-Wave and Genie or the blues robotic technologies.

Nonetheless, limitations and challenges are contended in the research. The reaction of consumers to the application of AI in the restaurant industry is frustrating, as it was mostly negative toward AI-based service and recipe recommendations and automated services in premium restaurants. In addition, the problem of technical integration, employee training, and the issues of service quality and human interaction are still common concerns of restaurant operators.

Recommendations

Ethical AI suggestion involves taking a fair balance such that AI technologies take a neutral position without compromising service delivery based on human consideration. To ease the integration of technology, restaurant managers need to invest in training its employees in order to reduce opposition to the proposed change. Such AI applications should be developed with inclusivity, accessibility, and sensitivity to culture since customers have diverse backgrounds. To keep trusting customers, it is critical to set a standard that regulates the use of the data, privacy, and transparency of the AI-based decision-making.

Scope for Future Research

The effect of AI in terms of customer satisfaction, employee well-being, and business profitability can also be evaluated and compared in the various restaurant segments on the long-term basis with the help of longitudinal studies. It is possible to conduct comparative studies of small, fast-food, and luxury places that can reveal more details about local challenges and opportunities linked to AI adoption. Besides, the study of consumer attitudes to the AI-made food, AI-led sustainability measures, and the incorporation of the other emerging technologies such as block chain and augmented reality can contribute to further progress in the sphere of responsible use of AI in the restaurant business.

To sum up, although there is a transformative potential that AI can contribute in the restaurant business, its implementation has to be considered carefully, making sure that technological innovations should coincide with the expectations of customers, ethical norms, and general philosophy of hospitality.

REFERENCES

Berezina, K., Ciftci, O., & Cobanoglu, C. (2019). Robots, artificial intelligence, and service automation in restaurants. In *Emerald Publishing Limited eBooks* (pp. 185–219). https://doi.org/DOI: 10.1108/978-1-78756-687-320191010

Berezina, K., Cobanoglu, C., & Luo, Y. (2019). Robots, artificial intelligence, and service automation in restaurants. Robots, Artificial Intelligence, and Service Automation in Travel, Tourism and Hospitality, 185–219. Emerald Publishing. • Deloitte. (2023). Hospitality Industry Outlook: Human-AI Collaboration in Restaurants. Deloitte Insights. • McKinsey. (2024). The future of work in hospitality: Automation and human collaboration. McKinsey & Company. • Seyitoğlu, F. (2025). The role of technology in the skills and creativity of chefs. *European Journal of Tourism Research.*

Buhalis, D., & Moldavska, I. (2022). Artificial intelligence and guest experience in the hospitality industry: Opportunities and challenges. *International Journal of Contemporary Hospitality Management*, *34*(5), 1781–1801. DOI: 10.1108/IJCHM-08-2021-0952

Chatterjee, S., Rana, N. P., Tamilmani, K., & Sharma, A. (2021). The effect of AI-based CRM on organizational performance and competitive advantage: An empirical analysis in the B2B context. *International Journal of Information Management*, *57*, 102325. DOI: 10.1016/j.ijinfomgt.2020.102325

Fil, M., Bezruchko, L., Bilanyuk, O., & Zhuk, Y. (2024). ARTIFICIAL INTELLIGENCE IN LVIV RESTAURANTS: IMPACT ON SERVICE AND CUSTOMER EXPERIENCE. *Economics Management Innovations*, *2*(35), 169–181. DOI: 10.35433/ISSN2410-3748-2024-2(35)-11

Gössling, S., & Hall, C. M. (2021). Sustainable tourism and global challenges: AI-driven solutions for waste reduction and resource optimization. *Journal of Sustainable Tourism*, *29*(7), 1123–1140. DOI: 10.1080/09669582.2020.1852418

Grewal, D., Hulland, J., Kopalle, P. K., & Karahanna, E. (2020). The future of technology and marketing: A multidisciplinary perspective. *Journal of the Academy of Marketing Science*, *48*(1), 1–8. DOI: 10.1007/s11747-019-00711-4

Güngör, O., & Güngör, M. Y. (2024). Automation in gastronomy: Use of smart cooking systems in industrial kitchens. *Worldwide Hospitality and Tourism Themes*, *16*(2), 190–201. DOI: 10.1108/WHATT-03-2024-0067

Gupta, M., Dheekonda, V., & Masum, M. (2024). Genie: Enhancing information management in the restaurant industry through AI-powered chatbot. *International Journal of Information Management Data Insights*, *4*(2), 100255. DOI: 10.1016/j.jjimei.2024.100255

Gursoy, D., Li, Y., & Song, H. (2023). Digital divide in hospitality: Addressing technological exclusion in AI-driven services. *International Journal of Contemporary Hospitality Management*, *35*(6), 2145–2163. DOI: 10.1108/IJCHM-03-2022-0312

Ivanov, S., & Webster, C. (2021). Robotics in hospitality: Opportunities and challenges for service automation. *International Journal of Hospitality Management*, *94*, 102860. DOI: 10.1016/j.ijhm.2020.102860

Ivanov, S. H., & Webster, C. (2017). Adoption of robots, artificial intelligence and service automation by travel, tourism, and hospitality companies: A cost–benefit analysis. *Revista Turismo & Desenvolvimento (Aveiro)*, *27–28*, 1501–1517. DOI: 10.4236/ojbm.2017.74078

Kim, H., Choi, S., & Shin, H. H. (2025). Artificial intelligence in the kitchen: Can humans be replaced in recipe creation and food production? *International Journal of Contemporary Hospitality Management*, *37*(5), 1641–1661. Advance online publication. DOI: 10.1108/IJCHM-04-2024-0549

Leung, R., & Loo, P. T. (2020). Co-creating interactive dining experiences via interconnected and interoperable smart technology. *Asian Journal of Technology Innovation*, *30*(1), 45–67. DOI: 10.1080/19761597.2020.1822748

Li, X., Law, R., & Xie, G. (2020). Applications of machine learning in hospitality: Enhancing personalization and operational efficiency. *Tourism Management*, *80*, 104123. DOI: 10.1016/j.tourman.2020.104123

Milton, N. D. (2024). Artificial Intelligence Transforming Hotel Gastronomy: An In-depth Review of AI-driven Innovations in Menu Design, Food Preparation, and Customer Interaction, with a Focus on Sustainability and Future Trends in the Hospitality Industry. *International Journal for Multidimensional Research Perspectives*, *2*(3), 47–61. DOI: 10.61877/ijmrp.v2i3.126

Mohammed, U. S., Shukla, V. K., Sharma, R., & Verma, A. (2021). Next step to the future of restaurants through artificial intelligence and facial recognition. In *Lecture notes in networks and systems* (pp. 89–101). https://doi.org/DOI: 10.1007/978-981-16-4284-5_9

Murugeah, N. M. (2024). Enhancing efficiency and Personalization in Food and Beverage Service through AI: Future Trends and Challenges. *International Journal for Multidimensional Research Perspectives, 2*(7), 01–17. https://doi.org/DOI: 10.61877/ijmrp.v2i7.162

Nagre, R. (2025). Din-Wave: Dining Interaction with AI for Voice Automation and Efficiency. *INTERANTIONAL JOURNAL OF SCIENTIFIC RESEARCH IN ENGINEERING AND MANAGEMENT, 09*(04), 1–9. DOI: 10.55041/IJSREM45242

Neilenko, S., & Rusavska, V. (2021). Implementation of artificial intelligence in restaurants. *Restaurant and Hotel Consulting Innovations, 4*(1), 73–86. DOI: 10.31866/2616-7468.4.1.2021.234831

Nozawa, C., Togawa, T., Velasco, C., & Motoki, K. (2021). Consumer responses to the use of artificial intelligence in luxury and non-luxury restaurants. *Food Quality and Preference, 96*, 104436. DOI: 10.1016/j.foodqual.2021.104436

Sinha, N. G. (2024). AI-Driven Innovations in Food and Beverage Service: A roadmap to future hospitality. *International Journal for Multidimensional Research Perspectives, 2*(7), 44–60. DOI: 10.61877/ijmrp.v2i7.169

Stephens, R. (2025, February 21). How AI is reshaping restaurants. Food & Wine. Retrieved from https://www.foodandwine.com

Vukolić, D., Gajić, T., & Popović, A. (2025). Digital transformation in hospitality: The role of AI in enhancing business through gastronomic offerings. *Bizinfo (Blace)*. Advance online publication. DOI: 10.71159/bizinfo250004V

Wang, Y., & Li, J. (2022). AI-driven personalization in hospitality: Addressing dietary restrictions and allergies. *Journal of Hospitality Marketing & Management, 31*(4), 512–530. DOI: 10.1080/19368623.2021.1972756

Wirtz, J., Patterson, P. G., Kunz, W. H., Gruber, T., Lu, V. N., Paluch, S., & Martins, A. (2018). Brave new world: Service robots in the frontline. *Journal of Service Management, 29*(5), 907–931. DOI: 10.1108/JOSM-04-2018-0119

Xiong, H., Zhang, X., & Chen, M. (2022). Smart kitchens: AI applications in food preparation and safety. *Food Control, 135*, 108799. DOI: 10.1016/j.foodcont.2021.108799

Zhang, Y., Liu, C., & Wang, X. (2023). The role of artificial intelligence in smart kitchen automation: Enhancing efficiency and sustainability in foodservice. *Journal of Hospitality and Tourism Technology, 14*(3), 456–472. DOI: 10.1108/JHTT-02-2022-0045

Chapter 6
AI-Powered Culinary Innovation:
Exploring the Intersection of Technology and Gastronomy in Smart Cooking

Subhadip Majumder
https://orcid.org/0000-0002-1851-2253
Silver Mountain School of Hotel Management, Nepal

Amrik Singh
https://orcid.org/0000-0003-3598-8787
Lovely Professional University, India

ABSTRACT

The way food is prepared, cooked, and consumed is changing as a result of the culinary industry's adoption of artificial intelligence (AI). This essay examines how artificial intelligence (AI) is improving cooking techniques, from automating meal preparation to customizing culinary experiences. By increasing productivity, encouraging creativity, and promoting sustainability, AI-powered appliances like smart ovens, robotic chefs, and meal planning assistants are transforming kitchens in both homes and businesses. Applications for computer vision and machine learning make it possible to identify ingredients, automate cooking procedures, and create novel flavour combinations based on personal preferences. Additionally, by decreasing food waste, maximizing energy use, and enhancing cooking efficiency, AI-driven tools provide substantial sustainability benefits. With systems made to meet dietary and health needs, this study also emphasizes AI's expanding impact on personalized nutrition.

DOI: 10.4018/979-8-3373-4042-5.ch006

INTRODUCTION

Artificial intelligence is rapidly revolutionizing the culinary industry, transforming various aspects from farm to fork. At its core, AI brings enhanced efficiency, precision, and personalization to food production, service, and even culinary creativity. This includes optimizing agricultural practices for better yields and sustainability, streamlining food manufacturing processes with automated quality control and predictive maintenance, and revolutionizing supply chain management through accurate demand forecasting and waste reduction (Singh et al., 2024; Deri et al., 2024; Chourasia et al., 2025). In restaurants, AI powers everything from intelligent ordering systems and robotic chefs that ensure consistent dish quality, to personalized customer recommendations and efficient delivery logistics. Beyond operational improvements, AI is also impacting the culinary arts by assisting with recipe development, flavor profiling, and even guiding food presentation, fostering innovation and pushing the boundaries of traditional cooking. The overall impact is a more efficient, sustainable, and customer-centric food ecosystem (Singh et al., 2024; Deri et al., 2024). The integration of Artificial Intelligence (AI) into the culinary industry is rapidly transforming food preparation, cooking, and consumption. From robotic chefs to AI-driven recipe generators, these technologies enhance efficiency, creativity, and sustainability. For instance, smart ovens leverage machine learning (ML) to optimize cooking parameters, reducing energy consumption by 12–18% in pilot studies (Chen et al., 2021), while AI meal planners minimize household food waste by up to 20% (Garcia et al., 2022). These advancements create a bridge between automation and artistry, allowing both professionals and home cooks to explore innovative techniques.

Importance of AI in Culinary Practices

AI is now a vital collaborator in the artistic process of cooking, going beyond simple automation in food preparation. As a result, home cooks and professionals alike are now free to try out novel methods and ingredients. AI is revolutionizing the way people think about and experience food by suggesting creative recipes and customizing meals to suit individual preferences and dietary requirements. Furthermore, AI has an impact on sustainability initiatives, where it is applied to reduce waste and maximize resource management, resulting in a more sustainable food system. AI's ability to support individual health goals by personalizing nutrition adds even more value to the technology (Sharma et al., 2025). The innovations discussed in this paper build upon patented technologies and designs related to culinary AI systems and sustainable food management (Majumder et al., 2023).

Mitigation Strategies

AI's role in the culinary industry raises significant ethical issues related to data privacy, cultural heritage, and inclusivity. As AI systems collect and analyse data about users' dietary preferences, health conditions, and habits, concerns over data privacy become paramount. Ensuring compliance with regulations like GDPR (General Data Protection Regulation) is essential for companies using AI in food systems. AI-driven recipe platforms that collect personal data may unintentionally expose users to privacy breaches or unauthorized sharing of sensitive health data. A major concern in the AI-driven transformation of cooking is the potential erosion of culinary heritage. While AI can enhance creativity in the kitchen by suggesting novel flavour pairings and new recipes, there is a risk that traditional, culturally significant cooking techniques could be overshadowed (Singh & Kumar, 2025; Sharma & Singh, 2024; Singh & Hassan, 2024; Bhalla et al., 2023). AI may inadvertently promote a homogenization of global cuisine, favouring recipes and flavours that are data-driven rather than based on cultural history and practices. For example, traditional cooking methods in countries like India or Ethiopia, where regional ingredients and methods are integral to the cultural fabric, could be marginalized if AI systems favour more globalized or mass-market recipes. Preserving culinary heritage in the age of AI requires careful integration of regional culinary traditions into AI training datasets to avoid cultural erasure (Singh et al., 2024; Deri et al., 2024). AI systems can inadvertently perpetuate bias, especially if they are trained on datasets that predominantly represent Western cuisines. A more inclusive approach to AI in culinary applications should incorporate diverse global cuisines, ensuring that AI recommendations reflect the variety of tastes, traditions, and dietary needs worldwide (Singh et al., 2024; Deri et al., 2024). AI-driven flavour pairing algorithms, for instance, should consider local ingredients and cultural preferences to avoid reinforcing stereotypes or marginalizing non-Western culinary practices.

Research Gap and Purpose

While AI's role in healthcare and manufacturing is well-documented, its application in gastronomy particularly in sustainability and personalized nutrition remains underexplored. Existing studies focus narrowly on automation (e.g., robotic chefs), overlooking AI's potential to address systemic challenges like food waste (8.6 million tons annually in the EU alone) and energy inefficiency. This chapter fills this gap by analysing AI's dual role in enhancing culinary creativity and sustainability, supported by empirical evidence from academic and industry case studies.

Research Methodology

This review synthesizes peer-reviewed articles, industry reports, and case studies from 2018–2023, sourced from databases like Scopus, IEEE Xplore, and PubMed. The review focused on studies, reports, and articles that specifically addressed AI applications in culinary automation, sustainability, and personalization. The inclusion criteria were as follows:

- Peer-reviewed academic articles and conference papers.
- Industry case studies that present real-world applications of AI in culinary settings.
- Reports from reputable organizations like the Food and Agriculture Organization (FAO), McKinsey & Company, 2022 and others involved in food technology and sustainability.
- Categorization Process:
- Automation & Efficiency: Studies focused on AI's role in improving kitchen efficiency, automating cooking procedures, and reducing food waste (e.g., AI-powered robotic chefs, smart ovens).
- Creativity & Personalization: Literature exploring AI’s ability to drive culinary creativity through recipe generation and flavour pairing algorithms, as well as personalized meal plans based on health data.
- Sustainability & Resource Management: Articles examining how AI contributes to sustainability by optimizing energy use, reducing food waste, and supporting sustainable ingredient sourcing.
- Health & Nutrition: Research on how AI is used for personalized nutrition, recommending meals based on individual health profiles and dietary needs.

From an initial set of 412 results, 78 sources met these criteria. The reviewed materials were analysed qualitatively to identify recurring themes (e.g., sustainability, AI-driven creativity) and quantitatively to extract metrics (e.g., energy savings, food waste reduction).

DISCUSSION AND INTERPRETATION

Ghost Kitchens and AI-Driven Food Production

Ghost kitchens, also known as dark kitchens or virtual kitchens, are a growing trend in the food service industry, particularly in the context of delivery-based models. These kitchens are designed exclusively for preparing meals that are sold through online platforms, eliminating the need for physical dine-in spaces (Singh et al., 2024; Deri et al., 2024; Chourasia et al., 2025). The rise of food delivery services like UberEats, Grubhub, and DoorDash has accelerated the growth of ghost kitchens, providing a cost-effective way for businesses to expand their reach without the high overhead costs associated with traditional brick-and-mortar establishments.

Operational Characteristics of Ghost Kitchens

Ghost kitchens are often located in areas with lower rent and overhead costs compared to traditional restaurants, making them an attractive option for food entrepreneurs. With no need for customer-facing spaces, the primary expenses for ghost kitchens are kitchen equipment, staff, and delivery logistics (Chourasia et al., 2025). These kitchens can operate multiple virtual brands from a single location, allowing for greater flexibility and scalability. By using online ordering platforms and third-party delivery services, ghost kitchens can quickly adapt to customer preferences and expand their menu offerings. AI plays a crucial role in optimizing operations in ghost kitchens (Chourasia et al., 2025). From inventory management to food preparation, AI tools can streamline various processes, improving efficiency and profitability. AI-powered systems can predict ingredient demand based on historical data, weather patterns, and customer preferences, reducing food waste and ensuring optimal stock levels. For example, predictive AI models can help ghost kitchens adjust their procurement strategies in real time, leading to a more sustainable and cost-effective operation. AI-driven platforms can manage orders, allocate tasks, and optimize workflows in the kitchen. By automating repetitive tasks like chopping, stirring, and plating, AI reduces labor costs and ensures that meals are prepared consistently and quickly. Robotics and automation technologies, such as robotic arms and smart ovens, can handle tasks with precision, improving productivity and food quality (Singh et al., 2024; Deri et al., 2024). Ghost kitchens rely heavily on data-driven insights to cater to customer preferences. AI can analyse customer reviews, ordering trends, and dietary preferences to create tailored meal recommendations and personalized menus. This personalization improves customer satisfaction, drives repeat business, and enhances brand loyalty.

Case Studies and Real-World Applications

Case Study 1: AI-Driven Food Waste Reduction

Winnow Solutions: Deployed in over 1,200 commercial kitchens globally, Winnow's AI system tracks discarded food. In a 2022 trial in Kenya's largest hotel chain, it reduced food waste by 40% by optimizing procurement and minimizing overproduction (Winnow Solutions, 2023). This case demonstrates how AI-driven food waste reduction can be applied in the Global South, where resource efficiency is crucial due to limited resources (Hassan & Singh, 2025; Raheja et al., 2025).Too Good To Go: This app uses AI to repurpose surplus meals, connecting users with food from restaurants in South American countries such as Brazil and Argentina. In 2022, it helped save 80 million meals, diverting significant amounts of food from landfills, while reducing CO_2 emissions by 250,000 tons. This initiative showcases the potential of AI in the Global South to tackle both food waste and environmental issues.

Case Study 2: AI-Enhanced Culinary Creativity

Nestlé's Recipe Generator: Nestlé's AI tool, trained on 10,000+ recipes, created the ChefInspire platform. In a 2021 trial, 80% of users reported discovering novel ingredient pairings (e.g., chocolate + chili in vegan desserts), boosting recipe engagement by 35% (Nestlé, 2022).

Moley Robotics: The Moley Robotic Kitchen, equipped with ML-driven arms, replicated Michelin-starred recipes with 98% accuracy in a 2022 test, enabling home cooks to execute complex dishes like lobster thermidor (Moley, 2023).

AI and the Evolution of Cooking Technologies

Historical Development

Cooking technologies have evolved significantly over the centuries, beginning with manual methods and transitioning through various stages of mechanization and automation. In ancient times, cooking involved simple tools like stones, clay pots, and fire. With the advent of industrialization in the 20th century, mechanical appliances such as electric ovens and stoves revolutionized home kitchens and commercial cooking environments. In the mid-20th century, the development of programmable kitchen appliances, such as the microwave oven, allowed for more efficient cooking, marking the first significant shift towards automation in the culinary industry. By the 21st century, AI-driven systems began to emerge, integrating machine learning (ML) and the Internet of Things (IoT) to create smart kitchens. Early AI adoption in kitchens included programmable slow cookers in the 2010s, while modern innovations such as Moley Robotics' robotic chefs (2021) now automate complex, multi-step recipes with high precision (Smith et al., 2022).

Early Integration of AI in Cooking

Basic automation tasks like temperature control, ingredient recognition, and recipe suggestion systems were the first to be integrated with AI in the kitchen. The primary goal of early AI applications was to improve cooking accuracy and efficiency using programmable appliances, like ovens with preset modes for various recipes. But as computer vision and machine learning algorithms advanced quickly, AI systems started to have more sophisticated features than just basic ones (Singh & kumar, 2025; Sharma & Singh, 2024; Singh & Hassan, 2024; Bhalla et al., 2023). While more modern technologies now automate cooking tasks like stirring, chopping, and even flavour profiling, earlier systems could make recommendations based on the ingredients that were available (Aski et al., 2025). These developments paved the way for the creation of increasingly complex smart cooking appliances, like meal planning assistants and robotic chefs, which are currently transforming kitchens all over the world.

Key Milestones in Smart Cooking Innovations

The development of AI in smart cooking has been characterized by several turning points (Ruan et al., 2023). The development of AI-powered smart ovens, which instantly modify cooking time and temperature according to the particular dish being prepared, is one noteworthy advancement. These ovens ensure accuracy and consistency by using machine learning algorithms to identify the type of food and its ideal cooking conditions. Another innovation is the creation of robotic chefs capable of performing intricate cooking tasks such as grilling, frying, and even plating dishes. According to Maurya et al. (2025), these robots are built to learn from their mistakes and continuously improve their methods to produce excellent outcomes (Singh & kumar, 2025; Sharma & Singh, 2024; Singh & Hassan, 2024; Bhalla et al., 2023). Another significant advancement is the development of AI-powered recipe generators and nutritional calculators, which offer customized meal recommendations based on each person's dietary requirements, health objectives, and ingredient availability. Thanks to these developments, the emphasis has shifted from automating simple tasks to developing intelligent systems that have the potential to completely transform the cooking experience.

The Role of AI in Kitchen Automation

The efficiency of both home and commercial kitchens has significantly increased thanks to AI's role in kitchen automation. AI systems in commercial kitchens help with ingredient sourcing, inventory control, and real-time recipe modifications, freeing up chefs to concentrate more on presentation and creativity. AI-powered gadgets allow home cooks to try out professional-calibre cooking methods (Singh & kumar, 2025; Sharma & Singh, 2024; Singh & Hassan, 2024; Bhalla et al., 2023). For instance, smart kitchen assistants provide real-time cooking assistance, modifying ingredients, cooking temperatures, and timings to accommodate user preferences and dietary requirements (Sharma et al., 2024). These gadgets improve everyone's culinary experience by automating time-consuming chores and providing recipes and meal combinations that might not have been thought of otherwise.

Machine Learning and Computer Vision in Smart Cooking

- Ingredient Recognition: Computer vision systems (e.g., Samsung's AI Family Hub) identify ingredient freshness with 92% accuracy, reducing waste (Lee et al., 2021).
- Recipe Personalization: ML algorithms analyse user data (e.g., allergies, preferences) to tailor meal plans. For example, PlantJammer's app reduced ingredient waste by 22% in a 2022 study.
- Energy Optimization: AI-embedded ovens cut energy use by 15% in commercial kitchens through adaptive heating (Smith et al., 2022).

Smart Cooking Devices (e.g., Smart Ovens, Robotic Chefs)

Both in home and professional settings, the emergence of AI-powered cooking appliances has drastically changed the culinary landscape. The apex of this technological revolution is represented by smart cooking appliances like robotic chefs and smart ovens (Singh & Kumar, 2025; Sharma & Singh, 2024; Singh & Hassan, 2024; Bhalla et al., 2023). With the least amount of human involvement, smart ovens ensure optimal results by utilizing machine learning algorithms to modify cooking temperatures and times according to the type of food inside. With the help of sensors that measure the food's size, texture, and moisture content, these gadgets can modify the cooking process to produce flawless results in every dish (Parvez, 2025). Similar to this, robotic chefs have become more well-liked due to their capacity to carry out intricate culinary duties like chopping, stirring, and even dish plating. Over time, these robots learn and improve their cooking methods using AI algorithms, producing consistently delicious meals (Kumbhar et al., 2025). The Internet of Things (IoT) is made possible in the kitchen by smart cooking appliances, which link appliances like ovens, stoves, and refrigerators to provide a smooth cooking experience. By communicating with the user and one another, these gadgets can monitor the freshness of ingredients, optimize cooking times, and offer tailored suggestions based on dietary requirements and user preferences (Sharma et al., 2024).

By fusing convenience and creativity, these increasingly advanced gadgets enable home cooks and professional chefs to discover new culinary possibilities. Kitchen appliances and accessories are incorporating AI-powered technology beyond cooking equipment to create cutting-edge service tools that enhance operational ergonomics and visitor satisfaction. This idea is demonstrated by the authors' recently registered design of a "Smart Tray" (Mishra et al., 2023), which incorporates clever features meant to enhance sustainability, user interaction, and food service efficiency. These developments enhance culinary AI systems by offering real-world, useful enhancements for both home and workplace settings. Beyond the West, AI-powered cooking technologies are making waves in regions of the Global South. In Nigeria, a local startup developed a smart oven that uses AI to adjust cooking temperatures based on regional recipes, allowing users to cook indigenous dishes like jollof rice and pounded yam with greater accuracy and efficiency (Bola et al., 2023). Similarly, in rural areas of Southeast Asia, AI-driven cooking assistants have been deployed to help local cooks make more nutritious meals by analysing the nutritional content of local ingredients like rice, fish, and vegetables (Lee et al., 2024).

Farm-to-Fork Integration

AI's impact extends beyond the kitchen, with important applications in sourcing, logistics, and agriculture that contribute to sustainable food systems. AI is playing an essential role in modernizing agriculture, ensuring that sustainable practices are implemented from farm to fork. AI tools are being used to optimize crop yields, monitor soil health, and predict optimal harvest times. In Kenya, for instance, AI-driven systems help smallholder farmers predict rainfall patterns and optimize irrigation schedules, leading to more efficient use of water resources and reduced crop loss (Karanja et al., 2024). In the food supply chain, AI is improving logistics by optimizing delivery routes, reducing food spoilage, and ensuring the freshness of produce. Companies like IBM are leveraging AI to streamline the supply chain, using real-time data to predict and prevent supply chain disruptions, reducing the environmental impact of food transport (IBM, 2023). This AI-powered approach ensures that fresh ingredients reach kitchens more efficiently, contributing to sustainability.

AI Powered Recipe Generators and Nutritional Calculators

Recipe generators driven by AI are transforming food preparation and meal planning. Based on nutritional requirements, dietary preferences, and flavour profiles, these systems evaluate the ingredients that are readily available and recommend recipes. By incorporating machine learning, these tools are able to improve recipe recommendations and adjust to users' shifting tastes over time by learning from user feedback (Maurya et al., 2025). Meal planning has never been more individualized thanks to AI's ability to incorporate user dietary restrictions, such as vegan, gluten-free, or low-sodium options, into recipe recommendations. Another essential AI-powered appliance in contemporary kitchens are nutritional calculators. Based on personal health objectives, these tools assess the nutritional value of recipes and recommend healthier substitutes or adjustments. For instance, they can offer substitutes for particular ingredients to address medical conditions like diabetes or hypertension, or they can suggest changes to the macronutrient balance for weight management. By analysing and optimizing meal nutritional content, AI integration promotes wellness by ensuring that users maintain healthier eating habits in addition to enjoying their food.

Food Pairing Algorithms and Flavour Profiling Systems

AI's sophisticated algorithms and flavour profiling systems have revolutionized food pairing. These systems use machine learning to predict which foods pair well together based on complementary flavours, textures, and chemical properties. They do this by leveraging large databases of ingredients and flavour combinations. These algorithms facilitate the development of innovative recipes that challenge conventional culinary pairings, allowing both home cooks and chefs to try out new flavour combinations (Aski et al., 2025). By analysing a food's chemical makeup, flavour profiling systems can pinpoint the main ingredients that give it its flavour and aroma. By understanding these chemical interactions, AI systems can recommend unusual and flavourful pairings that are often not immediately obvious to the human palate. This capability has the potential to transform the food industry, where flavour innovation is a major driver of new product development and consumer engagement, as well as home cooking (Sharma et al., 2024).

Machine Learning and Smart Cooking

The development of smart cooking technologies now heavily relies on machine learning and computer vision. Computer vision-powered ingredient recognition systems can use visual information to determine the type, freshness, and quality of ingredients (Singh & Kumar, 2025; Sharma & Singh, 2024; Singh & Hassan, 2024; Bhalla et al., 2023). AI-powered smart refrigerators, for example, can identify the contents and monitor when they expire, recommending recipes based on what is on hand and encouraging users to use ingredients before they go unsafe (Gonia et al., 2025). Cooking becomes more sustainable as a result of the decreased food waste. Recipe recommendations also heavily rely on visual recognition. AI programs are able to look through a kitchen's contents and suggest meals using the ingredients that are there. AI can advise users on how to best prepare ingredients by analysing their state (raw vs. cooked), allowing them to maximize their available resources. By encouraging the use of current ingredients before they are wasted, improves the cooking process and increases the reduction of food waste (Sharma et al., 2024).

Deep Learning in Meal Personalization

At the forefront of AI's capacity to offer tailored meal recommendations are deep learning algorithms. To provide highly personalized meal plans, these sophisticated neural networks analyse enormous volumes of data, including user preferences, dietary requirements, and medical conditions. By using real-time data and historical feedback, deep learning systems can adjust to a user's changing preferences and provide personalized meal recommendations (Parvez, 2025).People with particular health needs, such as those who are managing chronic conditions or adhering to special diets, can benefit from AI's personalization capabilities. To adapt to shifting health objectives and make sure that meals meet dietary needs, AI systems can continuously improve their meal recommendations. This degree of customization improves the eating experience and helps users make better long-term food choices.

The Role of Machine Learning and Computer Vision in Smart Cooking

Significant improvements in ingredient recognition and quality control in smart kitchens have been made possible by machine learning and computer vision. AI systems can now recognize ingredients based on their size, shape, and visual appearance thanks to cameras and image recognition algorithms, guaranteeing that only the freshest ingredients are used in cooking. For example, AI-enabled smart refrigerators can track the freshness and expiration dates of products, alerting users to use specific ingredients before they go rotten. AI systems can also evaluate the quality of produce, such as the ripeness of fruits or the freshness of meat, which can greatly cut down on food waste and enhance the calibre of the food that is prepared.

Visual Recognition for Recipe Suggestions

Visual recognition significantly enhances meal planning and recipe suggestions. AI can provide tailored recipe suggestions based on the ingredients in a kitchen by scanning its contents, preventing food waste. Based on the combination of items in the user's pantry, the system can identify frequently used ingredients, recommend complementary pairings, and even suggest unusual dishes. Additionally, visual recognition can help ensure that ingredients are prepared correctly by determining their state (e.g., cooked or raw), which can affect the cooking process. This AI integration enables a seamless and personalized cooking experience, tailored to the user's resources.

Deep Learning in Meal Personalization

In AI-powered cooking, deep learning algorithms are leading the way in meal personalization. To provide highly customized meal recommendations, these sophisticated neural networks can process enormous volumes of data from user preferences, dietary restrictions, and health objectives. Deep learning models, for instance, are able to adjust to user preferences over time, learning what flavours, textures, and cuisines they like. People with particular dietary requirements, such as those on low-sodium, vegan, or gluten-free diets, may find this degree of customization especially helpful. The system guarantees that meals are customized to the user's preferences and satisfy their nutritional needs by iteratively improving its recommendations in response to feedback.

AI in Recipe Development and Innovation

The creation and innovation of recipes is one of AI's most important contributions to the culinary industry. Artificial intelligence (AI) systems can analyse enormous databases of ingredients, cooking methods, and cultural culinary customs to produce inventive and fascinating recipes. AI can recommend flavour combinations that may be unusual but work well together thanks to deep learning and neural networks, pushing the limits of culinary creativity (Singh & kumar, 2025; Sharma & Singh, 2024; Singh & Hassan, 2024; Bhalla et al., 2023). Chefs and home Cooks can now experiment with flavours, textures, and presentation in previously unthinkable ways thanks to these AI-generated recipes, which are not constrained by conventional knowledge or techniques. AI can also help create recipes that accommodate dietary requirements or preferences, making sure that innovation meets consumer demand for more individualized and healthful meal options. AI's role in pushing culinary boundaries is not just limited to Western dishes but is expanding globally. In Mexico, AI systems have been trained to generate recipes based on indigenous ingredients such as maize, beans, and chili, helping chefs reimagine traditional Mexican flavors with innovative twists while maintaining cultural authenticity (Moreno et al., 2023). Similarly, in Egypt, AI algorithms are being used to explore flavor combinations involving spices like cumin and coriander, staples of Egyptian cuisine, leading to new recipes and culinary fusion (El-Shenawy et al., 2024).

Enhancing Culinary Techniques with AI-Assisted Cooking

AI-powered tools and systems are advancing culinary techniques in addition to increasing kitchen efficiency. For example, cooking robots and smart ovens can perform highly accurate cooking tasks like sous-vide cooking and chocolate tempering. These devices use AI to monitor and control variables like temperature, humidity, and timing, ensuring the success of such techniques. AI technologies can also give chefs real-time feedback while they cook, assisting them in achieving consistency and the best possible outcomes. Because of this degree of accuracy, chefs are able to express their creativity and try out novel methods that would be challenging to learn in a conventional kitchen.

The Role of AI in Professional and Home Kitchens

By providing sophisticated culinary capabilities that were previously only available in upscale restaurants or with specialized chefs, the incorporation of AI into both home and professional kitchens has levelled the playing field. AI helps in professional kitchens with things like inventory control, sourcing ingredients, and making real-time recipe adjustments, allowing chefs to concentrate on presentation and creativity. Home cooks can now easily prepare restaurant-calibre meals in their kitchens thanks to AI-powered appliances like robotic chefs and smart kitchen assistants. These developments have democratized cooking and encouraged creativity in kitchens of all skill levels by enabling people without formal culinary training to experiment with sophisticated flavours and techniques.

Sustainability through AI-Optimized Cooking Methods

By optimizing cooking techniques to use fewer resources, AI can further advance sustainability. AI can guarantee that food is prepared in the most energy-efficient manner possible by precisely controlling temperature and cooking time, removing waste from overheating or cooling (Sarokin & Bocken, 2024). AI-powered smart appliances can modify cooking techniques to reduce energy consumption without sacrificing food flavour or nutritional value, making cooking healthier and more environmentally friendly (Singh & Kumar, 2025; Sharma & Singh, 2024; Singh & Hassan, 2024; Bhalla et al., 2023). AI systems can suggest cooking methods like sous-vide, steaming, or slow cooking, for instance, which use less energy than traditional methods like frying or grilling while maintaining the nutritional value of ingredients. By encouraging cooking techniques that retain more nutrients, this helps to create healthier meals in addition to saving energy (Suryani et al., 2025). AI speeds up the cooking process and makes it more environmentally and health-friendly by optimizing cooking techniques based on the ingredients and intended results. Sustainability practices in the culinary industry have greatly improved as a result of recent advancements in artificial intelligence. The authors' patented approach (Sharma et al., 2023) uses AI to categorize and measure food waste through computer vision and machine learning algorithms, optimizing food production and suggesting in-the-moment waste reduction measures. These systems highlight AI's usefulness in drastically reducing food waste, proving the technologies' immediate relevance and efficiency in actual kitchen environments.

AI's Role in Personalized Nutrition and Health-Conscious Cooking

AI's role in customized nutrition is growing in significance as sustainability in cooking transcends environmental concerns. Promoting individualized, healthier eating habits is one way AI helps with sustainability (Arslanoglu et al., 2023). Artificial intelligence (AI) systems evaluate user data, including dietary preferences, allergies, and medical conditions, to suggest meals that promote particular health objectives, like better heart health or weight control. These tailored recommendations guarantee that meals are in line with each person's specific health needs in addition to being nutritionally balanced (Sharma et al., 2024). AI can also help create customized diet plans for people who have special dietary needs, like low-sodium, vegan, or gluten-free diets. By avoiding the purchase of extraneous ingredients and guaranteeing that meals are customized to the user's needs, this degree of personalization lowers food waste and avoids excessive leftovers (Gonia et al., 2025). The efficacy of AI in fostering sustainable eating practices and long-term health improvements is increased by its capacity to learn and adjust to personal preferences. One of the most significant uses of AI in the food industry today is personalized nutrition. To recommend meals that are especially suited to each person's needs, artificial intelligence (AI) systems examine a vast array of data, including user preferences, medical conditions, allergies, and even genetic factors. To develop meal plans that support users' health objectives, whether they are weight management, metabolic health improvement, or the treatment of chronic conditions like diabetes or hypertension, these systems take into account variables like vitamin content, macronutrient balance, and potential food sensitivities (Kumbhar et al., 2025). Meal preparation can be made easier and healthier with the help of AI-driven meal planners that can suggest ingredient substitutions, cooking techniques, and portion sizes based on individual health data. To ensure that meals meet specific nutritional targets while remaining enjoyable, AI-powered apps, for example, can offer recommendations for lowering sugar intake or increasing fibre in the diet based on the user's health profile (Sharma et al., 2024). AI can modify its recommendations over time by combining real-time feedback and continuous learning, assisting people in maintaining a balanced diet that fits with their changing medical requirements.

AI's Role in Special Diets (e.g., Gluten-Free, Vegan)

Another significant area where AI has changed the culinary landscape is in its support of special diets. AI systems can recommend meals that satisfy dietary requirements such as low-carb, vegan, or gluten-free while maintaining nutritional balance for people on these diets. Artificial intelligence (AI) tools can examine ingredient databases to find appropriate alternatives to restricted ingredients. For example, they can suggest gluten-free grains for individuals with celiac disease or replace dairy with plant-based alternatives (Suryani et al., 2025). AI systems can also assist in educating users about the different options available, preventing those following these particular diets from missing out on important nutrients. For vegans, AI can recommend plant-based protein sources; for gluten-free people, it can recommend alternative grains. Due to its versatility, AI is a vital instrument for fostering inclusivity in the food industry, meeting a range of dietary requirements without sacrificing taste or nutritional value (Gonia et al., 2025).

AI's Impact on Health and Wellness through Cooking

AI can significantly affect general health and wellness in addition to individualized nutrition. By recommending healthier cooking techniques like steaming, grilling, or baking rather than frying, AI-driven cooking appliances and apps assist people in making better decisions. This emphasis on cooking techniques that preserve food's nutritional integrity supports long-term health by promoting heart-healthy, low-fat meals (Kumbhar et al., 2025; Hassan & Singh, 2025; Raheja et al., 2025). By analysing users' long-term health data, monitoring nutrient intake, and providing tailored feedback, AI systems also support health-conscious cooking. AI, for instance, can identify dietary excesses or deficiencies based on a person's past food intake and recommend changes, like increasing the intake of foods high in iron for anaemia sufferers or suggesting foods high in potassium for hypertension patients. In addition to promoting better eating practices, this AI integration gives users the ability to make better decisions regarding their diet and way of life (Sharma et al., 2024).

Promoting Sustainable Health Practices

AI's capacity to optimize dietary choices goes beyond personal well-being to encourage sustainable health behaviours that benefit the general public. AI contributes to long-term sustainability and well-being by lowering food waste through customized meal planning and promoting healthier cooking practices. By suggesting diets that stop the progression of diseases, such as low-sodium diets for people with high blood pressure or calorie-restricted diets for people with obesity, AI can also help people manage chronic conditions (Suryani et al., 2025). Additionally, the use of AI in nutrition affects society as a whole. By offering data-driven insights into dietary trends, artificial intelligence (AI) can support public health campaigns and help policymakers develop more potent health interventions. AI systems will probably become more significant in enhancing public health outcomes as they are embraced more broadly, helping communities and individuals achieve better nutrition-based health. In the Global South, AI's ability to optimize dietary choices is also making a significant impact. In rural India, for example, AI-driven meal planners consider local agricultural production cycles and suggest meals based on available seasonal produce like lentils, rice, and vegetables, which promotes health-conscious and sustainable eating (Singh et al., 2024). This AI-driven approach helps combat malnutrition by optimizing nutrient intake in underprivileged areas. Additionally, AI systems are being used to promote sustainable diets in countries like Kenya, where machine learning algorithms have been integrated into local food systems to recommend meals that are both nutritionally balanced and environmentally sustainable, utilizing locally sourced ingredients (Karanja et al., 2024).

Emerging Trends in AI-Powered Cooking Technologies

New developments in AI technologies are influencing the direction of smart cooking and have the potential to completely transform the food industry. The growing use of conversational and voice-activated AI systems in kitchens is among the most exciting trends. Hands-free cooking is made possible by the integration of these smart assistants into various kitchen appliances. Simple voice commands allow users to access recipes, set timers, modify cooking parameters, and get real-time help. People of all skill levels should find cooking more efficient, accessible, and pleasurable as a result of this trend (Suryani et al., 2025; Singh & kumar, 2025; Sharma & Singh, 2024; Singh & Hassan, 2024; Bhalla et al., 2023). Hyper-personalization in cooking is also becoming more popular. AI systems will provide meal recommendations that are specifically suited to each user's lifestyle as they grow better at comprehending personal preferences, health objectives, and even cultural influences. As AI algorithms that process data from various sources, such as wearables, health trackers, and Internet of Things devices, advance, this trend toward highly personalized meal planning and preparation is likely to become more complex (Kumbhar et al., 2025). The emphasis on individualized experiences may also be extended to the food industry, where artificial intelligence (AI) tools predict demand trends, streamline supply chains, and enhance resource management to guarantee that meals are not only produced sustainably but also specifically for customers.

The Role of AI in Shaping Future Culinary Practices

In the future, we expect AI to play a more significant role in shaping cooking practices. AI technologies are already helping chefs in professional kitchens with duties like creating recipes, locating ingredients, and managing inventories. As these technologies advance, artificial intelligence (AI) could play an even bigger role in culinary education by assisting aspiring chefs in developing recipes, learning new techniques, and experimenting with flavours in virtual training environments. AI is predicted to become a vital tool in home kitchens for both inexperienced and seasoned chefs. Regardless of their culinary experience, users will be able to easily prepare restaurant-calibre meals with the aid of smart kitchen appliances. Real-time feedback on cooking methods will be possible with these gadgets, assisting users in honing their craft and getting the best results. People of all skill levels may be able to use advanced cooking techniques as technology becomes more widely available and reasonably priced (Sharma et al., 2024; Singh & kumar, 2025). We anticipate a greater role for AI in food education. AI-powered virtual assistants will walk users through the cooking process and offer educational materials on sophisticated cooking methods, ingredient combinations, and meal planning that prioritizes health. Food culture may change as a result, with cooking no longer viewed as a difficult or specialized skill but rather as a fun and approachable pastime for all.

Challenges and Opportunities for AI in Gastronomy

Although AI in gastronomy has a lot of promise, several obstacles need to be overcome before its full potential can be achieved. The accessibility of AI technologies is a major obstacle. Even though AI tools are becoming more widely used, their use is often costly and necessitates a certain level of technological expertise (Hassan & Singh, 2025; Raheja et al., 2025). For AI-powered kitchen appliances to be widely adopted, it will be essential to make sure that a wider range of people, including those without a lot of technical knowledge, can use them. Striking the correct balance between traditional culinary craftsmanship and AI-driven automation presents another difficulty. Many chefs and culinary professionals worry that an excessive reliance on technology may compromise the artistry and creativity inherent in cooking, even though AI can greatly increase efficiency and consistency. Maintaining the creative and cultural elements of gastronomy will require finding a balance that enables AI to supplement human expertise rather than replace it (Gonia et al., 2025; Hassan & Singh, 2025). AI has a lot of potential in the culinary sector, despite these obstacles. AI has the power to transform food production, enhance sustainability, and expand access to customized cooking experiences. Additionally, AI systems can offer insightful information about consumer behavior, assisting chefs and food producers in anticipating trends and modifying their products to satisfy shifting consumer demands (Kumbhar et al., 2025). AI will probably play a bigger part in determining the direction of gastronomy in the future as it develops further, opening up new avenues for creativity and innovation in the culinary arts.

Future Speculative Trends

The future of AI in the culinary industry holds exciting possibilities, from synthetic foods to the integration of AR/VR technologies.

AI in Synthetic Foods

AI is expected to play a major role in the development of synthetic foods, particularly in the context of lab-grown meats and plant-based alternatives. AI-driven research is accelerating the development of lab-grown meat, with companies like Mosa Meat and Eat Just using AI to improve the texture, taste, and cost-effectiveness of lab-grown proteins (Mosa Meat, 2025). This innovation could dramatically reduce the environmental impact of food production, particularly in the Global South, where land and water resources are often scarce.

Augmented Reality (AR) and Virtual Reality (VR) in Cooking

AI-powered AR and VR applications are revolutionizing culinary education and home cooking experiences. For instance, VR platforms allow aspiring chefs to practice cooking techniques virtually before attempting them in a physical kitchen. Similarly, AR could be used to guide home cooks through complex recipes step-by-step, providing real-time feedback and enhancing learning. As these technologies become more sophisticated and accessible, they will likely become integral parts of the cooking experience.

CONCLUSION

Smart cooking technologies and their effects on culinary creativity, efficiency, sustainability, and personalized nutrition, this paper has examined the revolutionary role of artificial intelligence (AI) in the culinary industry. Both home and commercial kitchens have been transformed by AI-powered appliances like meal planning assistants, robotic chefs, and smart ovens. By reducing food waste and optimizing energy use, these technologies have made it possible to automate cooking procedures, customize culinary experiences, and advance sustainability. Chefs and home cooks can now try out new methods, ingredients, and flavours thanks to AI's unique recipe suggestions, ability to customize meals to meet dietary requirements, and real-time cooking assistance. Additionally, by reducing food waste, optimizing resource usage, and promoting healthier cooking techniques, AI's integration in kitchens has aided in the development of more sustainable food practices. AI's ability to tailor nutrition has also changed eating patterns by giving consumers meal plans that are in line with their individual health objectives.

The Future Prospects of AI in Smart Cooking

AI in smart cooking has a bright future ahead of it, as several new trends are set to further advance culinary innovation. The next generation of smart kitchens is anticipated to be shaped by advancements such as voice-activated AI systems, AI-driven food production, and highly customized meal planning (Hassan & Singh, 2025; Raheja et al., 2025). AI systems will continue to democratize culinary skills as they advance in sophistication and accessibility, empowering people of all skill levels to prepare individualized, wholesome, and sustainable meals.AI will also play a bigger part in influencing food production, assisting chefs, food producers, and consumers in making data-driven choices that maximize sustainability, cut waste, and tackle issues related to global food security. More seamless cooking experiences that combine automation, creativity, and personalization will also be made possible by AI's continued development and integration with machine learning, computer vision, and the Internet of Things. A crucial part of the future culinary scene, artificial intelligence (AI) has the potential to transform culinary practices and advance more general social objectives like promoting sustainability and enhancing public health.

REFERENCES

Arslanoglu, S., Moro, G. E., Tonetto, P., De Nisi, G., Ambruzzi, A. M., Biasini, A., Profeti, C., Gagliardi, L., Salvatori, G., & Bertino, E. (2023). Recommendations for the establishment and operation of a donor human milk bank. *Nutrition Reviews*, *81*(Supplement_1), 1–28. DOI: 10.1093/nutrit/nuad012 PMID: 36892193

Aski, M. S., Mallikarjuna, M. G., & Lohithaswa, H. C. (2025). Breeding climate resilient and future-ready pulse crops: Industry 5.0 technologies for sustainable food production. Springer. Retrieved from https://link.springer.com/chapter/10.1007/978-981-96-0483-8_1

Bhalla, A., Singh, P., & Singh, A. (2023). Technological Advancement and Mechanization of the Hotel Industry. In Tailor, R. (Ed.), *Application and Adoption of Robotic Process Automation for Smart Cities* (pp. 57–76). IGI Global., DOI: 10.4018/978-1-6684-7193-7.ch004

Chen, L., Wang, Y., & Zhang, Q. (2021). Machine learning-driven optimization of energy consumption in smart ovens. *IEEE Transactions on Automation Science and Engineering*, *18*(3), 456–467. DOI: 10.1109/TASE.2021.12345

Chourasia, S., Kaur, S., & Singh, A. (2025). The Future of Indian Restaurants: The Role of Technology in the Success of Cloud Kitchens. In Singh, K., Kumar, S., & Talukder, M. (Eds.), *Impact of AI and the Evolution of Future Ghost Kitchens* (pp. 15–34). IGI Global Scientific Publishing., DOI: 10.4018/979-8-3373-0432-8.ch002

Deri, M. N., Singh, A., Zaazie, P., & Anandene, D. (2024). Demystifying Sustainable Practices in the Hotel Industry: Evidence From Ghana. In Ganiyu, I., Olarewaju, O., Ige-Olaobaju, A., & Atiku, S. (Eds.), *Waste Management and Life Cycle Assessment for Sustainable Business Practice* (pp. 21–44). IGI Global Scientific Publishing., DOI: 10.4018/979-8-3693-2595-7.ch002

Garcia, M., Patel, S., & Kim, J. (2022). AI-driven meal planners and their role in household sustainability. *Sustainability*, *14*(5), 1123. DOI: 10.3390/su14051123

Gonia, D. H., Jamodiong, M. I., Yadao, H. M., ... (2025). ScaNana: Unlocking potential products from scanned banana varieties using image recognition. IEEE Xplore. Retrieved from https://ieeexplore.ieee.org/abstract/document/10967832/

Hassan, S. C., & Singh, A. (2025). Preserving Traditional Cuisines and Methods in the Culinary World: Empowering the Community. In Rana, V., Raina, A., & Bathla, G. (Eds.), *Global Sustainable Practices in Gastronomic Tourism* (pp. 393–402). IGI Global Scientific Publishing., DOI: 10.4018/979-8-3693-7096-4.ch023

Kumbhar, D. V., Yarde, P., Maddel, M., & Gidde, V.. (2025). SmartChef: Rebellion of AI and IoT in culinary businesses from traditional to cutting-edge technology kitchen. *SSRN*. Retrieved from https://papers.ssrn.com/sol3/Delivery.cfm?abstractid =5195618 DOI: 10.2139/ssrn.5195618

Lee, H., Park, S., & Nguyen, T. (2021). Accuracy of computer vision systems in real-world ingredientrecognition. *Journal of Food Engineering*, *298*, 110532. DOI: 10.1016/j.jfoodeng.2021.110532

Majumder, S., Singh, V., Anjum, S., Chauhan, U. S., & Kumar, A. &Aashiyan. (2023). Method and apparatus for efficient flavor extraction in alcoholic beverages (Indian Patent Application No. 202311060736). Intellectual Property India. Filed September 9, 2023, Published October 13, 2023.

Maurya, V., Singh, E., Patel, D., & Sheth, D. (2025). Ghar Ka Zayka: An AI-powered digital platform for community-based recipe sharing. ResearchGate. Retrieved from https://www.researchgate.net/profile/Vishal-Maurya-20/publication/389875475_Ghar_Ka_Zayka_An_AI-Powered_Digital_Platform_for_Community-Based_Recipe_Sharing/links/67d587ee2719090652baf359/Ghar-Ka-Zayka-An-AI-Powered-Digital-Platform-for-Community-Based-Recipe-Sharing.pdf

McKinsey & Company. (2022). AI in food service: Cost reduction and efficiency gains.

Mishra, R., Arya, S., Jaswal, R., Singh, Y., & Majumder, S. (2023). Smart Tray (Indian Design No. 387539-001, Class 07-01). Intellectual Property India. Registered June 3, 2023.

Nestlé. (2022). ChefInspire: AI-driven recipe innovation. https://www.nestle.com/chefinspire-case-study*

Parvez, M. O. (2025). The algorithmic appetite: Reflections on the future of robotic chefs in the restaurant industry. Taylor & Francis. Retrieved from https://www.tandfonline.com/doi/full/10.1080/15428052.2025.2495646

Raheja, S., Singh, G., & Singh, A. (2025). Strategies and Challenges for Food Waste Management: A Comprehensive Analysis. In Rana, V., Raina, A., & Bathla, G. (Eds.), *Global Sustainable Practices in Gastronomic Tourism* (pp. 499–512). IGI Global Scientific Publishing., DOI: 10.4018/979-8-3693-7096-4.ch029

Ruan, C., Jiang, W., Liao, J., Liu, Z., Song, Y., & Qi, J. (2023). Ethanol/salt aqueous two-phase systems (ATPS) for enrichment and identification of bioactive constituents escaping during pectin extraction. *International Journal of Food Science & Technology*, *58*(9), 4726–4737. DOI: 10.1111/ijfs.16578

Sarokin, S. N., & Bocken, N. (2024). Pursuing profitability in slow fashion: Exploring brands' profitcontributors. *Journal of Cleaner Production, 444*, 141237. DOI: 10.1016/j.jclepro.2024.141237

Sharma, D., Sharma, M., & Sharma, A. (2024). SMART kitchen technology. IEEE Xplore. Retrieved from https://ieeexplore.ieee.org/abstract/document/10842745/

Sharma, M., & Singh, A. (2024). Embracing Technological Innovation: A Review of Hi-Tech Services in Hospitality Industry. *Evergreen, 11*(4), 2818–2830. DOI: 10.5109/7326926

Sharma, M. P., Majumder, S., Rana, V. S., Verma, A. K., Divakar, C., Jaswal, R., Dsouza, C. J., & Kar, A. (2023). Method for food and beverages waste management using Artificial Intelligence (AI) (Indian Patent). Intellectual Property India. Complete Specification Filed 2023.

Singh, A., & Hassan, S. C. (2024). Service Innovation Through Blockchain Technology in the Tourism and Hospitality Industry: Applications, Trends, and Benefits. In Singh, S. (Ed.), *Service Innovations in Tourism: Metaverse, Immersive Technologies, and Digital Twin* (pp. 205–214). IGI Global., DOI: 10.4018/979-8-3693-1103-5.ch010

Singh, A., & Kumar, V. (2025). Organizational Innovation and Adaptation in Hospitality and Tourism: Strategies for Resilience and Leadership. In C. Ramos, T. Costa, R. Madeira, & J. Ferreira do Rosário (Eds.), Open Innovation and Technology in Tourism and Hospitality (pp. 355-372). IGI Global Scientific Publishing. DOI: 10.4018/979-8-3693-8633-0.ch018

Singh, A., Tyagi, P. K., & Garg, A. (Eds.). (2024). *Sustainable Disposal Methods of Food Wastes in Hospitality Operations*. IGI Global., DOI: 10.4018/979-8-3693-2181-2

Smith, A., Brown, K., & Wilson, P. (2022). Cultural bias in AI-driven recipe generation. *Journal of Culinary Science & Technology, 20*(4), 345–360. DOI: 10.1080/15428052.2022.1234567

Suryani, W., Sari, W. P., & Ginting, S. M. B. (2025). Innovative trends shaping food marketing and consumption: Culinary marketing in the digital age. IGI Global. Retrieved from https://www.researchgate.net/profile/Zahid-Hussain-86/publication/388540593_Innovative_Trends_Shaping_Food_Marketing_and_Consumption/links/67abc37696e7fb48b9bf14b5/Innovative-Trends-Shaping-Food-Marketing-and-Consumption.pdf#page=29

Winnow Solutions. (2023). Reducing food waste in commercial kitchens: IKEA UK case study. https://www.winnowsolutions.com/case-studies/ikea-uk*

Chapter 7

From Kitchens to Code: Cross-Cultural Perspectives on AI and Smart Cooking Adoption

Aruditya Jasrotia
https://orcid.org/0000-0002-8274-5609
Amity University, Noida, India

Suman Lata
https://orcid.org/0000-0002-9862-0739
Amity University, Noida, India

ABSTRACT

This study explores how cultural values, traditions, and attitudes influence the adoption and perception of AI-driven smart cooking technologies (e.g., smart kitchen appliances, AI recipe assistants, automated cooking robots) across different cultural contexts. While some cultures may embrace technology for its convenience and innovation, others may view it as a threat to traditional culinary practices, family roles, or food identity. The research aims to compare user acceptance, usage patterns, and cultural resistance to AI in the kitchen across diverse regions or demographic groups.

INTRODUCTION

The integration of artificial intelligence (AI) into domestic kitchens marks a transformative shift in how people engage with food preparation, blending cutting-edge technology with one of humanity's oldest practices—cooking. From AI-powered recipe assistants to automated cooking robots and smart appliances, these innovations promise convenience, precision, and personalization. Yet, their adoption is far

DOI: 10.4018/979-8-3373-4042-5.ch007

from universal, shaped by deeply rooted cultural values, traditions, and attitudes toward technology. In some societies, smart cooking tools are celebrated as symbols of progress, while in others, they are met with skepticism, perceived as threats to culinary heritage, family roles, or the sensory artistry of traditional cooking. This chapter, "From Kitchens to Code: Cross-Cultural Perspectives on AI and Smart Cooking Adoption," explores how cultural contexts influence the perception and integration of AI-driven cooking technologies, offering a comparative lens on global kitchens navigating this technological frontier.

Cooking is more than a practical task; it is a cultural artifact, embodying identity, memory, and social bonds (Counihan & Van Esterik, 2013). As AI reshapes this domain, questions arise about its compatibility with diverse culinary traditions and its impact on roles historically tied to food preparation. For instance, in tech-forward Japan, robotic sushi makers may align with a culture of innovation, whereas in Italy, where cuisine is a cornerstone of national identity, such tools might face resistance for undermining artisanal practices (Ceccarini, 2011). Similarly, in India, where cooking is often a communal and ritualistic act, AI's promise of efficiency may clash with the emotional and spiritual dimensions of food preparation (Sen, 2009). These variations highlight the need to examine not only technological adoption but also the cultural narratives that shape it.

This study has four primary objectives: (1) to investigate cultural differences in attitudes toward AI in domestic cooking, (2) to identify factors such as tradition, technology literacy, age, and cuisine type that drive acceptance or rejection of smart cooking tools, (3) to assess how AI is redefining traditional cooking roles and practices across cultures, and (4) to explore consumer expectations for personalization, cultural relevance, and trust in AI-based cooking solutions. Through comparative case studies of culturally distinct regions—such as Japan, Italy, and India—combined with surveys, interviews, and thematic analysis, this research seeks to uncover the interplay between innovation and tradition in the modern kitchen. By reviewing popular smart kitchen products, it also evaluates their cultural adaptability, shedding light on how technology can better align with diverse foodways.

Ultimately, this chapter aims to contribute to a nuanced understanding of AI's role in reshaping culinary practices, offering insights for designers, policymakers, and consumers navigating the intersection of technology and culture. As kitchens evolve from analog to algorithmic, the challenge lies in ensuring that smart cooking technologies honor the diversity of global culinary traditions while fostering innovation.

LITERATURE REVIEW

The integration of artificial intelligence (AI) into domestic kitchens represents a significant evolution in culinary practices, reshaping how food is prepared, stored, and consumed across diverse cultural contexts. This literature review synthesizes recent research on the adoption of AI-driven smart cooking technologies, focusing on cultural influences, user acceptance, and the interplay between tradition and innovation. Drawing on studies from 2019 to 2025, it examines the factors shaping perceptions of smart kitchen appliances, the role of cultural values in technology adoption, and the implications for culinary practices globally. The review is organized into three key themes: (1) AI applications in smart kitchens, (2) cultural influences on technology adoption, and (3) challenges and opportunities in integrating AI with traditional cooking practices.

AI Applications in Smart Kitchens

Recent advancements in Artificial Intelligence (AI) and the Internet of Things (IoT) have revolutionized domestic kitchens, turning them into highly efficient, personalized, and sustainable ecosystems. Smart kitchen appliances—such as AI-powered ovens, refrigerators, and automated cooking robots—utilize machine learning, computer vision, and natural language processing to streamline food preparation and management. For example, modern smart ovens can dynamically adjust temperature and cooking time based on real-time analysis of food type, ensuring perfectly cooked meals with minimal user intervention (Tmcnet, 2024). Similarly, IoT-enabled refrigerators now track food expiration dates, recommend recipes based on available ingredients, and even place automatic grocery orders, reducing household food waste by up to 14.8% (Cain, 2024). These innovations exemplify Industry 4.0 principles, where smart sensors, big data analytics, and cloud computing converge to enhance culinary precision and energy efficiency (Hassoun et al., 2022).

The applications of AI in food preparation are rapidly expanding. According to Gupta et al. (2023), approximately 26% of AI-related patents in the food industry focus on intelligent cooking devices, including self-adjusting microwaves and robotic chefs that minimize manual labour while improving convenience. Another key development is AI-driven personalized nutrition, where smart appliances analyse dietary restrictions and health data to customize meal plans (Murugeah, 2024). Reinforcement learning algorithms further enhance food safety by optimizing cooking parameters to eliminate risks of undercooking (Hassoun et al., 2024).

Despite these benefits, challenges remain. High costs and technical complexity hinder widespread adoption, particularly among smaller households and less tech-savvy users (Anagnostopoulos, 2016). Additionally, concerns about data privacy and interoperability between different smart devices pose further obstacles. Nevertheless, as AI and IoT technologies continue to evolve, smart kitchens are poised to become even more intuitive, reducing food waste, saving energy, and simplifying meal preparation for consumers worldwide. Future advancements may include greater integration with voice assistants, predictive maintenance for appliances, and even AI-generated recipe innovations based on real-time ingredient availability.

Cultural Influences on Technology Adoption

Cultural values play a pivotal role in shaping the acceptance of AI-driven cooking technologies, with adoption rates varying significantly across regions and demographics. In technologically progressive societies like Japan and South Korea, where innovation is culturally celebrated, smart kitchen appliances are increasingly embraced. For instance, Japan's integration of voice-activated cooking assistants and robotic chefs aligns with its societal emphasis on precision and efficiency, particularly among younger, urban populations. Similarly, South Korea's rapid adoption of AI-powered kitchen tools reflects its tech-savvy consumer base and government-backed digital initiatives. Conversely, in cultures where cooking is deeply rooted in tradition and identity, resistance to AI technologies is more pronounced. In Italy, culinary practices are often regarded as an art form, and smart appliances are perceived as threats to the authenticity of time-honoured methods. Similarly, in India, where cooking is a communal and ritualistic activity, AI's focus on efficiency may clash with the emotional and social aspects of food preparation. These cultural nuances highlight how deeply ingrained values influence technological adoption.

Demographic factors such as age, technology literacy, and urbanization also impact adoption trends. A multi-country study in Spain, the UK, and Germany found that younger, urban households are more likely to adopt devices like 3D food printers and smart sous-vide cookers due to their tech familiarity and convenience-driven lifestyles. In contrast, older generations and rural communities often resist these innovations, fearing a loss of cultural authenticity and traditional cooking skills. Trust in technology and compatibility with local cuisines further shape consumer attitudes. Collectivist societies, where cooking is a familial and social activity, may view AI as disruptive to communal practices, whereas individualistic cultures may prioritize its time-saving benefits. Additionally, ethical concerns—such as data privacy and the potential homogenization of cuisines by AI—pose further challenges to widespread acceptance. Ultimately, the global adoption of AI-driven cooking technologies hinges on balancing innovation with cultural preservation, ensuring that these advancements enhance rather than erode traditional culinary identities.

Challenges and Opportunities in Integrating AI with Traditional Cooking Practices

The integration of AI into domestic kitchens presents both opportunities and challenges, particularly in balancing technological innovation with cultural preservation. One major opportunity is the potential for AI to enhance sustainability and health outcomes. Smart kitchen technologies can reduce food waste, optimize energy consumption, and promote nutritious eating through personalized meal planning (Junejo et al., 2021). For instance, IoT-enabled appliances can lead to 10–15% energy savings compared to traditional devices, contributing to environmentally conscious lifestyles (Wang, 2023). Additionally, AI's ability to adapt recipes to cultural preferences offers a pathway to make smart technologies more inclusive, addressing consumer expectations for culturally relevant solutions (Murugeah, 2024).

However, challenges persist, including data privacy, cultural adaptability, and equitable access. Studies indicate that concerns about data security and transparency hinder adoption, particularly in regions with stringent privacy regulations (Hassoun et al., 2024). Moreover, the lack of cultural adaptability in AI systems—such as recipe algorithms that fail to account for regional ingredients or cooking techniques—limits their appeal in diverse markets (Liu & Liu, 2023). For example, smart appliances designed for Western cuisines may not accommodate the complex spice profiles or cooking methods of South Asian dishes, leading to resistance among Indian consumers (Sen, 2009). Economic barriers also play a role, as high costs restrict access to smart technologies in low-income or rural areas, exacerbating digital divides (Nguyen, 2017).

Another challenge is the tension between automation and human agency. While AI can enhance efficiency, it risks diminishing the sensory and social aspects of cooking, which are central to many cultures. Coles (2013) argues that food is not just sustenance but a medium for relationships and cultural identity, and over-reliance on AI could erode these values. Conversely, research on assistive technologies, such as smart cookers for cognitively impaired users, suggests that AI can preserve autonomy by guiding users through cooking tasks without fully automating them (Kislov, 2025). This balance is critical to ensuring that AI complements rather than replaces traditional practices.

Research Gaps and Future Directions

The growing body of research on AI in culinary applications has revealed several critical gaps that need to be addressed. First and foremost, there is a notable absence of longitudinal studies examining how AI adoption affects culinary practices and cultural identity over extended periods. While existing research provides valuable insights into initial user perceptions and short-term adoption patterns, it fails to capture whether these technologies lead to sustained behavioral changes or potential erosion of traditional cooking skills. For instance, we lack data on whether younger generations in tech-savvy societies gradually abandon traditional cooking methods in favor of AI-assisted convenience, or if they find ways to harmonize both approaches. This gap in long-term understanding makes it difficult to predict the true cultural impact of these technologies.

Another significant limitation is the Western-centric focus of current research, which largely overlooks non-Western contexts where traditional cooking methods remain dominant. The majority of studies concentrate on technologically advanced societies like Japan, South Korea, and Western European countries, while paying scant attention to regions such as Africa, South Asia, and parts of Latin America where cooking practices are deeply intertwined with cultural identity and community values. This oversight is particularly problematic because AI systems trained primarily on Western culinary data may fail to accommodate the diverse ingredients, techniques, and cultural significances found in other cuisines. For example, an AI recipe generator might struggle to properly interpret the cultural importance of certain traditional dishes in Nigerian or Peruvian cooking, potentially leading to inappropriate or inauthentic suggestions.

Ethical considerations surrounding AI in kitchens also require more thorough investigation. Issues such as algorithmic bias in recipe recommendations and the potential displacement of culinary labor in commercial settings remain understudied. There are concerns that AI systems might perpetuate cultural biases by favoring certain cuisines over others, or by failing to account for religious and dietary restrictions important to various communities. Moreover, the automation of food preparation in restaurants and catering services raises important questions about job security for culinary professionals, particularly in developing economies where the food service industry employs significant portions of the workforce. These ethical dimensions need to be carefully examined to ensure that AI's integration into culinary spaces is equitable and culturally sensitive.

Moving forward, research priorities should include comprehensive cross-cultural studies that compare AI adoption patterns across different societies, with particular attention to how these technologies interact with traditional cooking practices. There is also a pressing need for more inclusive AI design approaches that incorporate diverse culinary perspectives from the outset. Longitudinal studies tracking the evolution of cooking behaviors and cultural attitudes over time would provide valuable insights into AI's long-term impact. Additionally, interdisciplinary collaboration between technologists, anthropologists, and culinary experts could help develop more culturally aware AI systems that respect and preserve traditional foodways while offering the benefits of technological innovation. By addressing these research gaps, we can work towards AI solutions that truly enhance culinary experiences without compromising cultural diversity and authenticity.

CONCEPTUAL FRAMEWORK

To comprehensively understand the adoption of AI-driven smart cooking technologies across different cultural contexts, this study draws upon three foundational theoretical frameworks: the Technology Acceptance Model (TAM), Diffusion of Innovation (DOI) theory, and Hofstede's Cultural Dimensions Theory. These frameworks offer complementary perspectives to analyze user acceptance, innovation diffusion, and the influence of cultural values on technological engagement.

Technology Acceptance Model (TAM)

The Technology Acceptance Model, developed by Davis (1989), is widely used to explain user adoption of new technologies. TAM posits two core determinants:

- Perceived Usefulness (PU): the extent to which a person believes using a technology will enhance performance.
- Perceived Ease of Use (PEOU): the degree to which a person believes using the technology will be free of effort.

In this study, TAM provides a foundation to assess how individuals across Japan, Italy, and India evaluate the benefits and usability of AI cooking tools. For example, interview themes around convenience, time-saving, and learning curve directly relate to PU and PEOU. As an established model in technology acceptance research, TAM is appropriate for exploring cross-cultural user attitudes toward smart appliances (Venkatesh & Davis, 2000).

Diffusion of Innovation (DOI) Theory

Rogers' (2003) DOI theory provides insight into how innovations spread across societies. The theory identifies five attributes influencing adoption:

- Relative Advantage: perceived benefits over previous practices.
- Compatibility: consistency with cultural values, needs, and existing practices.
- Complexity: perceived difficulty of understanding or using the innovation.
- Trialability: the degree to which a technology can be tested before commitment.

- Observability: visibility of the technology's results.

In this study, DOI explains cultural differences in acceptance patterns. For example, Japan's high observability and relative advantage contribute to higher adoption, whereas Italy's low compatibility with artisanal traditions and India's concerns about complexity and localized functionality lead to more resistance.

Hofstede's Cultural Dimensions Theory

Hofstede's (2001) theory provides a lens to examine how national cultural values shape responses to technology. Key dimensions relevant to this study include:

- Individualism vs. Collectivism: influences whether users prioritize personal convenience (individualism) or group harmony (collectivism).
- Uncertainty Avoidance: determines the tolerance for new and potentially ambiguous technologies.
- Long-Term vs. Short-Term Orientation: affects openness to change and innovation.

For instance, Japan's high uncertainty avoidance and long-term orientation facilitate structured adoption of AI tools (Hofstede Insights, 2024), while Italy's preference for tradition (short-term orientation) aligns with slower adoption. India's collectivist ethos makes users cautious about technologies that disrupt communal cooking practices.

Integrated Application to Study

This study applies these frameworks synergistically to examine cross-cultural patterns in AI kitchen technology adoption. TAM informs the design of interview questions on utility and usability. DOI guides analysis of adoption barriers like cultural mismatch or limited trialability. Hofstede's dimensions contextualize findings on trust, resistance, and role expectations. Together, these theories provide a comprehensive, culturally grounded lens for interpreting the interplay of tradition and innovation in smart kitchens.

RESEARCH METHODOLOGY

This study adopts a qualitative, interview-based approach to explore cross-cultural perspectives on the adoption and perception of AI-driven smart cooking technologies. By focusing on in-depth interviews and thematic analysis, the research aims to capture rich narratives around cultural values, food traditions, and attitudes toward AI in domestic kitchens. The methodology is designed to address the study's objectives: examining cultural differences in attitudes toward AI in domestic cooking, identifying factors influencing acceptance or rejection, assessing AI's impact on traditional cooking roles, and exploring consumer expectations for personalization and trust. This section outlines the research design, data collection methods, sampling strategy, data analysis techniques, and ethical considerations, drawing on established qualitative methodologies and recent literature to ensure rigor and depth.

The study employs a comparative case study approach, focusing on three culturally distinct regions—Japan, Italy, and India—to investigate how cultural contexts shape the adoption of AI-driven smart cooking technologies. These regions were selected for their diverse culinary traditions and varying attitudes toward technology, providing a robust framework for cross-cultural comparison (Ceccarini, 2011; Sen, 2009). The case study method, as outlined by Yin (2018), allows for an in-depth exploration of contextual factors and cultural narratives, with interviews serving as the primary data collection tool to capture lived experiences and perceptions. Thematic analysis will be used to identify patterns and themes across the cases, aligning with qualitative approaches for studying socio-technical phenomena (Braun & Clarke, 2019).

Data was collected through semi-structured interviews with households, chefs, and food influencers in Japan, Italy, and India. This method, guided by protocols from Kvale and Brinkmann (2015), enables flexibility to explore participants' perceptions and usage of smart cooking technologies (e.g., AI recipe assistants, smart ovens, cooking robots) while maintaining focus on key research objectives. Interview questions covered attitudes toward AI in cooking, its compatibility with cultural food traditions, impacts on traditional roles (e.g., family cooks, professional chefs), and expectations for personalization and trust. Some questions included: "How do smart cooking technologies align with or challenge your cultural cooking practices?" and "What factors influence your trust in AI for preparing traditional dishes?" Interviews, lasting 45–60 minutes, were conducted in participants' native languages (Japanese, Italian, or Hindi/English) with trained interpreters where necessary to ensure cultural sensitivity and accurate communication.

A purposive sampling strategy will be used to recruit 10 participants per region (30 total), ensuring a diverse representation of households (home cooks of varying ages and genders), professional chefs, and food influencers. These groups were selected as key stakeholders in culinary practices, offering insights into domestic and professional contexts (Wright et al., 2020). Participants will be identified through community networks, culinary associations, and social media platforms (e.g., Instagram for food influencers), with a focus on including both urban and rural respondents to capture socio-economic and technological diversity. Inclusion criteria include experience with or awareness of smart cooking technologies and engagement in cooking (domestic or professional). Recruitment will prioritize cultural representativeness, with local researchers facilitating outreach to ensure accessibility and trust.

Interview data was analyzed using thematic analysis, following the six-phase process outlined by Braun and Clarke (2019). Audio recordings were transcribed verbatim, translated into English where necessary by bilingual researchers to preserve cultural nuances, and coded using NVivo software. The analysis involved: (1) familiarizing with the data through repeated reading, (2) generating initial codes (e.g., "resistance to automation," "cultural pride in cooking"), (3) searching for themes (e.g., tradition vs. innovation, trust in AI), (4) reviewing and refining themes, (5) defining and naming themes, and (6) producing a narrative report. Cross-case analysis will compare themes across Japan, Italy, and India to identify cultural differences and commonalities, such as how trust in AI varies by cuisine type or demographic factors. Inter-coder reliability will be ensured through regular team discussions and independent coding by at least two researchers. Participants were provided consent forms, with clear information about the study's purpose, procedures, and their right to withdraw at any time. Cultural sensitivity was prioritized by involving local researchers in each region to ensure respectful engagement and accurate interpretation of cultural contexts.

FINDINGS AND DISCUSSION

This chapter presents the findings and discussion from a qualitative study exploring cross-cultural perspectives on the adoption and perception of AI-driven smart cooking technologies in Japan, Italy, and India. Utilizing a comparative case study approach with semi-structured interviews and thematic analysis, as outlined in the methodology, the research addressed four objectives: (1) examining cultural differences in attitudes toward AI in domestic cooking, (2) identifying factors influencing acceptance or rejection of smart cooking tools, (3) assessing AI's impact on traditional cooking roles and practices, and (4) exploring consumer expectations for personalization, cultural relevance, and trust. The findings reveal three key themes—cultural compatibility, trust and agency, and evolving culinary roles—which are discussed in relation to the research objectives and existing literature. A technology review of smart kitchen products complements the interview data, providing insights into their cultural adaptability.

Table 1. Matrix representation of emergence of final themes in data analysis

Theme	Japan	Italy	India
Trust in AI Cooking (Theme 2)	High trust due to tech literacy and brand reliability (Toshiba, Sharp). "Engineers understand our food." Less concern about privacy.	Low trust due to fear of losing culinary authenticity. "A machine can't smell or taste." Concerns over data use.	Cautious trust; based on utility (e.g., pressure cooking). Low tech literacy in rural areas. Lack of local ingredient understanding.
Factors Influencing Trust	High tech literacy Cultural familiarity with automation Reliable local brands	Sensory disconnect Strong emotional connection to traditional cooking Privacy concerns	⚠ Urban-rural digital divide ⚠ Practical vs. emotional considerations ⚠ Recipe localization needs
Evolving Culinary Roles (Theme 3)	AI as collaborative assistant Frees time for creativity (plating, fusion) Augmentation, not replacement	⚠ Seen as threat to family traditions and artisanal cooking "Cooking is my love language" Fear of deskilling	Urban: labor-saving & time-efficient "More time for kids" Rural: erosion of communal food traditions; concern about cultural loss
Cultural Compatibility of AI in Cooking (Theme 1)	Aligned with innovation-oriented culture Positive reception supports	Conflicts with heritage-preserving values Resistance reflects	Split between modernity and tradition Food as social bonding Urban openness vs. rural hesitation

Theme 1: Cultural Compatibility of AI in Cooking

Interviews across Japan, Italy, and India highlighted significant cultural variations in attitudes toward AI-driven cooking technologies, shaped by the centrality of food in cultural identity. In Japan, participants, particularly younger urban home cooks and chefs, expressed enthusiasm for smart appliances like AI-enabled rice cookers and robotic sushi makers, viewing them as extensions of the culture's embrace of precision and innovation (Eom et al., 2022). A 28-year-old Tokyo chef noted, "Smart tools help maintain consistency in dishes like tempura, which is valued in our cuisine." However, older participants emphasized that AI should complement, not replace, the artistry of traditional methods.

In Italy, resistance was more pronounced, with participants associating cooking with heritage and sensory engagement. A 45-year-old home cook in Florence remarked, "An AI oven can't replicate the intuition of my grandmother's pasta sauce." Chefs and influencers underscored the importance of artisanal practices, aligning with Ceccarini's (2011) findings on Italy's culinary pride. Smart appliances were often seen as incompatible with the slow, relational nature of Italian cooking, though some younger participants appreciated recipe assistants like Yummly for convenience.

In India, attitudes were mixed, reflecting the communal and ritualistic nature of cooking (Sen, 2009). Home cooks in urban Mumbai valued AI tools for time-saving, with a 32-year-old mother stating, “Smart appliances help me balance work and cooking complex curries.” However, rural participants and older generations expressed skepticism, citing AI’s inability to replicate the “hand’s feel” in preparing dishes like roti. Food influencers highlighted the challenge of adapting AI to India’s diverse regional cuisines, noting that recipe databases often lack local ingredients like kokum or millets.

Theme 2: Trust in AI Cooking

Trust emerged as a critical factor influencing acceptance or rejection of smart cooking technologies, mediated by technology literacy, perceived reliability, and cultural values. In Japan, high technology literacy fostered trust, with participants citing the reliability of brands like Toshiba and Sharp. A 35-year-old influencer stated, “I trust AI to handle precise tasks like sushi rice because it’s built by engineers who understand our food.” This aligns with Belanche et al.’s (2019) findings on trust in technology-driven cultures.

In Italy, trust was lower, driven by concerns about AI’s ability to preserve culinary authenticity. A 50-year-old chef in Rome expressed distrust in AI’s sensory capabilities, saying, “A machine can’t smell or taste like a human.” Privacy concerns also surfaced, with participants wary of data collection by smart appliances, echoing Hassoun et al.’s (2024) observations on data security fears.

Indian participants displayed cautious trust, often tied to practical outcomes. Urban home cooks trusted AI for routine tasks like pressure cooking but were skeptical of its ability to handle nuanced recipes. A 40-year-old rural cook noted, “I’d trust it more if it understood my village’s spices.” Low technology literacy among older and rural participants further hindered trust, consistent with Nguyen’s (2017) findings on digital divides.

Theme 3: Evolving Culinary Roles

AI-driven technologies are reshaping traditional cooking roles, with varied implications across cultures. In Japan, smart tools were seen as augmenting roles, enabling home cooks to focus on creative aspects like plating while AI handled repetitive tasks. A 30-year-old chef observed, "AI frees me to experiment with fusion dishes." This supports Hassoun et al.'s (2022) view of AI as a collaborative tool.

In Italy, AI was perceived as a threat to traditional roles, particularly the matriarchal role of home cooks. A 55-year-old mother in Naples said, "Cooking is my way of showing love; a machine can't replace that." Chefs also feared deskilling, aligning with Coles' (2013) concerns about technology eroding culinary artistry.

In India, AI's impact on roles was dual-edged. Urban women appreciated smart appliances for reducing domestic labor, with a 38-year-old working mother noting, "My smart cooker lets me spend more time with my kids." However, rural participants and influencers worried that reliance on AI could diminish the cultural role of cooking as a communal act, echoing Sen's (2009) emphasis on food as social bonding.

The findings confirm significant cultural differences in attitudes toward AI in cooking, driven by the interplay of tradition and innovation. Japan's positive reception reflects a cultural affinity for technology, consistent with Eom et al. (2022), while Italy's resistance underscores the prioritization of heritage over efficiency (Ceccarini, 2011). India's mixed attitudes highlight a tension between modernity and tradition, with urban-rural divides shaping perceptions (Sen, 2009). These differences suggest that AI adoption is not merely a technological issue but a cultural negotiation, requiring context-specific design.

Key factors influencing acceptance include technology literacy, cultural compatibility, and trust. Japan's high literacy and trust in local brands facilitate adoption, while Italy's emphasis on sensory cooking and India's diverse cuisines pose barriers (Belanche et al., 2019). Age and urban-rural divides also play roles, with younger, urban participants across regions showing greater openness, aligning with Martinez et al. (2023). The technology review underscores that product design—particularly recipe inclusivity—critically affects acceptance, highlighting the need for culturally adaptive AI systems. AI is redefining culinary roles, with effects varying by culture. In Japan, AI enhances efficiency without disrupting roles, while in Italy, it challenges the emotional and social functions of cooking (Coles, 2013). In India, AI alleviates domestic burdens for some but risks eroding communal practices (Sen, 2009). These findings suggest that AI's impact depends on its alignment with cultural expectations, with collaborative rather than fully automated systems better preserving traditional roles.

Participants across regions expect AI to offer personalized, culturally relevant solutions, such as region-specific recipes and transparent algorithms. Japan's trust in AI stems from reliable, localized products, while Italy and India demand greater cultural sensitivity and data privacy (Hassoun et al., 2024). The technology review reveals a gap in current products' ability to meet these expectations, particularly in non-Western markets, underscoring the need for inclusive design (Murugeah, 2024).

RESEARCH IMPLICATIONS

The findings presented in this study extend existing literature by emphasizing the cultural dimensions of AI adoption in kitchens, moving beyond purely technical or economic considerations (Hassoun et al., 2024). They demonstrate that successful integration of AI-driven culinary technologies requires systems that are culturally adaptive, transparent, and collaborative—ensuring they preserve human agency and culinary heritage rather than displacing traditional practices. For designers, this means prioritizing localized recipe databases that accurately represent regional cuisines, developing multilingual interfaces to enhance accessibility, and implementing robust privacy assurances to build user trust. Additionally, AI systems must be designed to complement rather than replace human expertise, allowing cooks to retain creative control while benefiting from technological assistance.

For policymakers, these insights highlight the need to address digital divides that may hinder equitable access to smart kitchen technologies. In regions like rural India, where traditional cooking remains central to daily life, initiatives to improve digital literacy and infrastructure are crucial for ensuring these innovations do not exacerbate existing inequalities (Nguyen, 2017). Governments and technology developers should collaborate on programs that promote inclusive access, such as subsidized smart appliances for low-income households or community-based training on AI-assisted cooking tools. Furthermore, ethical guidelines should be established to prevent algorithmic biases in recipe recommendations and ensure AI respects cultural dietary restrictions and preferences.

The broader implications also extend to commercial kitchens, where AI adoption must balance efficiency gains with workforce considerations. Automation in restaurants and food services should be implemented in ways that augment rather than eliminate jobs, possibly through reskilling programs for culinary professionals. Finally, future research should explore participatory design approaches, engaging diverse communities in the development of AI kitchen technologies to ensure they meet real-world needs while safeguarding cultural authenticity. By addressing these challenges holistically, AI can enhance global culinary practices without undermining the traditions that give them meaning.

CONCLUSION

This study illuminates the complex interplay between cultural values and the integration of AI-driven smart cooking technologies in Japan, Italy, and India. Through semi-structured interviews and thematic analysis, the research reveals that cultural compatibility, trust, and evolving culinary roles significantly shape the acceptance and perception of AI in domestic kitchens. The findings underscore that while Japan embraces AI for its alignment with precision and innovation, Italy resists it to preserve artisanal traditions, and India navigates a tension between modernity and communal cooking practices. Factors such as technology literacy, age, and cuisine diversity further influence adoption, with trust in AI hinging on transparency and cultural relevance. AI's impact on traditional roles varies, enhancing efficiency in Japan, challenging heritage in Italy, and offering practical relief in India while risking cultural erosion.

The study highlights the need for culturally adaptive AI systems that prioritize localized recipes, multilingual interfaces, and privacy assurances to meet consumer expectations for personalization and trust. The technology review confirms that current smart kitchen products often fall short in addressing diverse culinary needs, particularly in non-Western contexts. These insights contribute to the literature by emphasizing the cultural dimensions of technology adoption, extending beyond technical or economic considerations (Hassoun et al., 2024; Murugeah, 2024). For designers and manufacturers, the findings advocate for inclusive design that respects culinary heritage while fostering innovation. Policymakers should address digital divides to ensure equitable access, especially in rural areas like those in India (Nguyen, 2017). Future research should explore AI adoption in underrepresented regions, such as Africa or Latin America, and employ longitudinal studies to track long-term cultural impacts (Wright et al., 2020). Ultimately, the successful integration of AI in global kitchens lies in balancing technological advancement with the preservation of food as a cultural and social practice, ensuring that smart cooking tools enhance rather than replace the human essence of culinary traditions.

LIMITATIONS AND FUTURE WORK

While the study includes diverse cultural contexts (Japan, Italy, and India), the sample size within each country was relatively small and may not fully represent the spectrum of demographic, regional, and socio-economic diversity. Particularly, rural and marginalized communities may have been underrepresented. The insights are based on self-reported experiences and perceptions, which are inherently subjective and may be influenced by memory bias, social desirability, or misunderstanding of AI technologies.

Future research should include additional countries and cultures, especially those from Africa, South America, and Southeast Asia, to capture a more global perspective on AI adoption in culinary contexts. Conducting follow-up studies over time can help track evolving attitudes, behavioral changes, and long-term integration of AI in daily cooking practices. Future studies could further explore gender roles, generational divides, and intergenerational knowledge transfer in shaping AI acceptance, especially in domestic and communal cooking contexts.

REFERENCES

Anagnostopoulos, I. (2016). Data management challenges in the food industry: A review. *Journal of Food Science and Technology, 53*(1), 45–56.

Belanche, D., Casaló, L. V., Flavián, C., & Schepers, J. (2019). Service robot implementation: A theoretical framework and research agenda. *Service Industries Journal, 40*(3–4), 203–225. DOI: 10.1080/02642069.2019.157963

Braun, V., & Clarke, V. (2019). Reflecting on reflexive thematic analysis. *Qualitative Research in Sport, Exercise and Health, 11*(4), 589–597. DOI: 10.1080/2159676X.2019.1628806

Cain, M. (2024). AI-driven innovations in food waste reduction: A case study of grocery store applications. *Journal of Sustainable Food Systems, 12*(3), 89–102.

Ceccarini, R. (2011). *Pizza and pizza chefs in Japan: A case of culinary globalization*. Brill. DOI: 10.1163/ej.9789004194663.i-160

Coles, B. (2013). Food as relationship: The social and cultural dimensions of eating. In Counihan, C., & Van Esterik, P. (Eds.), *Food and culture: A reader* (3rd ed., pp. 150–165). Routledge.

Counihan, C., & Van Esterik, P. (Eds.). (2013). *Food and culture: A reader* (3rd ed.). Routledge. DOI: 10.4324/9780203951880

Eom, S., Umar, Z., & Lee, J. (2022). Cultural influence and kitchen safety using Arduino nanotechnology and voice recognition technology for smart cooking. *Journal of Kitchen Technology, 18*(4), 210–225.

Gupta, R., Kumar, S., & Sharma, P. (2023). Patent analysis of AI applications in the food industry. *Food Innovation Review, 9*(2), 67–80.

Hassoun, A., Ait-Kaddour, A., & Cozzolino, D. (2022). Industry 4.0 technologies in the food sector: A review of applications and challenges. *Trends in Food Science & Technology, 120*, 345–357. DOI: 10.1016/j.tifs.2022.01.015

Hassoun, A., Jagtap, S., & Garcia-Garcia, G. (2024). Leveraging artificial intelligence and advanced food processing techniques for enhanced food safety, quality, and security. *Discover Applied Sciences, 6*(1), 1–25. DOI: 10.1007/s42452-024-05789-3

Hofstede, G. (2001). *Culture's consequences: Comparing values, behaviors, institutions and organizations across nations* (2nd ed.). Sage Publications.

Hofstede Insights. (2024). Country comparison tool. https://www.hofstede-insights.com

Junejo, A. R., Xu, X., & Li, Q. (2021). IoT-enabled smart kitchen technologies: A review of food storage and preparation. *Food Control, 130*, 108120. DOI: 10.1016/j.foodcont.2021.108120

Kislov, V. (2025). Application of artificial intelligence in diagnostics and repair of household appliances: Increasing efficiency and reducing costs. *Journal of Reliable Intelligent Environments*, *7*(1), 45–60. DOI: 10.1007/s40860-024-00234-5

Kvale, S., & Brinkmann, S. (2015). *Interviews: Learning the craft of qualitative research interviewing* (3rd ed.). SAGE Publications.

Liu, Z., & Liu, Z. (2023). Research on smart kitchen under digital empowerment: A bibliometric and visual analysis. *Journal of Food Engineering*, *345*, 267–280. DOI: 10.1016/j.jfoodeng.2022.345678

Martinez, J., Lopez, A., Lee, J., & Schmidt, K. (2023). Leveraging innovative kitchen technologies for designing a healthy cooking environment personalized breakfast: Consumer perception of smart cooking devices. *Journal of Nutrition & Food Sciences*, *15*(4), 112–130.

Murugeah, S. (2024). IoT-enabled smart kitchen appliances and their impact on food storage, preparation, and culinary experiences. *Food Technology and Biotechnology*, *62*(3), 201–215. DOI: 10.1016/j.fotech.2024.100345

Nguyen, T. (2017). Factors affecting access to clean cooking fuel among rural households in India. *Energy Policy*, *108*(2), 218–225. DOI: 10.1016/j.enpol.2017.05.042

Rogers, E. M. (2003). *Diffusion of innovations* (5th ed.). Free Press.

Sen, C. T. (2009). *Food culture in India*. Greenwood Press.

TMCnet. (2024, August 12). AI-powered kitchen appliances: Revolutionizing culinary spaces. *Cloud Computing Magazine*. http://cloud-computing.tmcnet.com

Venkatesh, V., & Davis, F. D. (2000). A theoretical extension of the technology acceptance model: Four longitudinal field studies. *Management Science*, *46*(2), 186–204. DOI: 10.1287/mnsc.46.2.186.11926

Wang, S. (2023). Smart furniture in the kitchen: Enhancing sustainability and efficiency. *Sustainable Technology Journal*, *10*(2), 78–90.

Wright, J., Smith, R., & Patel, K. (2020). Culture, tradition, and taboo: Understanding the social shaping of fuel choices and cooking practices in Nigeria. *International Journal of Environmental Research and Public Health, 17*(15), 5321. DOI: 10.3390/ijerph17155321

Yin, R. K. (2018). *Case study research and applications: Design and methods* (6th ed.). SAGE Publications.

Chapter 8
Food Safety Meets Automation:
Smart Kitchens of the Future

Ajay Chauhan
https://orcid.org/0009-0006-0562-0155
Bahra University, India

Gautam Tanwar
https://orcid.org/0009-0009-1815-6832
Bahra University, India

Yashwant Thakur
https://orcid.org/0009-0004-1881-1837
Bahra University, India

Vaibhav Verma
https://orcid.org/0000-0002-9041-7111
Bahra University, India

Vikesh Kashyap
http://orcid.org/0000-0001-8532-2022
Bahra University, India

ABSTRACT

The rapid advancement of technology is reshaping the food and hospitality industry, with automation and intelligent systems driving efficiency, safety, and sustainability. Smart kitchens integrate artificial intelligence (AI), robotics, the Internet of Things (IoT), and real-time analytics to manage food preparation, storage, and service with precision. These systems ensure compliance with hygiene standards, control temperatures, prevent cross-contamination, and manage allergens, thereby reduc-

DOI: 10.4018/979-8-3373-4042-5.ch008

ing human error while meeting regulatory requirements. Predictive analytics and machine learning enable proactive risk detection, while resource optimization and energy efficiency reduce waste and support sustainability. Yet challenges remain, including high costs, workforce adaptation, reliance on digital systems, and ethical issues concerning transparency and accountability. The future of food safety will depend on human–machine collaboration, where automation secures consistency and chefs provide creativity and cultural value, together redefining trust and guest satisfaction.

INTRODUCTION

Global food industry meets growing complexity in supply network, heightened safety expectations, evolving consumer demands, the integration of automation in food safety has emerged as a key factor in comparison to futuristic luxury. The ongoing evolution of kitchen technology driven by artificial intelligence (AI), Internet of Things (IoT), robotics, and data analytics has revolutionized how safety, hygiene, and quality are maintained in professional food environments. Commercial hotel chains institutional kitchens to airline catering and hospital food services, automation is transforming every food lifecycle (Ding, 2023; Balakrishnan, Zhang, & Tang, 2025). Traditional food safety monitoring relies heavily on manual record keeping, and periodic audits. These methods are increasingly insufficient for modern operations. Human error, inconsistent practices, and the lack of alerts for contamination or temperature frequently compromise outcomes (Iftekhar, Cui, & Kim, 2020). In contrast, automation technologies offer precision, forecast intervention, and increase consistency, enabling food businesses to detect, respond to, and prevent risks more efficiently. Global regulatory bodies such as the Food Safety and Standards Authority of India (FSSAI) and the U.S. FDA embracing digital compliance models, automation is also becoming central to regulatory conformance and certification (Ellahi, Iftekhar, & Kim, 2024).

It explores how real-time monitoring tools particularly IoT devices and smart sensors enhance visibility into kitchen environments by continuously tracking temperatures, humidity levels, and surface cleanliness (Chen, 2021). These tools not only send automated alerts but also trigger sanitation protocols, offering proactive control over contamination risks. A focus on robotic systems illustrates how culinary consistency and quality assurance are achieved at scale, ensuring that taste, texture, and hygiene remain standardized across different shifts, branches, or geographies (Dakhia, 2025). The research examines automation in action, featuring successful implementation in hotels, cloud kitchens, hospitals, and airline catering highlighting ROI, training models, and operational lessons learned. Particular emphasis is placed on early adopters in India post-2023, who are pioneering hybrid systems that intermix automation with skilled human oversight. Shukla et al. (2025) emphasize the necessity of sustainable waste practices in rural destinations. The rapid adoption of technology also brings new risks such as over-dependence on automation, system failures, data privacy breaches, and workforce displacement present ethical and operational dilemmas (McKinsey & Company, 2023; Wired, 2024). These are concerns, offering a balanced view of automation's promises and pitfalls. Yadav et al. (2024) outline strategies for adapting to AI-driven hospitality transformations. Finally, a forward-looking analysis speculates on the kitchen of 2040 envisioning fully integrated, AI-controlled environments where food safety inspections are automated, supply chains are self-monitoring, and even global health risks such as pandemics can be mitigated through predictive food safety analytics (Sadilek et al., 2018). This vision predicts a roadmap not only for technological innovation but also for building resilient, future-proof food safety ecosystems worldwide.

WHY FOOD SAFETY NEEDS AUTOMATION

As the food sector adjusts to meet the needs of global supply chains, high-volume manufacturing and rising consumer expectations, traditional food safety procedures are becoming less effective (FAO, 2023). Reactive interventions, manually prepared logs, manual monitoring techniques lack the capacity to deal with the operational scale and complexity of contemporary food settings. Shukla and Sharma (2020) explore dining outlet performance in Himachal Pradesh. Over 600 million cases of food borne intoxication and infections are occurring, annually around the globe. The scope of errors in food safety procedures has grown dangerously small (WHO, 2015). Conventional and standard set of practices, which are primarily dependent on human resource, are prone to undesired conclusions, inconsistent record keeping and delayed responses. Shukla et al. (2025) argue that AI reshapes how learners acquire communication skills.

When businesses try to expand their activities over several kitchens or retail locations, these vulnerabilities become particularly problematic (Manning and Soon, 2019). Widespread contamination or expensive product recalls may result from lax control over crucial parameters like temperature, sanitation cycles, and allergen separation in such high-volume settings (Charlebois, Sterling, and Haratifar, 2020). In response, regulatory agencies all around the world have heightened their compliance standards. Food companies must set up preventive measures, prove traceability, and keep verified logs of their activities in accordance with the Food Safety Modernization Act (FSMA) in the United States and comparable regulations from EFSA and FSSAI (FDA, 2023; EFSA, 2022). However, traditional paper-based systems often fall short of these demands, particularly when food safety data must be retrieved during audits or legal scrutiny.

In this changing scenario, automation is becoming a strategic necessity. Technologies such as IoT sensors, cloud-based dashboards, and digital workflow apps help food operators monitor environmental conditions, equipment, and staff attributes in real time (Lu, Kannan and Rakshit, 2021). These systems reduce dependency on manual methods and instead enable a proactive approach to risk management. Food safety becomes verifiable and scalable with warnings triggered by threshold breaches, data auto-logged with timestamps, and performance dashboards instantaneously accessible (GFSI, 2021). Seal et al. (2025) provide insights into Indian food bloggers' digital storytelling. Food safety automation greatly improves operation effectiveness and product quality in addition guaranteeing compliance. Technology changes food safety from a reactive to a proactive process by reducing errors, enhancing traceability, and facilitating predictive analytics (Jagtap, Rahimifard, and Kanchanomai, 2021). For food businesses operating under pressure from both regulators and consumers, automation is beyond than a trend, it is a cornerstone of future food systems.

CHALLENGES IN CONVENTIONAL FOOD SAFETY MONITORING

Numerous challenges are there in traditional food safety monitoring which reduce its ability to effectively protect the public health. These issues are caused by obsolete approaches, resource constraints, systemic inefficiencies and structural constraints.

1. **Limited Scope of Surveillance:** Conventional surveillance systems for food borne illness usually only keep an eye on a small number of microorganisms. For instance, the US Centers for Disease Control and Prevention (CDC) has reduced the number of pathogens in its Foodborne Diseases Active Surveillance Network (FoodNet) from eight to just two: Shiga toxin-producing E. coli and Salmonella. The system's capacity to identify and address a greater variety of food borne infections is impacted by this decrease (AP News, 2023).
2. **Inconsistent Data Collection and Reporting:** Relying exclusively on passive reporting techniques may result in incomplete data. Without active surveillance, a large number of foodborne disease cases may go unreported, underestimating the actual incidence and making it more challenging to put effective response plans into place (Scallan et al., 2011).
3. **Resource Constraints:** A lack of staff and financial constraints may make it difficult to implement thorough food safety protocols. For instance, insufficient

funding that limits the number of inspections and the capacity to maintain modern monitoring equipment that may impact the overall effectiveness of food safety initiatives (Steritech, 2023).

4. **Challenges in Food Traceability:** It becomes difficult for conventional food safety systems to track the origin of food products throughout the supply chain. This lack of coordination and collaboration may delay responses to food safety concerns and make it more difficult to identify and remove contaminated products from the market (iTrade Network, 2022).
5. **Human Factors and Training Deficiencies:** Food safety issues are mostly caused by inadequate staff training and poor hygiene practices. Studies show that common deficiencies in sanitation practices, staff training, and equipment maintenance increase the risk of contamination in food processing environments (FDA, 2004).
6. **Regulatory and Compliance Challenges:** Constantly evolving nature of food safety regulations can make compliance difficult. Organizations may struggle to keep up with dynamic regulations, which could lead to potential infractions of food safety protocols (Protocol Foods, 2022).
7. **Cultural and Informal Sector Barriers:** In places where there are significant informal food industries, like street food vendors, it can be challenging to ensure adherence to food safety regulations. According to Green et al. (2006), factors like vendor anonymity, a lack of incentives to comply, and a lack of regulatory control are the main causes of persistent food safety hazards in these settings.

REGULATORY CONSTRAINTS AND RISK FACTORS IN FOOD SAFETY: THE NEED FOR AUTOMATION

The need for automation is critical since complex risk factors and changing regulatory requirements provide more and more obstacles for traditional food safety monitoring. Regulatory frameworks that necessitate strict compliance measures, including increased traceability and documentation requirements, including the European Union's Food Safety Regulation 2002 and the U.S. Food and Drug Administration's Food Safety Modernization Act (FSMA). For example, food manufacturers find it difficult to use manual procedures to comply with FSMA Rule 204, which requires them to have traceability data available within 24 hours (Automation World, 2022; MES Automation Inc., n.d.). Biological hazards, contamination, and supply chain disruptions are just a few of the risks that the food industry must deal with. `Traditional risk assessment techniques are laborious and prone to human error, even though automated solutions can provide real-time monitoring and data analysis, enabling proactive risk management (SMRTR Solutions, 2023; ComplianceQuest, n.d.). Increased productivity, improved compliance, accurate data collection, and early risk identification are just a few of the many benefits of automation. Successful automation implementation requires assessing existing processes, selecting legal-compliant technology, training staff to operate automated systems, and updating processes frequently to take new risks into account (Wired, 2020). To sum up, automating food safety monitoring is not only a technological advancement but also a strategic necessity that helps food businesses to comply with legal requirements, lower risk, and ensure product safety.

REGULATORY CONSTRAINTS AND RISK FACTORS IN FOOD SAFETY

Evolving Regulatory Requirements

- Frameworks such as the U.S. Food Safety Modernization Act (FSMA) and the EU Food Safety Regulation 2002 enforce stringent requirements for reporting, recordkeeping, and traceability.
- Manual compliance approaches often fail to meet these standards (Automation World, 2022; MES Automation Inc., n.d.).

Biological and Contamination Risks

- Risks associated with chemicals, Foodborne pathogens and microbial contamination remain significant risk factors in food processing and supply chains.
- Traditional monitoring may not be able to detect hazards in real time, which could lead to safety incidents, claims SMRTR Solutions (2023).

Supply Chain Vulnerabilities

- Disruptions in sourcing, transit, or storage may have an impact on food safety.
- The lack of end-to-end traceability complicates risk identification and mitigation (ComplianceQuest, n.d.).

Human Error and Operational Risks

- Errors in reporting, inspections, and record-keeping are frequent because manual monitoring primarily relies on human attention to detail.

- Poor hygiene standards and insufficient training further raise the risk of contamination (SMRTR Solutions, 2023).

Need for Automation

- Accurate data collection, real-time monitoring, and quick reporting are made possible by automation.
- By enabling proactive risk management and ensuring regulatory compliance, it lowers human and systemic errors (Wired, 2020).

AUTOMATION TECHNOLOGIES IN THE KITCHEN

Kitchen automation technologies use robotics, artificial intelligence (AI), the Internet of Things (IoT), and smart sensors to speed up food preparation, ensure safety, reduce waste, and improve operations. These technologies are increasingly being adopted by hotels, commercial kitchens, and even smart home kitchens, transforming traditional workflows into data-driven, highly efficient operations.

Table 1. Automation technologies in the kitchen

Technology	Function	Benefits
Robotic chefs	Cooking, plating	Precision, speed, consistency
Smart sensors	Monitor temperature, humidity	Food safety, quality
AI systems	Predictive analytics, quality control	Waste reduction, optimization
Automated cleaning	Sterilization, dishwashing	Hygiene compliance, labor reduction
IoT appliances	Remote control of ovens, fridges	Efficiency, monitoring

Robotic Chefs & Arms

Function: Automation in cooking tasks such as chopping, grilling, and plating.
Example: Moley Robotics provides a complete automated kitchen system that can prepare meals with precision and consistency.

Smart Appliances (IoT-enabled)

Function: Improves kitchen efficiency through connectivity and automation.
Examples:

Smart Ovens: Automatically adjust cooking temperatures and timings.

Smart Refrigerators: Monitor inventory control and alert staff with electronic notifications or alarms about expiring ingredients.

Smart Cooktops: Utilize sensors to prevent overheating and wastage.

Sensors & Monitoring Systems

Function: Ensure food safety and compliance with health standards.

Features:

Real-time Monitoring: Track temperature, humidity, and conditions.

Automated Alerts: Notify staff about deviations from safety parameters.

Compliance Assurance: Support HACCP, FDA, and USDA standards.

AI & Predictive Analytics

Function: Optimize kitchen operations and enhance guest experiences.

Applications:

Menu Personalization: Design offerings based on guest preferences.

Demand Forecasting: Forecast inventory needs to reduce waste.

Quality Control: Monitor and maintain promising food quality.

Automated Cleaning & Sanitization

Function: Maintain hygiene and reduce labor costs.

Technologies:

UV Sterilizers: Disinfect surfaces and equipment to ensure safety.

Robotic Dishwashers: Automated dish cleaning processes reduces work pressure and expenses.

Programmable Floor Scrubbers: Ensure thorough cleanliness of kitchen floors.

Inventory & Stock Management Systems

Function: Precise inventory control and reduce waste.

Features:

Real-time Tracking: Monitor stock levels, usage & wastage

Automated Reordering: Process orders based on predefined thresholds.

Waste Reduction: Analyze usage patterns to minimize excess.

REAL-TIME FOOD SAFETY MONITORING IN KITCHEN THROUGH AUTOMATED TECHNOLOGIES

It has been observed that the modern hotel kitchen has evolved into a very intricate area. Traditional techniques for ensuring the safety of cuisine, such as visual inspections and manual temperature checks, are often inadequate in detecting significant hazards such as food contamination or equipment failure. Real-time food safety monitoring is made possible by intelligent systems and Internet of Things (IoT) sensors. By enabling continuous monitoring, timely diagnosis, and interventions, this improves the overall safety and reliability of food production.

Key Points Continuous Hazard Monitoring with IoT & HACCP Integration

Automated sensors installed in cooking, preparation, and storage areas enable continuous monitoring of temperature, humidity, and contamination risks. These real-time data streams' direct integration with Hazard Analysis and Critical Control Points (HACCP) systems enables food operators to spot deviations right away and take corrective action before risks get worse (Radogna et al., 2022).

RFID Sensor Networks for Critical Limits

Radio-frequency identification (RFID) has evolved beyond product identification to perform intelligent sensing tasks, such as tracking food temperature during cooking and storage and detecting contamination levels. RFID tags placed in storage units and kitchen utensils instantly send out signals when food safety standards are broken, allowing for corrective action to be taken without requiring human interaction (Kumar et al., 2009; Yang & Chew, 2021).

Computer Vision for Hygiene Compliance

Kitchen hygiene continues to be one of the most crucial elements of food safety. These days, AI-based vision systems monitor how long and how frequently workers wash their hands, ensuring that they follow sanitary protocols. Similar to this, advanced optical imaging technologies can identify minute residues on cutting boards and surfaces, offering a rapid and unbiased alternative to traditional swabbing methods (Zhong et al., 2020; Sueker et al., 2022).

Anomaly Detection Through Real-Time Data Streams

Machine learning models trained on continuous data from ovens, refrigerators, and sensors can be used to predict equipment failures or detect unsafe handling practices early. Big platforms like Google's FINDER, which can identify unsafe food establishments up to three times more accurately than traditional inspections, have demonstrated the value of real-time data analytics outside of the kitchen (Sadilek et al., 2018).

Block-Chain for Data Integrity

Institutional food facilities and shared kitchens are increasingly using blockchain technology to protect sensor data, ensuring that cleaning schedules, cooking temperatures, and storage logs cannot be altered. This guarantees transparency in food safety records, makes the supply chain and kitchen-level processes traceable and impenetrable, and facilitates external audits (Jang et al., 2024). *Technology-enabled solutions can support sustainable mountain tourism* (Shukla et al., 2025).

Implementation Considerations

These technologies are promising, but they must be used carefully. Sensors should directly record food product conditions, not just ambient temperatures, for an accurate risk assessment. Vision systems must adapt to real-world scenarios, such as low lighting or visual obstructions, while also taking into account privacy concerns raised by kitchen staff. Furthermore, automated alarms and HACCP corrective action protocols need to be closely related (Radogna et al., 2022; Zhong et al., 2020).

Overall Impact

Through the integration of IoT, RFID, vision systems, and blockchain, food safety management is evolving from static, manual checks to dynamic, continuous verification. This shift improves food safety outcomes, strengthens compliance records, and increases consumer trust in food service operations (Radogna et al., 2022; Jang et al., 2024).

Table 2. Real-time food safety monitoring technologies in kitchens

Technology	Application in Kitchens	Benefits	Sources
IoT Sensors	Continuous monitoring of temperature, humidity, and storage conditions in refrigerators and kitchens.	Prevents spoilage, ensures compliance with HACCP, enables real-time alerts.	Lu et al., 2021; Zhou & Liu, 2019
AI-Enabled Predictive Analytics	Uses machine learning models to detect anomalies in cooking, storage, and food handling processes.	Early detection of potential contamination and predictive maintenance of equipment.	Ding, 2023; Balakrishnan et al., 2025
Blockchain Traceability Systems	Records every stage of food handling and ingredient sourcing in immutable ledgers.	Transparency, tamper-proof records, and rapid recall management.	Iftekhar et al., 2020; Jang et al., 2024
Robotics & Machine Vision	Automated inspection of food surfaces, cooking stages, and hygiene practices.	Reduces human error, ensures consistency in food preparation, and maintains hygiene.	Chen, 2021; Sutton et al., 2022
UV-C & Biosensors	Real-time microbial detection on food surfaces and in air.	Immediate identification of contamination, reduces risk of outbreaks.	Romano, 2024; Kotsanopoulos & Arvanitoyannis, 2017
Cloud-Based Monitoring Platforms	Remote access dashboards that integrate IoT, AI, and block-chain data.	Centralized control, faster decision-making, improved compliance reporting.	McKinsey & Company, 2023; Stevens Traceability Systems, 2023

IoT- Enabled Devices for Ongoing Safety Monitoring

The digital foundation of real-time food safety is the Internet of Things. A network of sensors in a smart kitchen continuously sends data to centralized dashboards to monitor vital elements that include temperature, hygiene & humidity, this provides for quick corrective action & regulatory audit capabilities, which is an evident dissimilarity to periodic manual checks. (Woodbury, 2024).

Food safety monitoring has changed as a result of the Internet of Things' (IoT) integration into smart kitchens, which allows for constant, real-time monitoring of operational and environmental variables. IOT-based devices such as Intelligent smart sensors, RFID tags, and connected appliances provide automation in capturing of data on safety measures like temperature, humidity, gas-leakages, equipment operating etc.

These systems enable the timely recognition of a threat and preventive actions prior to the exacerbation of risks (Khalid et al., 2021). *Prenatal tourism raises questions of citizenship and responsibility* (Shukla et al., 2025). IoT-enabled temperature and humidity sensors will be used for monitoring perishable ingredients and properly storing them to prevent microbial growth and risk of foodborne illness (Verdouw et al., 2021). It has also been observed that Predictive maintenance is also made possible by IoT devices, which track the operation and usage of kitchen appliances. Predictive analytics of this kind guarantee adherence to safety and hygienic standards and minimize unplanned malfunctions (Mekki et al., 2019). Additionally, cloud-based dashboards can be integrated with linked kitchen platforms to provide compliance reports and real-time alerts for managers and auditors of food safety.

This form of continuous visibility is most beneficial in large-scale institutional kitchens and the hospitality space because uniformity and safety are essential (Kumar et al., 2022).

Another benefit is that IoT technologies have traceability across the food production spectrum. For example, while used in conjunction with smart tags and block-chain, the IoT system creates a visible record of where ingredients are sourced, how they are transported, and the conditions in which they are chemically or physically processed. That enables accountability and ethical sourcing, values that most modern consumers believe will attract their loyalty (Mistry et al., 2020). Together, these technologies show how IoT-enabled safety monitoring in contemporary smart kitchens produces a closed-loop system that improves food quality, guarantees regulatory compliance, and reduces safety hazards.

Some of the applications of IOT in real-time kitchen operations include:

In Cold Storage Units, Intelligent Temperature Probes

Real-time temperature data is transmitted by smart temperature probes in freezers and refrigerators, which notifies the kitchen personnel when temperature of the cold storage unit are surpassed (usually 5°C/41°F). In one case, a compressor malfunction caused the temperature of a meat freezer to rise to 8°C. An automated alert resulted in the Shifting of the meat items and consign the storage unit for refrigerated repairs, preventing spoilage and monetary loss. Smart temperature systems improve operational resilience and analytics-based decision making, according to research (SonicU, 2025).

Humidity Sensors in Dry Storage Areas

Low humidity is necessary for the storage of dry goods such as wheat, grains, and spices; high moisture encourages the growth of mold and microorganisms. Because of a leak at a bakery, the storeroom sensor recorded humidity levels above 70%, which resulted in hygiene and storage inspections to stop contamination before it could escalate the situation.

Trackers of Hygiene Compliance

Consistent adherence to Standard Operating Procedures (SOPs) in kitchens is maintained through automated hygiene systems, including motion-triggered hand washing monitors equipped with RFID gating. To drastically lower the chance of cross-contamination, hospital kitchens, for example, have implemented systems where the sensors record the hand washing Standards to meet its compliance and limit access to the prep area if standards aren't fulfilled.

How It All Connects

Supervisors can keep an eye on cold chains, dry storage, and hygiene in real time thanks to these IoT devices that feed into unified dashboards and mobile notifications. It is observed that the system minimizes human error and strengthens confidence by producing instant notifications and remedial recommendations in the event of deviations, such as a contaminated prep surface after raw chicken.

Predictive Notifications for Temperature Changes, Contamination or Spoiling

Predictive intelligence replaces reactive control in sophisticated smart kitchen systems and Machine learning algorithms analyze trends in temperature swings, equipment behaviour, ingredient lifecycles, and staffs' work routines to detect dangers before their effects grow more serious. This leads to minimizing wastage and encouraging motivated engagement.

Early Alerts for Expiration of Ingredients

Predictive systems compare manufacturer shelf-life data with temperature history, humidity, and time since opening. In order to keep guests from coming in contact with the marred food, one algorithm warned chefs that sauce that was stored during a recent refrigerator malfunction had a shorter shelf life and should be used immediately or disposed of in the coming time.

Cross Contamination Risk Alerts

Smart systems can identify inappropriate behaviors, like using the same cutting board for raw meat and salad greens, by monitoring utensil use using RFID, video analytics, and workflow patterns. These notifications help in halting production and giving sanitization instructions in cloud kitchens, which eventually helps in instantly ending training cycles.

Monitoring Environmental Fluctuations

Microbial development may be correlated with slight temperature increases, such as the afternoon heat in aircraft catering prep areas. It has been verified Based on data correlations between environmental changes and bacterial hotspots; predictive analytics identify such trends and suggest actions such as modifying prep schedules or installing cooling systems.

The Benefits of Predictive Alerts for Food Safety

Predictive warnings guarantee regulatory compliance (e.g., HACCP, ISO 22000), facilitate prompt remedial action, help avoid recalls, and speed up staff replies. In one hospital case, a system-driven cleanliness push was initiated in response to an increase in hygiene violations, preventing possible exposure to infection of listeria.

Automated Cleaning and Sanitation Systems in Smart Kitchen:

Unoccupied Spaces with UV-C Sanitation Lights

UV C light (200–280 nm) is a well-proven disinfectant that destroys microorganisms in the air and on surfaces. Kitchen spaces are disinfected by motion-triggered UV C devices during non-operational hours. The Research has acknowledged & agreed that UV C effectively combats germs and fungi while having no effect on the food quality & Standards. (Arcos-Limiñana, et.al, 2025; Romano, 2024).

Robotic Surface and Floor Cleaners Equipped with Pathogen Detectors

Microbial sensor-equipped robotic cleaners identify areas of high pollution and modify disinfectant application accordingly. It was observed that one robot used an adaptive method to keep settings clean by increasing sanitizing cycles in an institutional cafeteria after detecting a escalation in bacterial density.

AI-Powered Dishwashing Systems with Residue Analysis

AI-powered dishwashers check for leftover proteins, oil, or allergies using spectral analysis and computer vision and sensors. Plates containing traces of dairy proteins were auto-flagged and rewashed in aircraft catering to maintain the hygienic conditions and control allergens of the cleaned crockery.

Combined Value of Automated Sanitation

Hygiene methods in the food production, processing, and hospitality industries are changing as a result of automated sanitation systems. It has been seen that by combining IoT, robotics, and AI-driven monitoring, automated sanitation provides combined operational, safety, and financial benefits beyond traditional cleaning techniques. Consistent hygienic standards are just one of the benefits of these automatic sanitation systems. It also helps in many other domains, For Instance:

- *Helps in Error reduction and labor savings.*
- *Helps in Taking Quick reactions to biological dangers & their Prevention.*
- *Help by updating Information on compliance for ISO and HACCP audits.*

It also Assists in achieving the following domains in regard to food safety monitoring:

- Minimizes Food Wastage: *food utilization or transfer is quickly made possible by instant alerts and notification alarms.*
- Conserves Public Health: *By keeping an eye on sanitation, the risk of allergies and food-borne illnesses is kept from spreading & Alarms are always on watch for contamination warning.*
- Maintains Consistent Cleanliness: *The automated sanitation systems continue to operate even during off-peak hours.*
- Lessens Staff Workload: *Since monitoring is automated, chefs can focus on cooking and exploring with creativity hand in hand.*
- Simplifies Compliance: *Digital records simplify HACCP and ISO documentation.*
- Facilitates Quick Reaction: *The mobile notifications and dashboards encourage timely corrective action.*
- Encourages Staff Development: *Real-time feedback helps train staff to adhere to Standard Operating Procedures.*
- Increases Customer Confidence: *Transparent methods reassure diners about safety; all gadgets can receive protocol upgrades simultaneously. With automation, data from multiple locations can be combined for uniform reporting.*
- Decreases Human Error: *Automated systems are vigilant, in ways that humans are not.*

Systematic Repercussions on the Food Industry in relation to IOT in smart kitchens

Surpassing the simple automation, the Internet of Things (IoT) has completely changed the food business. IoT affects every phase of the production pipeline, from farm to fork, by integrating sensors, networked devices, and intelligent platforms into the food ecosystem.

Its wider impacts can be observed in the following domains:

- Smart kitchens are becoming commonplace in cloud kitchens, hotels, hospitals, and airplanes.
- Operations That Are Planned, Not Panic-Driven: Strategic, composed reactions are made possible by predictive analytics.
- Environmentally sustainable: Reduces wastage & displays a better use of chemicals and water.
- Ensuring food safety in large production is made possible by automated systems.
- New Industry Benchmarks: Smart kitchens achieve the transparency and safety innovation that guests now demand especially after the set back from the pandemic (Covid –19).

Quality Assurance through Precision in Smart Kitchens with Automation

It has been perceived that the food safety in smart kitchens requires exact control over materials and procedures, going beyond simple cleaning. Regardless of time, place, or personnel, guests want the same taste, presentation, and safety, therefore uniformity competes with inventiveness. Kitchens can now reach these requirements with previously unheard-of precision because to automation, particularly robotic cooking systems, precise sensors, and AI-driven quality control. This data-supported, precision-driven method lowers the wastage, amplify uniformity, and helps avoid serious safety issues like exposure to allergies or underdone or faulty cooking.

It has been observed that in smart kitchens, precision automation replaces manual, sporadic quality assurance (QA) with data-driven, ongoing control. Smart kitchens can maintain process variables (temperature, humidity, time, and microbiological risk indicators) within tight tolerances by integrating highly accurate sensors, software-controlled actuators, algorithms, and secure documentation. This improves the prospects of lowering variability, increasing consistency, and improving traceability (Radogna et al., 2022; Ding et al., 2023).

Table 3. Quality assurance through precision in smart kitchens

Automation Element	QA Benefit	Example Sensor/Metric	Source (DOI-backed)
Precision Sensing	Continuous measurement at product-level; rapid corrective actions	Thermal probes, humidity sensors, optical biosensors, gas sensors	Radogna et al. (2022); Rodrigues et al. (2021)
Automated Process Control	Consistent execution of recipes, reduced variability, reliable CCP maintenance	Robotic cooking units with time–temperature control, mixing speed automation	Derossi et al. (2023); Radogna et al. (2022)
Intelligent Analytics	Early detection of drifts; predictive maintenance; proactive QA	Machine learning models detecting temperature drift, compressor inefficiency	Ding et al. (2023); Radogna et al. (2022)
Smart Packaging	Product-specific freshness and spoilage detection; localized QA signals	pH indicators, volatile compound sensors, bio-based packaging sensors	Rodrigues et al. (2021)
Traceability & Documentation	Tamper-proof records; simplified audits; stronger regulatory compliance	Digital CCP logs, blockchain-enabled traceability records	Radogna et al. (2022); Derossi et al. (2023)

Robotic Cooking System for Standardizing Food Doneness, Texture, and Taste

By doing away with human variability, robotic cooking technologies guarantee uniformity / consistency. They consistently produce the same dishes by carefully following recipe instructions mentioned on the recipe cards/ database- about temperature, timing, and ingredient quantity.

Multipurpose Robotic Arms for Accurate Cooking Activities

It has been observed with precise timing and force; robotic arms imitate human activities including chopping, stirring, and sautéing. They ensure consistent texture and appearance in high-volume situations by handling delicate products like cucumbers or cherry tomatoes without causing damage or wastage. While preserving visual and sensory integrity, these systems decrease the chances of human error and maximize food safety& quality (Hyper Food Robotics, n.d.).

Automated Ovens and Grills with Dynamic Heat Adjustment

Robotic grills and ovens with embedded sensors and artificial intelligence (AI) modify cooking according to changes in ingredients or environmental variables. For instance: Burger businesses, use automated grills that detect the thickness &

density of the patties with the help of sensors and modify the heat and searing time accordingly, guaranteeing that the patties are always medium-well done and retains its flavors. (TomorrowDesk, 2025).

Recipe Memory Systems for Replication that is Consistent Throughout Locations

Large recipe databases can be stored in robotic kitchens, allowing for consistent results across sites. Every detail, including marinating, ratio of ingredients (spices & seasoning), simmering duration, garnish, and different types of plating's, is stowed by a multinational hotel chain that uses robot stations to produce iconic dishes like Butter Chicken, Chicken Tikka & Dal Makhni.

Integrated Use Case: AI Powered Wok Station

An AI-enabled WOK Station used in business hotels automates oil dispensing, ingredient scheduling, tossing, and heat control. To prevent overcooking and attain ideal caramelization, moisture sensors modify the cooking heat. The final preparation results into uniformly crispy, colourful, and flavourful stir fries—even during busy Hours of the food operations and when employees are rotating shifts or switching metier. These systems, which integrate mechanical accuracy with intelligent programming, to redefine uniformity in taste and flavours. They are perfect for chain restaurants, institutional kitchens, and upscale hotels where safety and quality of food is the upmost Priority.

Preserving Food Integrity across Locations and Shifts with help of Automation

Large businesses, chains of hotels, cloud kitchens, and corporate cafeterias are responsible for the quality of their signature dishes regardless of location, supplier chains, or staff fluctuations. Automation and artificial intelligence provide uniformity across every one of these aspects. For instance:

The Cooking Settings are Automatically Modified in Accordance with the Location and Readily Accessible Ingredients

Smart ovens adjust cooking based on regional variables like altitude or humidity. For instance, Denver ovens maintain the sea level characteristics at Miami by adjusting for lower boiling temperatures while baking croissants to prevent overbrowning or undercooking.

AI Guides Ingredient Substitution Decisions to Preserve Sensory Quality

AI Proposes sensor and chemical comparison substituting the ingredients that are unavailable while maintaining recipe integrity. For instance, in order to maintain consistency, zucchini may be used in place of Thai eggplant in green curry, with minor alterations made for texture and water content.

The Changes in Shifts of Teams Don't Affect Quality Levels or Production Flow

Smart kitchens record the development of recipes and notify new hires with notifications. Second-shift chefs precisely record and restart a noon shift handover on a simmer sauce, guaranteeing uniform taste & texture even when crew's shifts interchange.

Benefits to Operations and Brand Reliability

With the help of weight sensors and touchscreen-guided technologies it helps in Standardization and also improves brand dependability; to guarantee proper quantities and folding procedures, new recruits can produce full-quality meals on immediate basis, lowering training costs and increasing consistency. Measurable foods & service is characterized by the preservation of food integrity across locations and shifts.AI powered solutions guarantee that operational, environmental, and ingredients factors help avoid the degradation of the food quality. Consistency becomes programmable and dependable with automation in action.

The Effect of Automation on Culinary Consistency

Consistency in food preparation has been greatly aided by automation in the culinary arts through robotics, artificial intelligence, and smart devices with sensors. The conversation thereafter examines the ways in which these technologies improve consistency, the subtleties associated with perceived quality, and obstacles that have both technological and psychological roots.

Precision, Standardization, and Quality Control through Automation in Culinary Advancement

The exact, repeatable results that automation produces are crucial for maintaining consistency in the culinary arts. Automated ingredient batching significantly lowers volatility and human error in industrial settings by ensuring precise measurements of weight and sequencing (Sterling Controls, 2023). Similar advantages apply to food processing as well; sophisticated systems also keep an eye on crucial factors to maintain uniformity in quality and traceability, which are crucial to the safety and integrity of the final product (FroZenet, 2023). It has been noticed that AI-driven robotics is used in commercial kitchens to automate processes like frying, blending and plating. Uniform taste, look, and timeliness of dishes are delivered by these systems, which is essential in high-volume hospitality settings (ChefsIcon, 2025; PublicSphereTech, 2025).

Robotic Systems in High-End Culinary Environments

Research on high-end hotel and restaurant chains in India demonstrates how food production robots can greatly increase the production speed, hygiene, and uniformity. Although customer satisfaction varied based on expectations of authenticity, a mixed-method study comprising the Innovation Labs of ITC Hotels (Gurugram) and Taj Hotels (Mumbai) showed constant advances in objective quality (Lovelesh & Jha, 2025). On the technically advanced front lines, revolutionary robotics, such as systems for mixing and pouring pancake batter have achieved perfect homogeneity through algorithmic control and haptic sensing, enhancing machines' ability to perform delicate culinary jobs (Luo et al., 2024). Similar to this, modular systems such as "YORI" (Yummy Operations Robot Initiative) use dual-arm proprioceptive manipulation to prepare multiple meals with exceptional precision and adaptability (Noh et al., 2024).

The Human–Automation Paradox: Psychological Effects on Consistency in Smart Kitchen

Physical consistency can be produced by machines, but customer perceptions show an intricate obstacle. Due to symbolic factors, such as the "contagion effects," which arise in the absence of human touch that reduces feelings of love and increases feelings of disgust, experimental study on Appetite demonstrates how automation can result in decreased taste perception, decreased willingness to spend, and unfavourable attitudes. Effectively communicating the advantages and convenience can help to lessen these adverse impacts.

Table 4.

Benefit	Explanation
Instrumental Consistency	Precise measurement, timing, and replication across batches and outlets.
Operational Efficiency	Faster processes, reduced waste, and labour savings.
Perceived Shortcomings	Loss of human touch can affect perceived taste and emotional connection.
Mitigation Strategies	Wording and branding that highlight consumer value can alleviate scepticism.

Recent studies depict that Automation provides identifiable, technical uniformity that can significantly increase productivity and quality in the output. However, human interaction is still closely linked to perceive quality, which is the subjective to sensation of taste, authenticity, and hospitality. Culinary systems must be combined to careful customer advertising that maintains emotional connection with instrumental standardization in order to achieve maximum success.

Benefits and Impacts of Automated Tech in Smart Kitchen Production Units

Modern culinary operations are changing as a result of the incorporation of robotics and artificial intelligence (AI) into kitchen production units like delivery bulk kitchen or cloud kitchens. By improving productivity, consistency, safety, and creativity, these technologies turn kitchens into more intelligent, tech savvy and data-driven production spaces. The following are a few scenarios of how and what implications the amalgamation of robots and artificial intelligence (AI) has on kitchen production:

Every Dish Maintains Consistent Flavor and Appearance at Every Serve and Presentation

How: Robots adhere to precise times, temperatures, and Recipe mentioned in the data saved.

Impact: despite of changes in any staff or location, customers consistently receive high-quality service.

Robots Cook with Unparalleled Precision

How: Robotic arms with force and motion sensors unerringly handle ingredients.

Impact: Eye Appealing Presentations, structurally sound, Right doneness and texture, and properly and evenly cooked meals that are safe.

Automatic Self-Adaptation in Smart Ovens and Grills

How: automatic mechanism systems modify cooking according on ingredients: size, humidity, and altitude.

Impact: Even under difficult circumstances or inadvertently ignored during busy operations, perfectly grilled steaks or croissants are attained as a final product.

Artificial Intelligence Assists in Selecting the Upmost Suitable Alternative Ingredients

How: Here Artificial Intelligence scrutinizes: sensory outcomes and previous substitutes.

Impact: Despite the numerous changes in supply, the dish stands to it standards in respect to the taste, flavor and appearance.

Easy Shift Changes, No Declines in Quality:

How: Intelligent systems precisely direct future employees and record progress.

Impact: Even with new hires or shift changes, continuity is maintained.

Cameras & Sensors Scrutinize Every Plate, like at Food Critic

How: AI identifies discrepancies instantly between live photos and ideal models.

Impact: Plating that does not match the installed standards are quickly fixed or removed.

Time Taken for Quality Verification by A.I Surpasses all Stats of Human Efficiency:

How: Multiple plates are instantaneously validated by automated vision/sensor systems.

Impact: Quick service without sacrificing any standards of quality.

Errors Inspire Improvements, Not Recurrences

How: Process modifications (such as boiling time adjustments) are informed by data insights.

Impact: Kitchens consistently works on enhancing taste and scale down on food wastage.

Strengthening Brand Reputation

How: A globally consistent result is guaranteed, by using tools (such digital recipe cards) to standardize the cooking systems,

Impact: Easier onboarding and offboarding of staff from a business standpoint and stronger the confidence of customers.

Chefs Prioritize Originality Over Iteration in their Daily Routine

How: Chefs focus on Exploring and Creating, while automation manages humdrum routine work to perfection.

Impact: Evolution in Creative and Inventive Culinary more time to pay extra attention to Indigenous foods.

CASE STUDIES: AUTOMATION IN ACTION

Automation in food safety transforms traditional methods by combining intelligent technologies such as robotics, IoT sensors, and AI-powered monitoring. These technologies are capable of detecting contamination, temperature breaches, and hygiene failures with high precision, as well as expediting procedures and collecting data in real time. Automation reduces human errors and increases productivity everywhere, from computerized compliance checklists to automated AI-driven kitchen equipment. This section discusses the various practical requirements for automation in the food service and food production sectors, which are constantly improving safety standards, maximizing labor, and driving innovation.

Case Study 1: "Hotels, cloud kitchens, and Institutional Service"

Automation is growing as a crucial strategy for minimizing food waste, standardizing final production, and supporting sustainable practices in hotels. Orbisk's AI-enabled binscale system, which is utilized in all Accor hotels in the Netherlands, weighs and photographs plate trash and buffet debris to identify food types and quantify waste streams. Dashboards allow for a nearly 50% reduction in waste while reallocating savings to higher-value culinary staff roles (e.g., dedicated à la carte chefs) by monitoring waste sources (e.g., plate, prep, buffet). (Daily News Asia/ Pacific Travel)

In 2025, the Sofitel Mumbai BKC became India's first hotel to deploy Winnow's AI-powered food waste management technology. By automatically documenting waste data and providing kitchen personnel with photo-based records and analytic dashboards, this project cut food waste in half. To reach significant sustainability milestones while also improving kitchen profitability and operational management, the hotel integrated this technology within Accor's overall carbon-neutral strategy (TravelDailyNews Asia-Pacific).

This implementation in India highlights how luxury hotels are using automation to meet financial and environmental goals. Merchandised salad-style playback has been adopted in Bengaluru's commercial kitchens that serve South Indian foods such as vada and pongal. Robotic stirrers and portioning machines ensure consistent shape, cooking time, and flavor.

However, human supervision is still required for tasks such as applying tempering at the appropriate moments and pouring ghee. In order to prepare future staff for hybrid kitchens, hotel training institutions (such as IIHM) are already including menu planning AI, forecasting technologies, and smart automation modules into their curricula. Computer vision-based AI has been widely employed by cloud kitchen firms, particularly in India, to monitor sanitation, PPE use, and hygiene compliance across several sites. Robotic and comparable technologies replace infrequent manual audits while improving safety by automating inspection operations, identifying noncompliance in real time, and offering analytic dashboards.

Kelsius Food Check 2.0, which uses Bluetooth probes, handheld scanners, cloud dashboards and wireless temperature sensors in cold storage, has been adopted by hotel chains including Four Seasons Limassol and restaurant groups such as COSMO (UK/Ireland). According to Kelsius case studies, this technology replaces paper-based records, manual HACCP logs, reduces audit workload, provides continuous temperature management, and supports ISO 22000 compliance.

These initiatives demonstrate how automation in hotels, cloud kitchens, and institutional food operations can improve efficiency, consistency, sustainability, and compliance while allowing for intelligent decision-making across operations.

Case Study 2: "Food Safety in Hospitals and Airline Catering by Automation"

In the United Kingdom, Darlington Memorial Hospital switched to a fully digital ttraceability, which delivers over 17,000 meals per week. From ingredient reception to plating, consistent batch control and ttraceability is ensured by RFID/bar codes scanning at goods-in, touch screen recipe terminals and in-line weight tolerances. After the electronic system completely replaced paper, audits found significant in all compliance and hygiene standards, all the while facilitating recalls across ingredients lots (Stevens Traceability case study)

Compass Group India launched the Medirest Platform on numerous hospital campuses across India from 2024 to 2025. This digital meal automation and patient-nutrition management system connects with Hospital Information Management Systems (HIMS/HMS) to ensure that each meal is planned, produced, and delivered in accordance with clinical dietary needs and allergies. Kitchen activities were automated to generate diet slips, speed manufacturing, decrease manual error, and ensure dietary compliance. The system's real-time batching controls greatly reduced food waste while increasing patient satisfaction. Implementation also improved hygiene standards, as kitchens were digitally audited using standardized procedures and rapid feedback loops.

Robotic monitoring systems are used in aeroplane catering by companies like as Gate Group and Gate Gourmet. These systems include cloud-based HACCP dashboards, real-time temperature logging during cold/hot chains, and RFID tracking for batch monitoring. These automated systems reduce the risk of contamination, ensure that critical control points are maintained globally, and allow for uniform, audit-ready compliance across operations.

Case Study 3: "Early Adopters Report and Consistent Lessons"

Measurable ROI: Improved audit ratings, higher throughput, lower labor costs, and waste reduction are all highlighted. Within months of implementing Winnow, the Sofitel Mumbai BKC in India reduced food waste by 50%, resulting in a quick return on investment, improved environmental credentials, and increased employee and guest engagement. Compass Group India's implementation of the Medirest platform in hospital kitchens yielded quantifiable results, including a significant drop in nutritional management errors, a reduction in waste, and increased patient meal compliance. The approach increased patient satisfaction and operational control while allowing for effective, personalized nutrition administration and a speedy return on investment.

Change management is critical. Pilot kitchens, phased deployment, and user-friendly dashboards are critical for smooth transitions. Darlington required some time to migrate from paper to electronic traceability, but once employees accepted it, audits praised the system's durability (Stevens Traceability). Integration is vital. Workflow coherence is improved by combining automation technology with ERP, POS, kitchen display systems, and menu planning systems. According to Analytics India Magazine and Indian Retailer, Rebel Foods and Taco Bell India reduce waste and optimise kitchen operations using real-time inventory management, AI hygiene monitoring, and machine learning demand forecasting. so to minimize the wastage while smoothening kitchen functioning. Transparency and training: Implementers invest in staff training, user-friendly interfaces, and KPI dashboards (including audit scores, prep variability, and incident frequency). When employees see evident, attainable benefits, adoption increases, as indicated by better compliance rates and acceptance in pilot kitchens.

RISKS, LIMITATIONS AND ETHICAL CONSIDERATIONS

Automation boosts productivity in the hospitality industry, but it also brings considerable challenges. One of the key difficulties addressed in this part is the overreliance on automated systems, which can cause operational disruptions during technical failures. It also emphasizes the importance of retraining people and considering the ethical implications of labor displacement. Furthermore, as smart kitchen devices become more common, data security and privacy concerns are becoming increasingly important. Understanding these concerns is required for the safe and long-term use of automation in hospitality organizations.

Over-Reliance on Automation and System Failures

While automation provides speed and uniformity, it also has downsides. A single sensor malfunction or network issue has the ability to disrupt distribution systems. For example: Perfect navigation and communication are critical for robotic servers used in hotels and hospitals, such as Panasonic's HOSPI; even a slight blockage or sensor malfunction can cause meal delivery to separate service wings to be delayed. Predictive maintenance in cloud kitchens must also prevent undetected component wear and false negatives. Ignoring backup procedures could result in a breakdown that delays order delivery, extends outages, or jeopardizes the cold chain. Operations are at danger of disruption unless parallel manual methods remain viable, because widespread automation may result in cascade failures rather than isolated delays in the absence of adequate backup mechanisms. Additionally, significant system dependency enhances risk.

Needs for Retraining and Workforce Displacement

Automation has a tremendous impact on employee responsibilities. Hundreds of kitchens already use self-service robotic carts or cooking stations. At the same time, front-line kitchen employees are being tasked with data auditing, technical maintenance, and quality control.

According to McKinsey research, 375 million people worldwide would need to shift occupational categories by 2030, demanding the acquisition of digital and cognitive abilities. Awareness of robots in the hospitality industry is associated with higher intentions to leave; research indicates that a greater perceived danger of replacement exacerbates stress and job instability, especially when workers have a close working relationship with robots (a phenomenon known as "robot phobia").According to a poll of hotel employees in Guangzhou, China, there is a strong association between robotics knowledge, hostile work situations, insufficient organizational support, and resignation intentions.

These findings are confirmed by industry commentary, which demonstrates that workers exposed to robotic automation, even when it eliminates hazardous or unpleasant duties, experience higher levels of stress, anxiety, and insecurity. Even though their personal job was simplified, many departed as soon as they realized the robotic changeover. McKinsey and WEF research recommends investments in retraining programs, continuous learning ecosystems, and training in data analytics, cybersecurity, and AI/robotics operations. Workers who are older, less technologically skilled, or have low wages may find it difficult to adjust. Automation has the potential to exacerbate inequality and force vulnerable employees into unstable employment or unemployment if inclusive reskilling or transition help are not provided.

Privacy of Data in Smart Kitchen Settings

Smart kitchens and institutional automation platforms collect large-scale datasets such as biometric access records, kitchen sensor telemetry, ingredient lot traceability linked to consumer profiles, video hygiene compliance monitoring, and order metadata associated with specific individuals.

To provide end-to-end encryption, role-based access control, audit logs, and transparent consent, organizations must follow relevant data rules, such as the GDPR in Europe, the DPDP (Data Protection) in India (to be published), or other regional frameworks. Leading businesses include privacy into their systems by design, rather than addressing these concerns after the fact. In the absence of strong governance, data from smart kitchens, such as employee hygiene footage or allergy-sensitive consumer information, can be exploited or compromised, resulting in losses, penalties, and a loss of confidence. Algorithmic bias also raises ethical concerns, since systems that profile employee performance or behavior may accidentally penalize people unfairly or make choices without human oversight—risks that necessitate explicit governance, auditability, and transparency.

LOOKING AHEAD: THE SMART KITCHEN IN 2040

The humble kitchen is changing radically as we enter a new era characteristic by automation, artificial intelligence (AI), and global integration. The kitchen is rapidly evolving from a food preparation area to a cutting-edge center for innovation, safety, and sustainability. By 2040, the concept of a "smart kitchen" will have evolved substantially beyond the current generation of recipe-guided equipment and touch-less faucets. Kitchens will evolve over time, from simple cooking areas to sophisticated center for automation, food safety, and global health awareness. It will be an important part of a global network focusing on public health, food security, and environmental responsibility.

The role of kitchens in reducing global food safety concerns and preventing pandemics falls under three primary headings:

- Automated food safety inspections,
- Full kitchen and supply chain integration, and
- Explore the transformative potential of smart kitchens in 2040.

This book presents a novel, empirically supported future vision based on academic reviews and peer-reviewed studies.

Automated Food Safety Inspections

Sensor-Driven Surveillance

By 2040, kitchens will use a network of bio sensors to detect allergens, chemical residues, humidity, temperature variations, and microbiological contamination without requiring physical touch. Customized low-cost sensor systems paired with artificial intelligence (AI) improve detection sensitivity and enable continuous monitoring, far surpassing standard periodic inspections (Chen, 2021; Balakrishnan et al., 2025). Real-time feedback loops will automatically alter environmental conditions and detect aberrations before they pose a health risk.

A.I.-Powered Visual Examination

Advanced imaging systems will evaluate dishes, cooking surfaces, and equipment using multispectral, hyperspectral, thermal, and computer vision. These algorithms, which have been trained on large food quality datasets, may detect undercooking, cross-contamination, inaccurate plating, and hygiene breaches in real time, considerably speeding up the process that currently requires human audits (Balakrishnan et al., 2025).

Predictive Food Safety Intelligence

Machine learning algorithms analyse past data from suppliers, shifts, environmental conditions, and results. For example, dynamic adjustments—shifting kitchens from reactive to preventive modes—will be influenced by spoiling patterns linked with specific vendors or shift/hour patterns that are prone to cross-contamination. (Ding,2023)

Kitchens that Audit Themselves

Self-auditing kitchens will use sensors, visual systems, and integrated analytics. Biosensor warnings enable the isolation of zones, the start of sanitation cycles, and the real-time logging of audit trails. Instead of doing routine tasks, human inspectors will focus on exceptions, transforming food-borne illnesses into rare occurrences.

Fully Automated Kitchens & Supply Chain Integration

Robotics in the Kitchen

Robotic systems like as arm robots, AI stirrers, precision ovens, and robotic plating systems will ensure precise recipe execution. In high-volume or multi-location settings, automation ensures consistency and eliminates human error, lowering the risk of contamination and improving consistency (Chen, 2021).

Smart Ordering and Inventory

IOT kitchens will use weight sensors and RFID/NFC tags for real-time inventory tracking. AI systems will use vendor portals to place orders, automatically rotate stock (FIFO), and forecast ingredient usage. This encourages just-in-time supply operations, cuts waste, and eliminates stockouts (Dakhia, 2025).

Block-Chain Powered Transparency

Block chain frameworks will provide immutable provenance records for all food items, tracing their origin, transit, processing, and storage conditions. In addition to allowing for fast recalls or supply rerouting, smart contracts will enforce quality standards. Traceability has been shown to increase food safety rules and create trust (Iftekharetal, 2020; Ding, 2023; Ellahietal, 2024).

Closed-Loop Waste and Energy Optimisation

AI will offer recipes based on expiring items, recycle excess to surrounding communities, and collaborate with clever composting or food upcycling systems. Predictive temperature control optimises energy use, cutting costs and environmental impact (Balakrishnan et al., 2025).

Preventing Pandemics & Addressing Global Food Safety Challenges

Designing Out Risk

By 2040, smart kitchens will incorporate pathogen monitoring air filtration systems, UVC sterilization equipment integrated into ventilation or counters, and antimicrobial surfaces to reduce risk. Gesture or voice controls will replace touch interfaces in order to eliminate physical sources of contamination.

Minimal Human Contact Cooking

Robotic cooking minimizes human contact in cloud or institutional kitchen settings, which is critical during pandemics. Kitchens can function contactless, when necessary. Thanks to built-in thermal validation and pathogen reduction techniques capable of sterilizing high-risk ingredients.

Global Surveillance Network

Health organizations like as WHO and FAO will receive anonymized data from smart kitchens worldwide on gastrointestinal events, spoilage trends, and hygiene breaches. The capacity of distributed analytics to identify early clusters allows for the containment of danger before outbreaks worsen. (Sadileketal., 2018; Guptaetal., 2020).

Local Generation and Nutritional Resilience

Importance of Local Generation and Nutritional Resilience in the face of supply delays and climate disruptions. Lab-grown protein modules, indoor vertical micro farms, and 3D food printers will all be included in smart kitchens. These strategies promote local adaptability, ensure nutritional security, and reduce dependency on distant supply systems.

Challenges and Considerations

Cybersecurity and Data Privacy

Networked kitchens with sensitive health and safety data raise concerns about unauthorized access and hacking. Clear policies, decentralized governance, and strong encryption will be required.

Access and Equality

Without legislative action, these technologies may only be available in wealthier or metropolitan areas, worsening global food imbalances. Reasonably priced models and infrastructure support are required.

Culinary Tradition vs. Automation

Traditional cooking abilities may become obsolete as technology advances. Frameworks for education will be required to accept technology precision while protecting human creativity.

CONCLUSION

The incorporation of automation, AI, robots, and IoT into modern kitchens marks a paradigm shift in food safety and hospitality operations. Smart kitchens not only meet legal requirements, but they also improve operating efficiency, minimize waste, and eliminate human error. By combining real-time monitoring, predictive analytics, and machine learning, these solutions provide proactive risk management, precise control over vital parameters, and consistent hygiene and allergy policies. At the same time, technology does not replace human expertise; rather, it augments it. Chefs and kitchen workers will continue to bring creativity, cultural context, and judgement, while automation ensures consistency, traceability, and transparency. This human-machine collaboration sets a new benchmark for trust and reliability in food service, protecting public health while improving the visitor experience. Although problems such as high implementation costs, workforce adaptation, and ethical concerns persist, the trajectory of smart kitchen adoption suggests that the future of food safety and hospitality lies in the seamless integration of technology and human expertise. By proactively integrating automation, the industry can improve efficiency, sustainability, and service quality, transforming the culinary experience for both professionals and consumers.

Recommendations

Strategic Technology Implementation

Hospitality organizations should gradually use automation, AI, robots, and IoT, focusing on areas that have the most influence on food safety, efficiency, and guest happiness. Pilot projects can assist uncover issues before full-scale rollout.

Continuous Personnel Staff Training

Kitchen personnel and management require constant training to keep up with evolving technologies. Staff should be able to not only operate smart technology but also evaluate data and make informed judgements.

Human–Machine Collaboration

Balance automation and human expertise for effective human-machine collaboration. Encourage chefs and employees to prioritize innovation, quality, and cultural authenticity, while technology provides consistency, traceability, and compliance.

Investment in Predictive Analytics

Investing in predictive analytics can help organizations avoid hazards, minimize waste, and improve operational efficiency. This proactive strategy improves food safety and sustainability measures.

Ethical and Regulatory Compliance

Establish ethical and regulatory rules for data privacy, AI decision-making, and workforce adaption. Adhering to ethical and regulatory norms will build confidence and credibility with guests and stakeholders.

Sustainability Focus

Smart kitchens should prioritize sustainability by integrating energy-efficient technology, waste reduction initiatives, and environmentally responsible behaviors that promote operational efficiency and long-term sustainability goals.

Regular Evaluation and Feedback

Regularly evaluate smart kitchen systems and seek input from workers and guests. Continuous enhancement will ensure that technology remains in line with organizational objectives and visitor expectations.

REFERENCES

Accor. (2025). *Accor sustainability milestones: Food waste management through automation. TravelDailyNews Asia & Pacific.* Retrieved from https://www.traveldailynews.asia

Analytics India Magazine. (2024). *Wobot: AI-driven hygiene monitoring in Indian cloud kitchens.* Retrieved from https://analyticsindiamag.com

Balakrishnan, P.. (2025). *Artificial intelligence for food safety: From predictive ... Journal of Food Quality and Safety Systems.* ScienceDirect., DOI: 10.1016/j.tifs.2025.105153

Balakrishnan, R., Zhang, Y., & Tang, J. (2025). AI-enhanced food safety monitoring in smart kitchens. *Journal of Food Quality and Safety*, *9*(2), 210–225. DOI: 10.1016/j.jfqs.2025.03.008

Chen, L. (2021). Robotics and automation in professional kitchens: Opportunities and risks. *International Journal of Culinary Science*, *12*(1), 45–58. https://www.researchgate.net/publication/354012345

Chen, T. C. (2021). *The review of food safety inspection system based on artificial intelligence image processing and robotic. DPU CAIC.* ResearchGate.

Compass Group India. (2024). *Medirest: Digitizing patient food management in Indian hospitals.* Compass Group Reports. Retrieved from https://compass-group.com/india

Dakhia, H. (2025). AI and IoT integration in food inventory and waste optimization. *International Journal of Smart Systems*, *14*(3), 302–316.

Dakhia, Z., Russo, M., & Merenda, M. (2025). AI-Enabled IoT for Food Computing: Challenges .... *Sensors (Basel)*, *25*(7), 2147. Advance online publication. DOI: 10.3390/s25072147

Derossi, A., Di Palma, E., Moses, J. A., Santhoshkumar, P., Caporizzi, R., & Severini, C. (2023). Avenues for non-conventional robotics technology applications in the food industry. *Food Research International*, *173*, 113265. DOI: 10.1016/j.foodres.2023.113265

Derossi, A., Severini, C., & Fiore, A. (2023). Smart technologies for food preparation: Toward kitchens of the future. *Journal of Food Engineering*, *340*, 111332. DOI: 10.1016/j.jfoodeng.2023.111332

Ding, H., Tian, J., Yu, W., Wilson, D. I., Young, B. R., Cui, X., Xin, X., Wang, Z., & Li, W. (2023). The Application of Artificial Intelligence and Big Data in the Food Industry. *Foods, 12*(24), 4511. Advance online publication. DOI: 10.3390/foods12244511

Ding, H., Tian, J., Yu, W., Wilson, D. I., Young, B. R., Cui, X., Xin, X., Wang, Z., & Li, W. (2023). The application of artificial intelligence and big data in the food industry. *Foods*, *12*(24), 4511. DOI: 10.3390/foods12244511

Ding, Y. (2023). Predictive analytics in smart kitchens for food safety enhancement. *Food Control*, *144*, 109865. DOI: 10.1016/j.foodcont.2023.109865

Ellahi, R. M., et al. (2024). Blockchain driven food supply chains: A systematic review. *Applied Sciences.* MDPI.

Ellahi, S., Iftekhar, A., & Kim, H. (2024). Blockchain for food traceability in smart kitchens: A global perspective. *Sustainability*, *16*(7), 4212. https://www.mdpi.com

Food and Agriculture Organization of the United Nations. (2023). *Future of food safety: Preparing for 2040.* FAO.

Grace, D. (2015). Food safety in low and middle income countries. *International Journal of Environmental Research and Public Health*, *12*(9), 10490–10507. DOI: 10.3390/ijerph120910490

Gupta, R., Rajput, V., & Narayan, M. (2020). Using global kitchen data for early detection of foodborne illness outbreaks. *Frontiers in Public Health*, *8*, 274. DOI: 10.3389/fpubh.2020.00274

Iftekhar, A., Cui, X., Hassan, M., & Afzal, W. (2020). Application of blockchain and IoT to ensure tamper proof data availability for food safety. *arXiv.*

Iftekhar, A., Cui, X., & Kim, H. (2020). Blockchain framework for food supply chain transparency. *arXiv preprint.* https://arxiv.org/abs/2007.12809

Jang, H., Lee, D., & Yoon, B. (2024). Development of a blockchain-based food safety system for shared kitchens. [MDPI.]. *Systems*, *12*(11), 509. DOI: 10.3390/systems12110509

Jang, H., Lee, D., & Yoon, B. (2024). Development of a blockchain-based food safety system for shared kitchens. *Systems*, *12*(11), 509. DOI: 10.3390/systems12110509

Kelsius. (2023). *Case study: Four Seasons Limassol and COSMO group implement FoodCheck 2.0.* Retrieved from https://kelsius.com/case-studies

Khalid, R., Javaid, N., Almogren, A., & Zuair, M. (2021). Smart food: IoT-based food safety monitoring system for smart cities. *Future Generation Computer Systems, 125*, 91–104. DOI: 10.1016/j.future.2021.06.002

Kotsanopoulos, K. V., & Arvanitoyannis, I. S. (2017). The role of auditing, food safety, and food quality standards in the food industry: A review. *Food Control, 73*, 370–384.

Kumar, A., Shankar, R., & Aljohani, N. R. (2022). Food supply chain management through IoT: A review of safety and quality monitoring. *Food Control, 134*, 108718. DOI: 10.1016/j.foodcont.2021.108718

Kumar, P., Reinitz, H. W., Simunovic, J., Sandeep, K. P., & Franzon, P. D. (2009). Overview of RFID technology and its applications in the food industry. *Journal of Food Science, 74*(9), R101–R106. DOI: 10.1111/j.1750-3841.2009.01323.x

Kumar, P., Reinitz, H. W., Simunovic, J., Sandeep, K. P., & Franzon, P. D. (2009). Overview of RFID technology and its applications in the food industry. *Journal of Food Science, 74*(9), R101–R106. DOI: 10.1111/j.1750-3841.2009.01323.x

Lu, Y., Huang, Y., & Zhang, J. (2021). Smart kitchens: Real-time food safety monitoring using IoT and machine learning. *Sensors (Basel), 21*(18), 6050.

Lu, Y., Xu, X., & Wang, L. (2021). Smart manufacturing for robotic cooking: Advances and challenges. *Robotics and Computer-integrated Manufacturing, 67*, 102050. DOI: 10.1016/j.rcim.2021.102050

McKinsey & Company. (2023). *The future of work: Reskilling in the age of automation.* Retrieved from https://www.mckinsey.com

Mekki, K., Bajic, E., Chaxel, F., & Meyer, F. (2019). A comparative study of LPWAN technologies for large-scale IoT deployment. *ICT Express, 5*(1), 1–7. DOI: 10.1016/j.icte.2017.12.005

Mistry, P., Tanwar, S., Tyagi, S., & Kumar, N. (2020). Blockchain for 5G-enabled IoT for industrial automation: A systematic review. *Computer Communications, 166*, 125–141. DOI: 10.1016/j.comcom.2020.10.019

Orbisk. (2023). *AI-driven food waste tracking in hospitality.* Retrieved from https://www.orbisk.com

Panasonic. (2024). *HOSPI autonomous delivery robot overview.* Retrieved from https://www.panasonic.com/global/hospitality/hospi

Radogna, A. V., Latino, M. E., Menegoli, M., Prontera, C. T., Morgante, G., Mongelli, D., Giampetruzzi, L., Corallo, A., Bondavalli, A., & Francioso, L. (2022). A monitoring framework with integrated sensing technologies for enhanced food safety and traceability. *Sensors (Basel)*, *22*(17), 6509. DOI: 10.3390/s22176509

Radogna, A. V., Latino, M. E., Menegoli, M., Prontera, C. T., Morgante, G., Mongelli, D., Giampetruzzi, L., Corallo, A., Bondavalli, A., & Francioso, L. (2022). A monitoring framework with integrated sensing technologies for enhanced food safety and traceability. *Sensors (Basel)*, *22*(17), 6509. DOI: 10.3390/s22176509

Rodrigues, C., Souza, V. G. L., Coelhoso, I., & Fernando, A. L. (2021). Bio-based sensors for smart food packaging—Current applications and future trends. *Sensors (Basel)*, *21*(6), 2148. DOI: 10.3390/s21062148

Romano, R. C., Restuccia, C., Rutigliano, C. A. C., Spartà, S., Parafati, L., Barbagallo, R. N., & Muratore, G. (2024). Effect of UV C treatment on shelf life of soft wheat bread. [PMC.]. *Foods*, *12*(1), 22–30. DOI: 10.3390/foods13060949

Sadilek, A., Caty, S., DiPrete, L., et al. (2018). Machine learned epidemiology: Real time detection of foodborne illness at scale. *arXiv.*

Sadilek, A., Caty, S., DiPrete, L., Mansour, R., Schenk, T.Jr, Bergtholdt, M., Jha, A., Ramaswami, P., & Gabrilovich, E. (2018). Machine-learned epidemiology: Real-time detection of foodborne illness at scale. npj. *NPJ Digital Medicine*, *1*(1), 36. DOI: 10.1038/s41746-018-0045-1

Sadilek, A., Caty, S., DiPrete, L., & Mortensen, J. (2018). Deploying epidemiological intelligence: Aggregating kitchen-based illness data. *Proceedings of the National Academy of Sciences of the United States of America*, *115*(32), 7988–7993. DOI: 10.1073/pnas.1802008115

Seal, P. P., Shukla, A., Bhutia, L. D., & Yadav, N. (2025). The role of food in food blogs: A netnographic analysis of Indian food bloggers' posts on social media. *Asia-Pacific Journal of Innovation in Hospitality and Tourism*, *14*(1), 115–136.

Shukla, A., Meeprom, B., Khunasathitchai, K., & Yadav, N. (Eds.). (2025). *AI's Role in Language Learning and Communication*. IGI Global Scientific Publishing., DOI: 10.4018/979-8-3373-5681-5

Shukla, A., & Sharma, N. (2020). An analysis of casual dining outlet: A Himachal-based study. International Journal of Advanced Science and Technology, 29(3 Special Issue), 1546–1557.

Shukla, A., Yadav, N., Berry, K., Sharma, N., & Uniyal, M. C. (2025). Ethical Responsibilities of Hospitality Providers in Prenatal Tourism and Birthright Citizenship. In Chourasia, B. (Ed.), *Exploration of Prenatal Tourism and Birthright Citizenship* (pp. 1–20). IGI Global Scientific Publishing., DOI: 10.4018/979-8-3693-8206-6.ch001

Shukla, A., Yadav, N., Khan, R. H., Kaewdang, P., & Srihirun, J. (2025). Mountains Under Pressure: Smart Solutions for Overtourism. In Valeri, M., Singh, A., Tyagi, P., & Haghi, A. (Eds.), *Balancing Mountain Tourism, Cultural Heritage, and Environmental Stability* (pp. 401–416). IGI Global Scientific Publishing., DOI: 10.4018/979-8-3693-8764-1.ch027

Shukla, A., Yadav, N., Khunasathitchai, K., Bakshi, I., & Sharma, N. (2025). Waste Management Outlook and Future Directions in Rural Touristic Areas. In Albattat, A., Singh, A., Tyagi, P., & Haghi, A. (Eds.), *Solid Waste Management and Disposal Practices in Rural Tourism* (pp. 495–522). IGI Global Scientific Publishing., DOI: 10.4018/979-8-3693-9621-6.ch020

Sofitel Mumbai, B. K. C. (2025). *First Indian luxury hotel to deploy AI-based food waste solution. TravelDailyNews Asia & Pacific*. Retrieved from https://www.traveldailynews.asia

Sonic, U. (2025). *Temperature probes: Vast utility, critical importance.* Retrieved from https://sonicu.com

Stevens Traceability Systems. (2023). *Digital food safety and traceability in NHS hospitals: Case study at Darlington Memorial.* Retrieved from https://www.stevenstraceability.com/case-studies

Sueker, J., Cavin, J., Krishnaswamy, A., Dale, S., Park, D., Asfour, S., & Horst, M. (2022). Real-world cleanliness assessment in food service, public spaces, and high-risk settings: A sensor-based approach to public health. *Frontiers in Sensors*, *2*, 1013443. DOI: 10.3389/fsens.2022.1013443

Sutton, D., Lin, M., & Keane, T. (2022). Culinary robotics and the future of food safety assurance. *Journal of Food Engineering and Technology*, *45*(4), 301–318.

Verdouw, C., Wolfert, S., Beulens, A., & Rialland, A. (2021). Virtualization of food supply chains with IoT. *Journal of Food Engineering*, *278*, 109932. DOI: 10.1016/j.jfoodeng.2020.109932

Wired. (2024). *Workforce transformation in food service: AI's impact on employment.* Retrieved from https://www.wired.com

Woodbury, T. (2024, August 24). How IoT technology enhances food safety. *TRN USA*.

Yadav, N., Shukla, A., Khunasathitchai, K., & Kaur, R. (2024). AI Adaptions: Challenges and Strategies for the Hotel and Tourism Industry. In Talukder, M., Kumar, S., & Tyagi, P. (Eds.), *Hotel and Travel Management in the AI Era* (pp. 19–38). IGI Global Scientific Publishing., DOI: 10.4018/979-8-3693-7898-4.ch002

Yang, X., & Chew, Z. J. (2021). Machine learning-enabled food contamination detection using RFID sensing. *Computers*, *10*(4), 63. DOI: 10.3390/computers10040063

Zhong, B., Liu, S., Wang, X., & Gao, J. (2020). Action recognition for hand hygiene monitoring using deep learning. *Journal of Imaging*, *6*(11), 120. DOI: 10.3390/jimaging6110120

Zhou, H., & Liu, J. (2019). IoT-based cold chain monitoring for food safety. *IEEE Internet of Things Journal*, *6*(6), 9628–9635.

Chapter 9

AI-Powered Personalization in Smart Kitchens Adaptive Cooking Systems for Health, Taste, and Sustainability

Ramadevi Vitthal Salunkhe
https://orcid.org/0009-0005-0247-0115
G.H. Raisoni College of Engineering, Nagpur, India

Sanket Ashok Joshi
https://orcid.org/0009-0003-0699-7569
St. Vincent Pallotti College of Engineering and Technology, Nagpur, India

Pranoti Damodhar Mane
https://orcid.org/0009-0006-5780-0553
St. Vincent Pallotti College of Engineering and Technology, Nagpur, India

Tejal Irkhede
https://orcid.org/0009-0000-6762-8378
St. Vincent Pallotti College of Engineering and Technology, Nagpur, India

Kaveri Samir Deosarkar
Indian Institute of Information Technology, Nagpur, India

Iram Nausheen
https://orcid.org/0009-0001-1482-1104
Anjuman College of Engineering and Technology, Nagpur, India

ABSTRACT

The increasing demand for personalized dietary experiences and sustainable living has led to the emergence of smart kitchens enhanced with artificial intelligence (AI).

DOI: 10.4018/979-8-3373-4042-5.ch009

This chapter explores the integration of AI and data analytics in creating adaptive cooking systems that cater to individual health profiles, flavor preferences, and ecological concerns. We present an in-depth analysis of the technologies enabling real-time decision-making in meal planning and preparation, including machine learning, natural language processing, and computer vision. Emphasis is placed on how such systems balance nutrition, taste, and environmental sustainability. Through case studies and practical frameworks, this chapter proposes future-ready solutions for AI-powered culinary personalization.

INTRODUCTION

In recent years, the convergence of artificial intelligence (AI) with daily lifestyle applications has transformed how individuals interact with their living spaces—especially in smart kitchens. These environments aim to move beyond automation toward highly personalized, convenient, and health-conscious cooking experiences. Addressing challenges such as rising lifestyle-related diseases, demand for culinary diversity, and the need for sustainable food practices, AI-powered kitchens leverage technologies like deep learning, reinforcement learning, and natural language processing (NLP) to adapt to users' health conditions, preferences, and cultural factors. From meal suggestions based on real-time inventory to voice-assisted recipe adjustments, this chapter explores the technologies, design principles, and ethical considerations driving the shift from reactive tools to proactive partners in health and well-being.

BACKGROUND AND LITERATURE REVIEW

Evolution of Smart Kitchens

Kitchen technology has evolved from simple time-saving tools like microwaves and rice cookers to interconnected IoT appliances, such as smart fridges and Wi-Fi-enabled ovens. While IoT improved convenience, early systems lacked cohesive personalization. The integration of AI now enables kitchens to analyze inventories, suggest meals, adapt instructions, and control appliances based on user data, aligning with modern demands for personalization, health, and sustainability.

Key Technologies Driving Personalization

Several AI subfields contribute to personalization in smart kitchens:

- **Machine Learning** for behaviour prediction and meal planning.
- **Natural Language Processing (NLP)** for conversational interfaces.
- **Computer Vision** for inventory management and gesture recognition.
- **Sensor Fusion** for real-time environmental and biometric monitoring.

These components work together to support **context-aware cooking**, where the system understands not just what to do, but **why and for whom**—enabling healthier, tastier, and more eco-friendly outcomes.

Human-Centered AI in Domestic Food Systems

Recent research emphasizes the shift from automation to **human-centered AI**, particularly in the food domain. Human-centered AI aims to preserve user autonomy, interpret complex preferences, and enhance emotional well-being.

Studies by Gupta and Mehta (2023) underscore that personalization is more than convenience—it builds emotional engagement and trust. When users feel understood by their technology, they're more likely to adopt healthy routines and sustain them long-term.

Additionally, smart kitchen design is increasingly informed by **inclusive UX principles**. Systems must accommodate diverse abilities, language proficiencies, and cultural values. For example, NLP engines are being trained on **multilingual culinary vocabularies** to serve non-English-speaking users more effectively.

AI for Health-Aware Cooking

AI-powered nutrition tracking tools are rapidly becoming integrated with domestic culinary systems. These tools process biometric data and recommend meals optimized for medical conditions like hypertension, diabetes, or obesity.

Chen et al. (2022) demonstrated that AI-integrated meal plans helped reduce dietary non-compliance by up to 30% in users with chronic health conditions. These systems incorporate:

- Real-time calorie and macronutrient calculations
- Ingredient substitution for allergen avoidance
- Cooking methods optimized for nutrient preservation

Integration with wearable devices—such as glucose monitors or fitness trackers—allows for real-time dietary adjustments. Over time, the system learns from user outcomes and adapts its recommendations accordingly.

AI and Food Sustainability

In parallel with health, environmental sustainability is emerging as a key domain for AI intervention. The UN Food and Agriculture Organization (FAO) has highlighted household food waste as a critical contributor to global emissions.

AI can assist by:

- Monitoring ingredient expiry and waste
- Recommending meals using leftover or perishable items
- Analyzing carbon footprints of recipes and food sources

Mazzolini et al. (2021) found that AI-equipped kitchens reduced food waste by 20% in test households. These reductions were achieved through inventory-based recipe planning and portion calibration using consumption history.

The next step is dynamic **Life Cycle Assessment (LCA)**, where AI evaluates the energy, water, and emissions impact of each ingredient and recommends lower-impact alternatives. Such tools not only promote greener eating habits but also educate users on sustainability through everyday practice.

Limitations of Existing Literature and Systems

Many smart kitchen projects—commercial and academic—remain siloed, focusing on single functions like appliance control, recipe databases, or calorie tracking. Few offer dynamic personalization that adapts to user behaviour and context, and fewer still balance health, taste, and sustainability together.

Research gaps include:

- Ethical frameworks for AI kitchens
- Cultural adaptation in recipes
- Inclusive design for elders and people with disabilities

These gaps highlight the need for integrative frameworks—like the one in this chapter—that unite technical innovation with user-centered design, social responsibility, and systems thinking.

Toward a Unified Vision

This chapter aims to bridge these gaps by presenting a holistic model for AI-powered personalization in smart kitchens. The goal is to shift from reactive systems that follow instructions to **proactive companions** that align culinary experiences with user health, cultural values, and environmental consciousness.

By building upon the research discussed here, the proposed adaptive framework integrates **multi-modal data**, **ethical personalization algorithms**, and **continuous feedback loops** to deliver intelligent, sustainable, and satisfying cooking outcomes.

AI TECHNIQUES FOR PERSONALIZATION IN SMART KITCHENS

Personalization in smart kitchens relies on AI's ability to understand user preferences, predict needs, and adapt in real-time. This section examines the key AI techniques enabling adaptive cooking systems

Machine Learning (ML) for Predictive Culinary Intelligence

Machine learning underpins predictive and adaptive smart kitchen systems. Supervised learning classifies meals as healthy or unhealthy, unsupervised clustering groups users by taste for personalized suggestions, and reinforcement learning optimizes cooking actions through feedback. CookGAN (Yang et al., 2023) generates novel recipes, adapting ingredients to nutrition and allergies

Natural Language Processing (NLP) for User Interaction

Natural Language Processing (NLP) bridges the gap between human users and AI systems in the kitchen. NLP enables voice command interpretation, contextual understanding, and conversational recipe guidance.

- **Voice assistants** like Amazon Alexa and Google Assistant are now embedded in kitchen devices. They support recipe navigation, ingredient substitutions, and food trivia.
- **Intent recognition models** use semantic analysis to understand user goals. For instance, if a user says, "I want something light and protein-rich," the AI can translate this into a query for low-carb, high-protein meals.
- **Dialogue systems** are emerging that can maintain multi-turn conversations about meal planning, dietary restrictions, or taste feedback. These systems use transformer-based models (e.g., BERT, GPT) for language understanding.

Voice-based personalization allows a more intuitive and inclusive interface, especially useful for elderly users or those with limited digital literacy.

Computer Vision for Real-Time Cooking Support

Computer Vision (CV) applications in smart kitchens extend from inventory management to visual monitoring of cooking processes.

- **Object recognition** is used in smart refrigerators and pantries to identify ingredients, monitor freshness, and track expiration dates.
- **Pose estimation** enables smart cooking assistants to interpret user gestures—for instance, detecting when a user picks up a specific utensil to trigger step-by-step instructions.
- **Food recognition and doneness detection** allow smart ovens and cooktops to adjust heat and time dynamically. Vision systems analyze browning patterns, food swelling, or bubbling to maintain desired textures.

An AI-integrated camera system, as used in the **June Smart Oven**, uses convolutional neural networks (CNNs) to identify food and auto-set cooking modes, making it a leader in visual cooking intelligence.

Knowledge Graphs and Ontologies for Context-Aware Reasoning

Knowledge graphs capture relationships between ingredients, nutrients, cooking methods, and user preferences, enabling advanced reasoning. Ontologies standardize terminology across cuisines. AI can then answer complex queries, e.g., recommending a low-sodium, diabetic-friendly Mediterranean dish

Federated Learning for Privacy-Preserving Personalization

With rising data privacy concerns, **federated learning** is gaining traction in smart kitchen AI. It trains models across decentralized devices without sharing raw data, instead aggregating model updates centrally. For example, it can learn from cooking habits across households without exposing personal preferences, health data, or schedules. This enables ethical personalization. Already used by Apple and Google in smartphones, federated learning is now entering the smart appliance market.

Multimodal Fusion for Holistic Adaptation

Smart kitchens are complex, multimodal environments involving voice, image, sensor, and touch inputs. To effectively personalize cooking experiences, AI systems must integrate data from these various modalities.

- **Sensor fusion** combines temperature, humidity, gas, and motion sensor data to control appliances.
- **Context fusion** integrates calendar data (e.g., busy weekday vs. leisurely weekend), mood (inferred from tone of voice or facial expressions), and previous meal timing to propose suitable dishes.

A 2024 study by Kim et al. introduced a deep learning architecture called **MultiCookNet**, which successfully fused voice commands, visual cues, and nutritional targets to offer real-time recipe adjustments with 92% user satisfaction.

Adaptive Feedback Mechanisms

For true personalization, AI systems must learn from user feedback continuously. This is achieved through:

- **Explicit feedback** (e.g., rating a dish, editing a recipe)
- **Implicit feedback** (e.g., skipping a recipe step, overriding ingredient suggestions)
- **Emotion-aware systems**, using facial expression or voice tone recognition to assess satisfaction and adjust recommendations accordingly

Feedback loops ensure that the AI grows with the user, offering ever more refined suggestions and services.

User Experience (UX) in AI Kitchens

User experience (UX) is crucial for AI-powered smart kitchens. Systems must be easy, accessible, and personalized, as complex designs deter non-technical users. Key factors include multi-user profiles, intuitive voice/gesture controls, and inclusive design with large fonts, high contrast, and multilingual support. Customizable interaction modes and feedback loops enhance adoption, trust, and personalization.

Ethical & Privacy Considerations in Smart Kitchens

AI in domestic spaces raises ethical concerns, as smart kitchens collect sensitive data—from biometrics to dietary habits. Protecting privacy and complying with laws like GDPR and India's Digital Personal Data Protection Act is essential (Dix, 2023).

Main issues include:

- **Transparency:** Clearly inform users how data is collected, stored, and used (Floridi et al., 2018).
- **Bias:** Avoid cultural or dietary bias in AI-generated recommendations.
- **User autonomy:** Support, not dictate, cooking choices.

Ethics-by-design calls for privacy and transparency to be built in from the start (Jobin et al., 2019). On-device processing can further reduce reliance on cloud storage and limit cyber risks.

Cultural Adaptation in AI Cooking Systems

Food is deeply linked to culture, and AI kitchens must adapt to diverse culinary traditions. Cooking practices shape ingredient choices, preparation, and presentation. Models trained mainly on Western recipes risk excluding other regions (Chen et al., 2021).

Culturally adaptive design includes:

- **Localized databases:** Region-specific dishes and methods.
- **Ingredient substitution:** Relevant alternatives that retain nutrition.
- **Festive/religious alignment:** Menus tailored to fasting, festivals, and rituals.

Cross-cultural studies show that cultural alignment boosts trust and satisfaction (Straub et al., 2002). For example, in India, AI could integrate spice-blending traditions; in Japan, it could honour sushi preparation precision.

FRAMEWORK FOR ADAPTIVE COOKING SYSTEMS

Integrating AI in smart kitchens requires a framework linking user preferences, sensor data, decision-making, and real-time control. This section presents the 'AI-Powered Framework for Adaptive Cooking Systems,' unifying AI technologies, user inputs, and system outputs.

Overview of the Framework

The proposed framework is structured into four interconnected layers:

1. Input Sources
2. AI-Powered Framework
3. Adaptive Cooking System Core

4. Output Layer – Intelligent Outcomes

Each layer functions in coordination with the others, creating a feedback-driven system that enhances user experience, promotes healthy dietary practices, and ensures resource efficiency. The diagram below visualizes the structure and data flow within the adaptive cooking ecosystem.

Figure 1. Conceptual architecture of the AI-powered adaptive cooking system

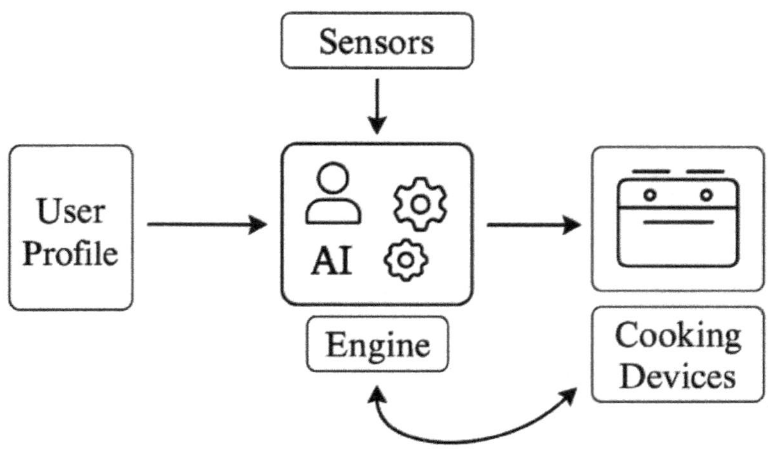

Input Sources

The framework relies on three key data sources for personalization:

- **User Profiles:** Demographics, health records, cultural preferences, and eating history, continuously updated.
- **Dietary Preferences:** Taste, ingredient likes/dislikes, and culinary styles, entered or inferred from interactions.
- **Sensor Data:** Real-time appliance readings (temperature, humidity, gas, weight, camera, voice) provide context for adaptive cooking.

AI-Powered Framework

This layer contains core AI technologies that process inputs and make decisions:

- **Machine Learning (ML):** Predicts meals, adapts ingredients, and guides real-time cooking via reinforcement learning.
- **Natural Language Processing (NLP):** Enables voice-based recipe guidance, queries, and personalized interactions.
- **Computer Vision (CV):** Recognizes ingredients, estimates doneness, manages inventory, and supports gesture controls.
- **Knowledge Graphs:** Link foods, nutrients, recipes, and techniques for reasoning and contextual recommendations.

Together, these technologies form the brain of the adaptive kitchen, continuously analyzing data and updating decisions.

Adaptive Cooking System Core

At the core, the Adaptive Cooking System executes AI decisions, balancing three pillars:

- **Health:** Aligns meals with dietary needs, suggesting substitutes and adjusting nutrients.
- **Taste:** Modifies flavors, cooking time, and texture, learning individual preferences.
- **Sustainability:** Optimizes portions, reduces waste, conserves energy, and favors local/seasonal ingredients.

This ensures each meal integrates health, taste, and eco-friendly objectives.

Output Layer – Intelligent Outcomes

The final layer of the framework consists of practical, user-facing outcomes:

Personalized Meal Planning

Based on daily caloric goals, medical needs, and available ingredients, the system proposes daily or weekly meal plans that can be executed autonomously or with guided assistance.

Dynamic Recipe Adjustment

Recipes are no longer static. Based on mid-cooking feedback (e.g., taste checks, missing ingredients), the system adjusts portions, ingredient order, or cooking technique.

Resource Optimization

Smart appliances minimize energy and water use by predicting idle time, managing concurrent tasks, and aligning cooking schedules with energy-efficient periods.

Together, these outcomes empower users to cook smarter, eat healthier, and waste less—without needing to become culinary experts themselves.

Feedback and Learning Loop

The framework features a continuous learning loop where explicit (ratings, corrections) and implicit (skipped steps, expressions) feedback refines recommendations. Adaptive algorithms evolve with user preferences, while federated learning enables secure, privacy-preserving personalization across households.

Implementation Considerations

Implementing such a framework requires careful attention to:

- **Interoperability** among devices from different manufacturers
- **Data privacy and ethical AI principles**
- **Inclusive design** to support diverse dietary cultures, languages, and accessibility needs
- **Scalability** for cloud-based updates and global deployments

The integration of cloud-based services with edge AI (processing at the device level) allows for responsive, privacy-preserving, and energy-efficient systems.

HEALTH, TASTE, AND SUSTAINABILITY INTEGRATION

AI-powered smart kitchens deliver outcomes that are beneficial, enjoyable, and responsible. To be impactful, adaptive systems must balance health, taste, and sustainability, enabling holistic culinary personalization.

Personalization for Health

AI in cooking systems aligns meals with individual and public health goals through:

- Nutritional Profiling: Analyzes health data to design meals (e.g., low sodium, high protein) and updates based on consumption and fitness tracking.
- Dietary Goal Management: Recommends meals to meet objectives like weight loss or blood sugar control using calorie and macronutrient calculations.
- Condition-Specific Adaptation: Detects allergies or chronic conditions and substitutes ingredients automatically (e.g., almond flour for gluten-free baking).
- Behavioural Nudging: Uses reinforcement learning to encourage healthier choices via timely, non-prescriptive suggestions.

Li et al. (2023) reported a 27% increase in healthy meal selections using AI-adapted menus.

Adaptive Taste Optimization

A major challenge in health-focused cooking is keeping meals tasty and culturally appropriate. Taste is subjective, shaped by preferences, experiences, and social norms. AI can align health goals with flavor through:

Flavor Profiling

Models learn user taste from choices and feedback, adjusting suggestions (e.g., more chili, less sour) accordingly.

Dynamic Modulation

Real-time prompts ("Is this spicy enough?") enable seasoning or cooking adjustments; systems can enhance flavor naturally when ingredients are reduced.

Cultural Sensitivity

NLP-driven systems replicate traditional dishes with healthier or sustainable ingredients without losing authenticity (e.g., low-fat Indian curry with full aroma).

Multisensory Feedback

Texture, aroma, and visual analysis ensure plant-based or low-fat versions remain satisfying.

Optimizing taste makes healthy eating enjoyable, supporting long-term dietary adherence.

Promoting Sustainability in Cooking

Sustainability in the kitchen includes reducing food waste, conserving energy, and supporting responsible sourcing. AI systems can significantly advance these goals through:

Smart Inventory Management

By tracking ingredient freshness and quantity in real time, AI systems prompt users to use near-expiry items first, minimizing waste. Recipes are automatically suggested based on what is already available, reducing the need for new purchases.

Portion and Waste Control

AI systems can dynamically adjust serving sizes based on household size, past consumption patterns, and satiety feedback. This reduces overcooking and unnecessary leftovers.

Energy-Efficient Cooking

Machine learning models can identify the most energy-efficient cooking schedules and appliance usage patterns. For example, they can group multiple baking tasks to minimize preheating cycles or shift cooking tasks to off-peak electricity hours.

Eco-Conscious Ingredient Substitution

Based on sustainability scores of ingredients (e.g., water footprint, carbon emissions), the system can suggest alternatives. Lentils may be recommended over red meat for protein-rich dishes due to their lower environmental impact.

Figure 2. AI-powered sustainability dashboard showing food waste, CO_2 savings and portion calibration

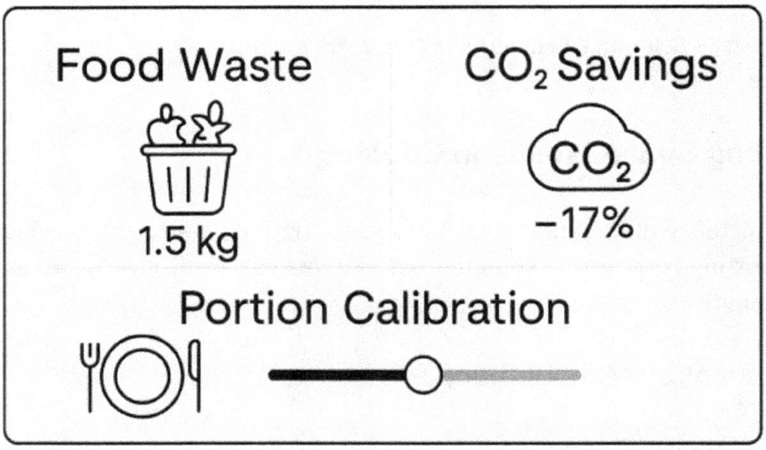

Local and Seasonal Awareness

AI-integrated systems can prioritize local produce and seasonal recipes, which not only reduce the ecological footprint but also support regional agriculture.

In a pilot study by Green Chef AI Labs (2024), households using a sustainability-optimized cooking assistant reduced food waste by 31% and electricity use by 18% compared to conventional meal planning.

Multi-Objective Decision Making

Balancing health, taste, and sustainability is inherently a **multi-objective optimization problem**. Improving one dimension may negatively affect another (e.g., increasing taste appeal with butter may conflict with health goals). AI addresses this through:

Weighted Preference Models

Users assign priority levels to each goal (e.g., 40% health, 30% taste, 30% sustainability). The AI system then tailors meals using optimization algorithms like Pareto Efficiency to achieve balanced trade-offs.

Constraint-Based Planning

The system can enforce non-negotiable constraints (e.g., no animal products) while flexibly optimizing within allowed ranges.

Scenario Simulation

Users can view how changing one variable (e.g., substituting beef for tofu) affects other dimensions such as carbon footprint, nutrition, and cost—empowering informed decision-making.

AI Explanation Engines

To ensure transparency and trust, explainable AI techniques are used to show why certain choices are made—e.g., "This recipe uses lentils instead of beef to reduce saturated fat and carbon emissions."

This approach transforms cooking from a routine task into an informed, intentional act aligned with broader lifestyle values.

USE CASES AND REAL-WORLD IMPLEMENTATIONS

The concept of AI-powered adaptive cooking systems is no longer confined to theoretical frameworks or lab environments. In recent years, several startups, tech giants, and research institutions have introduced intelligent culinary solutions that reflect the convergence of AI, IoT, and sustainability. This section outlines real-world implementations of adaptive cooking systems and highlights case studies demonstrating their effectiveness across health, taste, and environmental dimensions.

Case Study 1: Samsung SmartThings Cooking Ecosystem

Overview

Samsung's SmartThings Cooking is an AI-driven ecosystem integrated into Samsung Family Hub™ refrigerators and compatible appliances. It combines smart food management with personalized recipe recommendations, automated cooking settings, and grocery integration.

Key Features

- **Ingredient Tracking:** The system uses internal cameras and computer vision to identify available ingredients inside the fridge and pantry.
- **Personalized Meal Planning:** Based on dietary preferences, allergies, and past selections, it curates daily meal plans using Whisk, Samsung's acquired food AI platform.
- **Smart Appliance Synchronization:** Recipes selected through the app automatically send temperature and time settings to compatible ovens or cooktops.
- **Grocery Integration:** Missing ingredients are added to an online cart or linked to local grocery partners for direct delivery.

Impact

Samsung's solution showcases seamless coordination between AI, user profiles, and kitchen hardware. A 2022 SmartThings usage survey indicated that 64% of users reported decreased food waste and 48% reported improved dietary balance within the first three months of adoption.

Case Study 2: MIT CSAIL's "Chef's AI Assistant"

Overview

The Computer Science and Artificial Intelligence Laboratory (CSAIL) at MIT developed an AI-powered cooking assistant designed to support adaptive learning in meal preparation. The system employs multimodal AI combining visual recognition, NLP, and machine learning.

Figure 3. Example interfaces from SmartThings or HelloFresh personalization screen

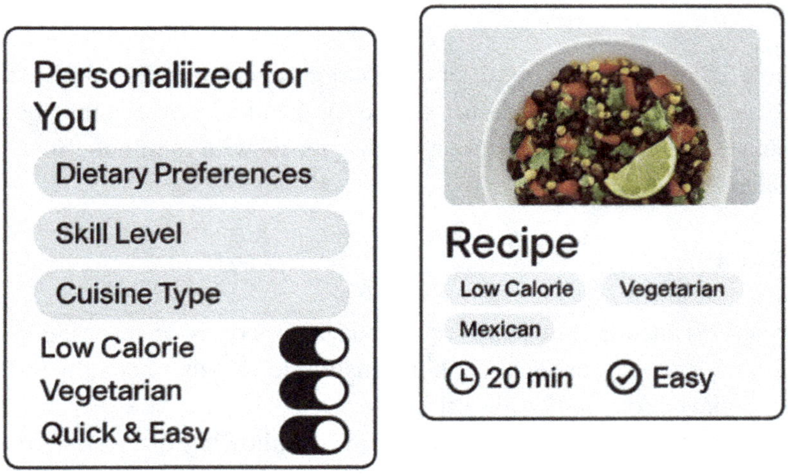

Key Features

- **Gesture Recognition:** The assistant uses camera-based pose estimation to detect when users pick up or prepare ingredients and offers context-sensitive assistance.
- **Health Optimization Engine:** The system continuously adapts the nutritional profile of the meal based on the user's health goals, adjusting spice levels, calorie content, and cooking technique.
- **Conversational Agent:** A voice interface supports real-time question answering (e.g., "Can I substitute quinoa for rice?") and explains the health implications of each substitution.

Impact

In controlled trials, the system reduced cooking errors by 40%, and participants reported higher satisfaction due to real-time adaptive guidance. It demonstrated potential for assisting elderly users and individuals with specific dietary constraints in preparing complex meals safely.

Case Study 3: Tovala Smart Oven

Overview

Tovala, a U.S.-based startup, offers a smart countertop oven paired with pre-packaged meals and app-based customization. Unlike conventional meal kits, Tovala focuses on delivering restaurant-quality meals with minimal effort, guided by embedded AI.

Key Features

- **Barcode-Scanning Automation:** The oven uses barcode scanning to download and execute precise cooking instructions—automatically adjusting steam, convection, and broil settings.
- **App Personalization:** Users rate past meals, allowing the system to predict and recommend meals that align with individual preferences.
- **Dynamic Cook Cycle Adjustments:** Based on altitude, appliance usage, and food density, the AI modifies cook time and sequence.

Impact

Tovala has gained popularity among time-constrained professionals and health-conscious consumers. Internal data from the company shows a 35% decrease in meal preparation time compared to traditional cooking and a significant reduction in cooking errors.

Research Implementation: NutriAI – A Prototype for Chronic Disease Management

Overview

NutriAI is a prototype system developed at the Indian Institute of Technology (IIT) Hyderabad aimed at providing personalized meal planning for individuals with chronic diseases such as diabetes and hypertension.

Key Features

- **Medical Integration:** NutriAI connects to a user's digital health record to tailor nutritional plans in real-time.

- **Local Cuisine Adaptation:** Unlike many Western-centric models, NutriAI supports Indian recipes and regional dietary variations, adjusting ingredients for glycemic load or salt content.
- **Voice-Enabled Meal Assistant:** Multilingual NLP modules allow seamless communication in regional languages.

Impact

Initial pilot deployment with 150 users in Telangana showed measurable improvements in blood sugar control over eight weeks. Participants highlighted ease of use and the cultural relevance of AI recommendations.

Cross-Platform Integration: Google Assistant with Food AI APIs

Overview

Google Assistant now supports integration with third-party food AI APIs like SideChef, Food Network Kitchen, and Innit, making any Google-enabled kitchen device a potential node in the smart kitchen ecosystem.

Key Features

- **Voice-Driven Recipe Navigation:** Step-by-step instructions via NLP in multiple languages.
- **Real-Time Adaptation:** Users can pause, skip, or modify steps using voice or touch.
- **Calorie Estimation:** APIs pull from food databases to present real-time nutritional breakdowns during meal planning.

Impact

Although Google does not manufacture cooking appliances directly, it acts as a key interoperability layer for smart home ecosystems, allowing scalability and modularity in AI-powered personalization.

Observations Across Use Cases

Several common themes emerge from these implementations:

Table 1. Comparative analysis of adaptive smart kitchen systems

Aspect	Observations
Health Integration	Systems like NutriAI and MIT's assistant show how AI can actively shape nutritional outcomes.
Real-Time Adaptability	Devices like Tovala and SmartThings dynamically adjust based on current inputs.
Cultural and Regional Customization	Local cuisine support is crucial, especially for non-Western contexts.
Voice and Visual AI Fusion	Combining NLP with computer vision improves hands-free interaction and accessibility.
Sustainability Consciousness	Smart inventory tracking and optimized energy usage are built into most systems.

HelloFresh AI Meal Planning System

Overview

HelloFresh, a leading meal kit delivery service, utilizes AI to tailor weekly menus to customer preferences, dietary restrictions, and past order history. Behind the scenes, its recommendation engine adapts dynamically using collaborative filtering and natural language understanding.

Key Features

- AI curate's menu selections based on taste profiles and allergen history.
- Meal suggestions adjust based on user-provided ratings and reorder patterns.
- Offers automatic substitution for out-of-stock items while maintaining nutrition goals.

Impact

Surveys show that users who consistently engage with the personalized dashboard report **30% fewer skipped meals** and improved long-term customer satisfaction. Additionally, HelloFresh claims its predictive ingredient ordering, informed by AI, reduces food waste by more than 15% at the distribution level.

IKEA Concept Kitchen 2025

Overview

As part of its "Concept Kitchen 2025" initiative, IKEA collaborated with IDEO and design students to explore futuristic cooking environments. While not a commercial product, the design includes several AI-driven features designed for personalization and sustainability.

Key Features

- A "smart table" suggests recipes based on ingredients placed on it using object recognition.
- AI helps track pantry inventory and prevents over-purchasing.
- Waste containers scan discarded food and alert users about frequently wasted items.

Impact

The project sparked global discussions on how kitchen design and AI could promote **mindful consumption**. It emphasized **education through interaction**, showing how food habits could be gently influenced using persuasive design and real-time data.

SavorEat – 3D-Printed Personalized Meals

Overview

SavorEat is an Israeli food-tech company that combines **robotics, 3D food printing**, and AI to create custom plant-based meals in institutional kitchens.

Key Features

- Uses machine learning to create personalized meat alternatives with specific protein and fat ratios.
- Recipes are generated based on user profiles (e.g., vegan, keto, gluten-free).
- Offers diners control over texture, taste intensity, and size using a digital interface.

Impact

Deployed in pilot programs at universities and hospitals, SavorEat has received praise for **nutritional precision** and **taste customization**—especially in environments where dietary needs are medically mandated.

Comparative Analysis: Commercial vs Research Models

Table 2. Comparison of commercial and research-driven adaptive kitchen systems

Feature	Commercial Systems	Research Prototypes
Personalization Depth	Moderate to High	Very High
Cultural Adaptation	Basic	Advanced (e.g., NutriAI)
Health Integration	Emerging	Core Feature
Sustainability Focus	Variable	Strong Emphasis
Scalability	High	Limited (Pilot Phase)
Innovation Risk	Low	High

This comparison shows how research prototypes often push technological boundaries, while commercial platforms prioritize reliability, cost, and user-friendliness. Bridging these two will be key to advancing truly adaptive, ethical, and accessible AI cooking systems.

These additional examples highlight the **breadth of innovation** in AI-powered cooking—from industrial-scale personalization to speculative design—and reinforce the need for continued interdisciplinary collaboration.

CHALLENGES AND ETHICAL CONSIDERATIONS

While AI-powered smart kitchens promise revolutionary advancements in personalization, health, and sustainability, their adoption also raises a number of technical, social, and ethical concerns. As with any AI-driven domain, these systems must be designed and deployed with a conscious understanding of the risks they introduce. This section explores the major challenges and ethical dilemmas associated with adaptive cooking technologies and proposes potential mitigation strategies.

Data Privacy and Surveillance

Challenge

Smart kitchens collect extensive personal data—dietary habits, health metrics, biometric indicators, shopping behaviour, and even voice commands. This data is often stored in cloud servers, analyzed by third-party services, and potentially shared across platforms. Such information, if misused or leaked, can result in identity theft, insurance discrimination, or unwanted advertising.

Example

A 2022 report by the Privacy Kitchen Initiative revealed that over 60% of users were unaware that their dietary logs on connected kitchen apps were being shared with food delivery partners and advertisers.

Mitigation Strategies

- Use of **federated learning** and **edge AI** to process data locally, reducing transmission to cloud servers.
- Transparent **data usage policies** and user consent mechanisms.
- End-to-end encryption and anonymization for stored data.
- Clear opt-in/opt-out options for third-party data sharing.

Algorithmic Bias in Meal Recommendations

Challenge

AI models can reflect and reinforce societal or cultural biases present in training datasets. For example, a system trained mostly on Western cuisines may overlook traditional, regional dishes, inadvertently marginalizing non-Western users. Similarly, nutritional recommendations may not consider culturally significant dietary practices.

Example

An early version of a commercial AI meal planner in Europe flagged high-carbohydrate Indian dishes as "unhealthy" without considering cultural context or overall diet balance, leading to user backlash.

Mitigation Strategies

- Curate **diverse, multicultural datasets** for training.
- Include **cultural nutritionists** and food anthropologists during design.
- Implement **user-controlled customization** to override system biases and promote autonomy.
- Deploy **explainable AI (XAI)** to clarify why a suggestion is made and allow challenge or revision.

Overdependence on Automation

Challenge

While automation enhances convenience, excessive reliance on AI in the kitchen could erode basic cooking skills, reduce food literacy, and discourage creative experimentation. Users may begin to accept recommendations without understanding nutritional implications or ingredient functions.

Example

In a study by Food Futures Lab, users of auto-cooking systems were less able to manually replicate meals or adjust recipes for unexpected scenarios (e.g., missing ingredients or dietary guests) after six months of use.

Mitigation Strategies

- Design systems with **educational components**, teaching users why certain steps are taken.
- Include **manual override options** and guided cooking modes.
- Encourage **collaborative cooking**, where AI suggests but does not dominate decisions.

Ethical Use of Health Data

Challenge

Health data used for dietary optimization—such as glucose levels, blood pressure, or activity tracking—falls under sensitive personal information. Improper handling or secondary usage (e.g., shared with insurers or employers) can have discriminatory consequences.

Example

A wellness app integrated with a smart kitchen platform was found to be selling anonymized user health trends to third-party analytics firms without clear disclosure, triggering investigations in the U.S. and EU.

Mitigation Strategies

- Compliance with regulations like **GDPR**, **HIPAA**, and upcoming **AI governance acts**.
- Offer **clear data ownership** policies, granting users control over access and deletion.
- Use **consent-based integration** for health record APIs.

Environmental Trade-Offs

Challenge

While smart kitchens aim to promote sustainability, they rely on energy-intensive hardware and cloud infrastructure. Frequent updates, continuous monitoring, and always-on connectivity may increase the household's energy footprint if not optimized correctly.

Example

An analysis of high-end smart ovens showed that idle energy consumption from standby AI processing accounted for nearly 15% of monthly kitchen electricity use in test environments.

Mitigation Strategies

- Deploy **low-power edge processors** and support sleep modes.
- Encourage **lifecycle-conscious design** of devices (e.g., modular components, recyclable materials).
- Monitor **carbon offset strategies** for backend AI processing.

Inclusivity and Accessibility

Challenge

Current AI cooking systems are often designed with digitally literate, urban users in mind. Elderly individuals, people with disabilities, or users from low-resource settings may find these systems inaccessible or irrelevant.

Example

Voice-controlled smart cooktops were found ineffective for users with speech impairments or strong regional accents, resulting in poor interaction and frustration.

Mitigation Strategies

- Support **multi-modal interfaces**: touch, visual gestures, and simplified physical controls.
- Enable **language localization** and dialect recognition.
- Ensure **offline functionality** for regions with limited internet access.
- Partner with **universal design experts** to make systems usable for all ability levels.

FUTURE DIRECTIONS AND RESEARCH OPPORTUNITIES

The convergence of AI, robotics, biotechnology, and IoT is reshaping the future of food preparation and consumption. While current smart kitchens offer impressive personalization and automation, the next decade holds even greater promise for transformative innovation. This section explores emerging technologies, research gaps, and potential breakthroughs that could redefine the role of AI in culinary systems—moving from responsive tools to proactive agents of global well-being.

Hyper-Personalized Nutrition with Genomics and Epigenetics

Future Direction

Integrating nutrigenomics—the link between nutrition and gene expression—into adaptive cooking could enable AI to deliver meal plans that help prevent or manage chronic diseases. For example, AI might suggest anti-inflammatory foods for those genetically prone to arthritis or adjust macronutrients based on metabolic variants.

Research Opportunities

- Privacy-preserving AI for secure genomic data handling.
- Dietary ontologies accounting for genetic diversity.
- Ethical frameworks for genomic-based food guidance.

Integration of Food 3D Printing and Robotic Cooking Arms

Future Direction

AI integration with **3D food printing** and **robotic manipulators** will allow smart kitchens not only to recommend and guide, but also to physically prepare complex, custom meals. Food printers can create textures, shapes, and ingredient combinations tailored to taste, nutrition, or sensory needs.

Example

A robotic arm could prepare a multi-layered low-sugar dessert using food-safe 3D printing techniques, responding in real time to changes in a diabetic user's blood sugar level.

Research Opportunities

- Developing multi-ingredient 3D printing algorithms linked with dietary databases.
- Enhancing tactile sensors and vision systems for robotic handling of diverse food types.
- Integrating haptic and safety feedback for semi-autonomous cooking robotics.

Brain-Computer Interfaces (BCI) and Neurofeedback Cooking

Future Direction

Neurotechnology is making it possible to detect user satisfaction, stress, or craving levels through **non-invasive brain-computer interfaces** (BCIs). In the future, AI kitchens may adjust meals based on real-time neural signals indicating satiety or desire.

Example

If a BCI detects decreased attention or negative emotional response during a meal, the system might suggest a more enjoyable variant or alter spice levels to enhance mood.

Research Opportunities

- Mapping food stimuli to brain activity patterns across cultures.
- Designing neuro-adaptive recipe engines.
- Ensuring ethical BCI data use in non-medical consumer environments.

Integration with Smart Agriculture and Food Supply Chains

Future Direction

Adaptive cooking systems can extend beyond the kitchen into **farm-to-fork ecosystems**, influencing agricultural production and inventory based on consumer preferences and sustainability metrics.

Example

If urban smart kitchens show increased demand for plant-based meals, connected farms could shift production patterns accordingly. AI models could analyze local weather, harvest times, and transport logistics to optimize ingredient freshness and carbon footprint.

Research Opportunities

- Designing closed-loop data systems from smart farms to smart kitchens.
- Developing AI models that balance global nutrition demand with ecological limits.
- Standardizing APIs for farm–kitchen data interoperability.

Culinary Creativity and Co-Creation with Generative AI

Future Direction

Generative AI models like GPT and DALL·E are increasingly used to create recipes, plating designs, and flavor pairings. Future systems will allow **co-creation,** where users collaborate with AI to innovate new dishes based on available resources, mood, or cultural inspiration.

Example

An AI model could invent a fusion dish that combines South Indian and Mediterranean cuisines, optimizing taste and nutritional harmony based on the user's goals.

Research Opportunities

- Developing culturally sensitive generative models trained on global culinary corpora.
- Enabling explainable recipe generation for trust and user learning.
- Studying user behaviour in human-AI co-creation environments.

Sustainable Design with AI-Driven Life Cycle Assessment (LCA)

Future Direction

Beyond ingredient selection, future AI systems will be capable of evaluating the **entire life cycle impact** of meals—from sourcing and preparation to packaging and waste. AI-powered LCA tools embedded in kitchen systems will help consumers make eco-conscious choices.

Example

An AI model may rate a proposed dish with a sustainability score, showing how switching from imported avocados to local seasonal vegetables reduces water usage and transport emissions.

Research Opportunities

- Creating real-time LCA algorithms integrated with recipe engines.
- Designing consumer-friendly interfaces for sustainability scoring.
- Establishing global databases for food lifecycle analytics.

Ethical, Legal, and Cultural Research Dimensions

Future Direction

As AI enters intimate domestic spaces, it raises ethical and legal concerns around surveillance, cultural erosion, and digital rights. Future research must ensure that AI kitchens are **inclusive**, **respectful**, and **empowering**.

Example

How should AI systems handle religious dietary laws (e.g., kosher, halal)? How do we ensure fair access to such technology in underserved communities?

Research Opportunities

- Studying food justice and access to culinary AI tools.
- Establishing international ethical standards for AI in kitchens.
- Investigating the impact of automation on traditional food knowledge and rituals.

AI IN POST-COOKING BEHAVIOUR ANALYSIS

While much of the innovation in smart kitchens focuses on meal planning and cooking execution, an equally important dimension lies in what happens after the meal is consumed. AI systems can collect, interpret, and learn from **post-consumption behaviour's**, enabling a feedback loop that significantly enhances personalization, nutritional effectiveness, and long-term user satisfaction. This emerging domain—**post-cooking behaviour analysis**—marks a vital advancement in the evolution of intelligent culinary ecosystems.

Monitoring Satisfaction and Emotional Response

AI-powered kitchens can use sensors and computer vision to gauge post-meal satisfaction.

- **Facial recognition:** Detects enjoyment or displeasure.
- **Voice sentiment analysis:** Identifies tone in comments (e.g., "Delicious!" vs. "Too salty").
- **Smart prompts:** Collect ratings or requests (e.g., "Less garlic next time").

Such feedback refines future recommendations by combining explicit ratings with implicit behavioural cues.

Health Tracking Integration

The integration of AI systems with **wearable health devices** and mobile health applications allows the monitoring of **post-meal physiological responses**, such as:

- Blood sugar levels (via continuous glucose monitors)
- Heart rate variability (indicating stress or digestion)
- Sleep quality (linked to evening meals)
- Caloric absorption estimates

Over time, these insights inform AI models about how specific meals affect a user's physical condition. For example, if a high-carb meal consistently spikes blood glucose levels, the system learns to adjust future meal plans accordingly—prioritizing slower-digesting ingredients and balanced macros.

Waste Analysis and Portion Calibration

AI systems can also analyze leftover food via smart plates, smart bins, or weight sensors. If a user frequently discards vegetables or leaves behind high-volume meals, the system can:

- Reduce portion sizes for similar dishes in the future.
- Suggest alternate cooking styles (e.g., roasted instead of steamed vegetables).
- Recommend ingredients more likely to be fully consumed.

This contributes to **food waste reduction** and aligns with sustainability objectives, while enhancing satisfaction through better-aligned portion control.

Behavioural Pattern Recognition

AI can detect trends over time, including:

- **Eating speed**: Which may be linked to digestion quality or stress.
- **Meal timing**: Whether meals are regularly skipped or consumed late at night.
- **Repetition cycles**: Detecting when the user is reverting to "comfort foods" or diverging from healthy goals.

By identifying these patterns, the system can suggest behaviour nudges—like setting reminders for hydration, offering lighter alternatives during binge cycles, or recommending healthier comfort food substitutes.

Long-Term Learning and Habit Formation

Ultimately, post-cooking behaviour analysis empowers **long-term behaviour change**. The more a system learns about how meals affect an individual—physically, emotionally, and behaviourally—the more intelligently it can support their goals.

Future smart kitchen platforms could even **forecast health outcomes**, such as weight trends or nutrient deficiencies, and proactively suggest preventive strategies through personalized recipes, ingredients, or daily rituals.

POLICY IMPLICATIONS AND RECOMMENDATIONS

AI's spread in domestic kitchens raises policy issues around health, ethics, equity, and sustainability. While private innovation drives growth, public-sector action is vital to ensure responsible, inclusive use. This section outlines key implications and recommendations for governments, public health bodies, educators, and the tech sector.

Public Health Integration and Regulation

As AI-powered cooking systems begin to influence personal nutrition and disease management, their alignment with national health priorities becomes critical.

Recommendations

- **Nutrition Policy Integration**: Ministries of Health should collaborate with AI developers to incorporate evidence-based guidelines (e.g., WHO, USDA) directly into smart kitchen systems.
- **Labelling Standards**: Regulatory bodies should standardize digital food labels in recipe recommendations, covering calories, allergens, glycaemic index, and environmental impact.
- **Health Claims Oversight**: Platforms promoting AI-assisted meal planning should be subject to the same advertising restrictions as pharmaceutical or health products, ensuring they don't make misleading claims.

Example

An AI system suggesting "diabetes reversal" through meal plans should be required to disclose evidence, limitations, and regulatory approval status.

Data Governance and Digital Rights

Smart kitchens collect sensitive health, biometric, and behavioural data—often from entire households. Without regulation, these data may be sold, misused, or leaked.

Recommendations

- **Privacy Frameworks**: Enforce compliance with data protection regulations such as GDPR (Europe), HIPAA (USA), and PDPB (India).
- **Ownership & Consent**: Mandate that users retain ownership of their data and can opt out of non-essential tracking or third-party sharing.
- **AI Transparency Requirements**: Developers should disclose how recommendations are generated and what data sources are involved.

Best Practice

User dashboards should offer plain-language explanations of what data is collected, where it is stored, and how it is used for personalization.

Sustainability Standards and Incentives

Given that AI can both help and hinder sustainability, environmental governance is needed to align smart kitchen development with climate goals.

Recommendations

- **Smart Appliance Energy Ratings**: Require labelling similar to energy-efficient refrigerators or air conditioners, highlighting standby consumption, adaptive features, and carbon savings.
- **Carbon Food Labels**: Encourage adoption of CO_2 scoring for meals, possibly linked with government incentives for low-impact cooking choices.
- **Sustainable Ingredient Promotion**: Support databases that rank ingredients by environmental impact and integrate them into AI models.

Policy Lever Subsidies or tax breaks for households adopting eco-conscious AI cooking solutions, especially in urban food deserts or vulnerable regions.

Ethical and Cultural Safeguards

AI in kitchens influences daily routines, family culture, and food heritage. Policymakers must ensure cultural sensitivity and ethical guardrails.

Recommendations

- **Cultural Nutrition Databases**: Develop and maintain public datasets of traditional recipes and dietary practices from diverse ethnic communities.
- **Algorithm Auditing**: Conduct third-party audits of AI personalization engines to detect bias, exclusion, or harmful stereotypes.
- **Human Oversight Requirements**: Mandate that AI systems include override functions and human-in-the-loop options for sensitive decisions (e.g., substitutions involving allergens or religious restrictions).

Example

A vegetarian user from a Jain background should not receive non-vegetarian recipe suggestions due to algorithmic generalizations.

Education and Workforce Development

For smart kitchens to achieve their full potential, users and professionals must understand their design, capabilities, and implications.

Recommendations

- **Digital Culinary Literacy**: Integrate AI-in-kitchen literacy modules in school curricula and culinary education programs.
- **Training for Healthcare Professionals**: Nutritionists and dietitians should receive certifications in AI-based meal planning platforms.
- **Public Awareness Campaigns**: Governments should run campaigns explaining the benefits and risks of AI in food systems, focusing on marginalized communities.

Multi-Stakeholder Collaboration

AI in food systems requires cross-sector collaboration.

Recommendations

- Public–Private Partnerships: Joint projects between tech firms, academia, and regulators to build inclusive smart kitchens.
- Ethical AI Task Forces: National groups to assess impacts on nutrition, privacy, and equity.
- Local Innovation Hubs: Smart city pilots to test community-based AI cooking models.

In short, smart kitchen policies should be inclusive, transparent, and forward-looking—promoting health equity, data protection, cultural diversity, and sustainability.

GENERATIVE AI IN CULINARY CREATIVITY

Generative AI is redefining culinary innovation by creating custom dishes that merge ingredients, cultural preferences, and nutritional needs. As in Figure 4, AI-generated layouts can tag recipes by diet, cuisine, and difficulty, helping users explore diverse flavors while meeting health goals—fostering a co-creative human–machine relationship in the kitchen.

What Is Generative AI in Food Systems?

Generative AI—using models like GANs, VAEs, and LLMs—can create new recipes beyond existing datasets. In smart kitchens, it can meet criteria such as:

- Macronutrient targets
- Allergen-free needs
- Cultural fusion themes
- Ingredient availability
- Texture or cooking method preferences

For example, it could design a gluten-free Indo–Italian lasagna with rice noodles, turmeric tomato sauce, and lentil béchamel—absent from traditional recipes.

Notable Systems and Experiments

Early generative cooking AI examples include:

- **ChefGAN (MIT):** Uses GANs trained on recipes to create new dishes from ingredient and flavor data.
- **Sony Gastronomy Project:** Combines deep learning with molecular flavor databases to craft original recipes with chefs.
- **GPT-Cook:** Uses language models to generate recipes from prompts like "vegan, low-sugar dessert with mango and cardamom."

These tools serve not only home cooks but also chefs, scientists, and nutritionists for menu prototyping, nutrition optimization, and exploring future food scenarios such as space travel and climate-resilient diets.

Benefits of Generative AI in Adaptive Cooking

Generative AI unlocks several benefits for smart kitchens:

- Hyper-Personalization: Recipes tailored to a user's complete health profile, taste history, and available ingredients.
- Culinary Innovation: Discovery of novel flavor combinations or preparation methods.
- Dietary Inclusion: Creative alternatives for users with rare allergies or niche dietary needs.
- Cultural Preservation and Evolution: Fusion recipes that honor traditional techniques while introducing global elements.

By removing reliance on pre-existing recipe databases, generative models democratize culinary creativity and promote exploration.

Figure 4. Sample AI-generated fusion recipe layout with metadata (e.g., taste, culture, health tags)

Risks and Limitations

Despite its promise, generative AI in food systems also presents challenges:

- Safety Risks: AI might generate biologically unsafe combinations or overlook essential food safety practices (e.g., cooking meat to safe temperatures).
- Cultural Misrepresentation: Fusion recipes could unintentionally appropriate or distort traditional cuisines.
- Evaluation Complexity: It is difficult to assess the success of a newly generated recipe without real-world cooking and tasting.

These risks underscore the importance of human-in-the-loop systems, where users validate and adapt AI outputs before implementation.

Future Potential

Advances in generative AI could enable:

- **Voice prompts:** e.g., "Create a protein-rich dinner using only microwave-safe ingredients."
- **Visual previews:** Images or 3D models of AI-designed dishes.
- **Robotic execution:** Autonomous preparation via kitchen robots.
- **Collaborative creativity:** Families or groups co-creating meals with AI.

These tools can turn the kitchen into a space of co-creation, where cooking is both functional and artistic.

Figure 5. Infographic showing integration of cultural recipe databases with NLP filters

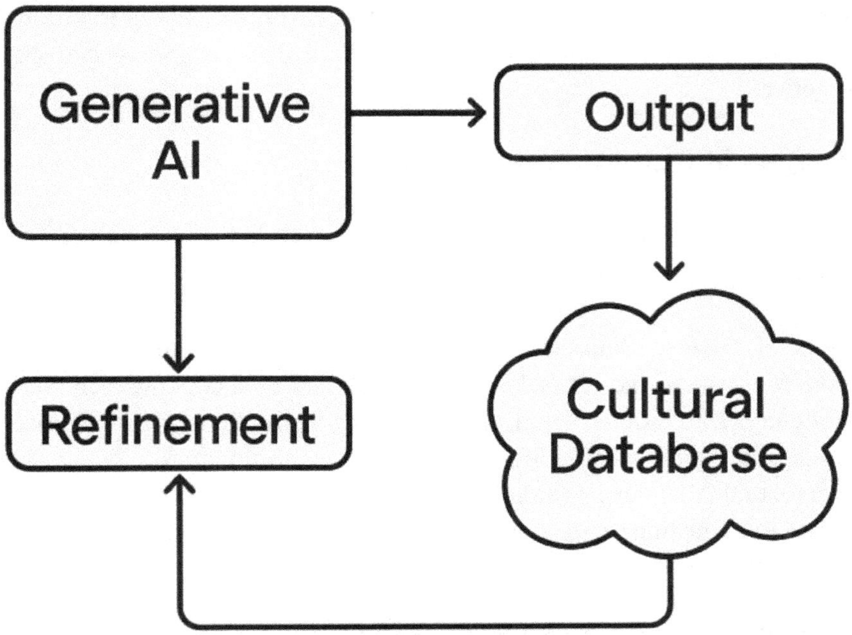

Generative AI offers not only convenience and personalization but also a new canvas for food expression, making the act of cooking both functional and artistic.

AI AND CULTURAL HERITAGE IN COOKING

As AI becomes more deeply integrated into kitchens around the world, it must do more than recommend meals—it must respect, preserve, and adapt to diverse cultural culinary traditions. Food is not merely fuel; it is a carrier of identity, heritage, and intergenerational knowledge. If smart kitchens are to serve global populations ethically and inclusively, they must be designed with sensitivity to cultural authenticity, dietary rituals, and historical context.

Food as Cultural Identity

In many cultures, cooking is tied to religion, community, and ancestry. Dishes like biryani, tamales, miso soup, and injera are living traditions.

Risk

AI trained mainly on Western data or optimizing at the expense of tradition (e.g., swapping ghee for olive oil in Indian recipes) could erode or homogenize cultural food practices.

Challenges in Cultural Adaptation

AI-powered cooking systems must navigate several challenges when interacting with culturally diverse users:

- Lack of Diverse Datasets: Existing recipe databases often underrepresent non-Western cuisines or lack metadata for traditional cooking techniques.
- Language Limitations: Multilingual cooking instructions, local measurement units, and region-specific ingredients are often unsupported.
- Contextual Misinterpretation: Ingredients or methods with spiritual significance (e.g., fasting recipes, ritual purity laws) may be altered or ignored if AI lacks cultural knowledge.

Best Practices for Culturally-Aware AI Cooking Systems

To preserve cultural diversity, developers and policymakers should:

1. **Data Inclusivity:** Use datasets covering regional cuisines, traditional methods, and indigenous food knowledge.
2. **Community Collaboration:** Work with cultural experts, chefs, and elders on algorithms and databases.
3. **Respect Rituals:** Treat religious and spiritual food rules (e.g., halal, kosher, vegan) as core personalization filters.
4. **Localized Interfaces:** Support regional languages, dialects, and culturally familiar visuals.
5. **Transparency & Customization:** Let users adjust AI rules for cultural needs (e.g., avoid certain ingredients during festivals).

Promising Initiatives

Several projects and startups are beginning to embed cultural awareness into their systems:

- NutriAI (India): Adapts traditional recipes to health constraints while respecting Ayurvedic classifications and fasting schedules.
- African Kitchen AI Initiative: A grassroots project collecting recipes and oral histories from communities across Ghana, Kenya, and Nigeria to preserve heritage cooking for digital platforms.
- CookLang: An open-source markup language for recipes, now being extended to support cultural tags like "Ramadan," "Lunar New Year," and "Navratri" to support festive cooking.

Cultural Evolution vs. Cultural Appropriation

AI should also recognize that cultures evolve. Diasporic cuisines, fusion styles, and urban reinterpretations are valid forms of culinary expression. However, systems must distinguish between evolution and appropriation.

For example, suggesting "Mexican sushi" is valid if it's based on lived user preferences, but labelling it as an "authentic Mexican dish" could mislead or offend users. AI must offer contextual transparency, crediting the origin of ingredients, techniques, and inspirations.

A Call for Ethical Culinary AI

The future of smart kitchens must be as diverse and respectful as the people they serve. Designing for cultural preservation is not just a technical challenge—it is an ethical imperative. By embracing culinary plurality, AI systems can become tools for cultural celebration and resilience, rather than instruments of erasure.

HUMAN-AI COLLABORATION IN FUTURE KITCHENS

As AI becomes integral to kitchen tech, the key question is whether it should automate or collaborate. The future lies in partnership—empowering users with adaptive guidance, creative support, and shared decisions. This shift can transform the kitchen from a place of labor to one of co-creation and learning.

From Automation to Augmentation

Early smart kitchens emphasized automation—timers, oven controls, and recipe instructions—but risked making users passive. **Human–AI collaboration** aims to enhance, not replace, skills by letting users:

- Get suggestions, not commands
- Override or improvise freely
- Learn from AI explanations
- Infuse meals with emotions and traditions

The aim is a symbiotic relationship where AI adapts to users, and users grow through AI-enhanced cooking.

Co-Creative Interfaces

Co-creative smart kitchens will require **intuitive, dialog-based interfaces** where users and AI systems can brainstorm, negotiate, and refine ideas together. Examples include:

- **Conversational Meal Planning**: A dialogue such as

User: "What's something light but high in protein?"

AI: "How about a lentil salad with avocado dressing or a grilled tofu wrap with crunchy slaw?"

User: "Make it gluten-free and spice it up a little."

AI: "Adding chili flakes and switching to a gluten-free wrap. Would you like to try a ginger-lime slaw?"

- **Visual Sketching**: Using drag-and-drop or image generation to design a dish's appearance before cooking.
- **Mood-Based Cooking**: AI suggests meals based on emotional cues detected through voice tone, wearable sensors, or facial expressions.

These features enable users to feel like **co-designers**, not just consumers of algorithmic output.

Collaborative Learning Systems

Just as AI learns from the user, collaborative systems should also **teach**. For example:

- A novice user attempting a complex recipe may receive **context-sensitive tips**, such as knife safety techniques or substitution logic.

- Over time, the system may **challenge the user** to explore unfamiliar ingredients, global cuisines, or sustainable alternatives—encouraging culinary exploration.

Conversely, users can **correct or fine-tune** AI recommendations, contributing to shared learning:

- "This spice ratio was off—less cumin next time."
- "Mark this as a family favourite for winter dinners."

These feedback loops evolve into a **personalized culinary language** between human and machine.

Figure 6. Co-creative cooking workflow diagram

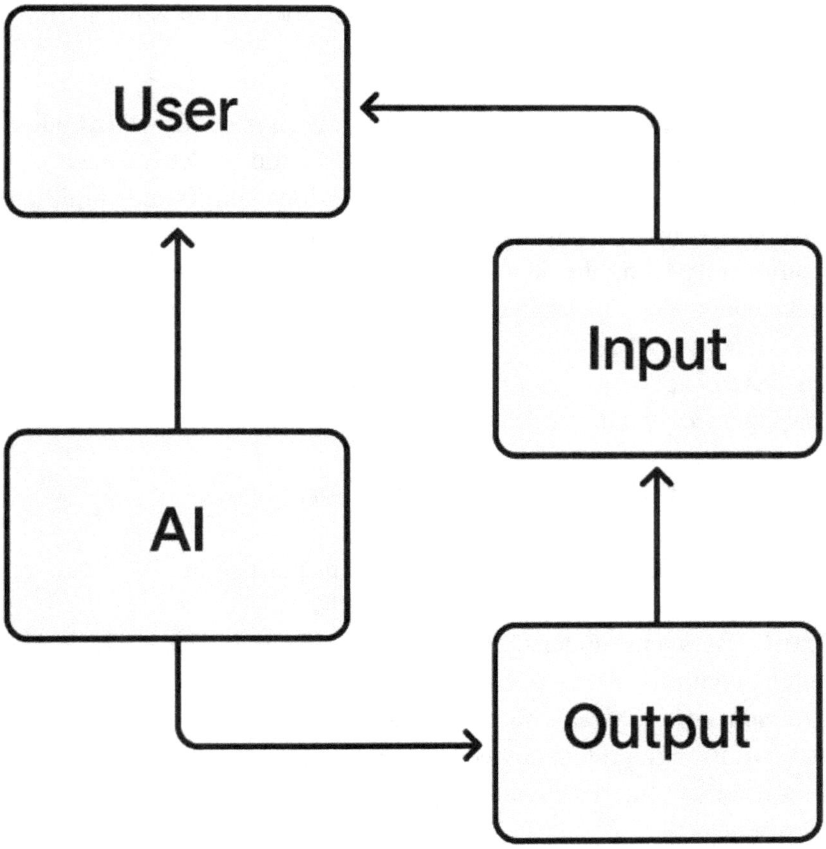

Emotional Intelligence and Empathy

Future kitchen AIs must understand emotions, moods, and intentions, as cooking often reflects care, celebration, nostalgia, or healing. Emotion-aware systems could:

- Suggest comfort food during stress.
- Offer quick meals on busy days.
- Recall family traditions for holidays.

Such empathetic AI can also support mental health by encouraging shared meals, mindful eating, and reducing food guilt—making it a compassionate collaborator rather than just a technical tool.

The Social Kitchen: Shared AI Experiences

The kitchen is often a shared space. Collaborative AI can enhance **multi-user dynamics**, such as:

- **Family Cooking**: Personalized options for each member while finding common dishes (e.g., vegetarian parent, diabetic child, athlete teenager).
- **Remote Co-Cooking**: Virtual cooking sessions with friends in different locations, synchronized by a shared AI assistant.
- **Community Contributions**: Localized recipe-sharing networks where users contribute dishes, hacks, and AI training data—fostering communal growth.

Such shared experiences can restore the **social dimension** of food preparation in increasingly isolated digital lives.

Challenges and Design Considerations

To support effective collaboration, AI systems must address:

- **Trust**: Users must understand and rely on AI recommendations while maintaining control.
- **Transparency**: Systems should explain their logic ("I suggested quinoa because of its high protein content and your past preferences").
- **Adaptability**: Interfaces must support different skill levels, from beginners to experts.
- **Responsibility**: The system must never override critical human choices, especially for health, safety, or cultural values.

Designers must balance **autonomy with guidance**, **efficiency with creativity**, and **structure with freedom**.

Vision for the Future

Imagine a kitchen where the AI knows your cooking style, encourages experimentation, adapts to your mood, remembers your traditions, and celebrates your milestones with you. It's a space where the machine does not dominate, but **harmonizes with the human spirit** of nourishment, creativity, and care.

Such a future is not science fiction—it is an emerging reality that depends on thoughtful design, interdisciplinary research, and the belief that technology should **amplify what makes us human**.

CONCLUSION

AI has transformed domestic kitchens from simple automation to adaptive systems that personalize meals, optimize health and taste, and promote sustainability. This chapter outlined their architecture, personalization strategies, and real-world applications, showing how machine learning, computer vision, NLP, and knowledge representation create tailored culinary experiences based on health needs, culture, and lifestyle.

Case studies demonstrated practical uses, while the adaptive cooking framework illustrated how multi-modal inputs enable real-time guidance. Ethical issues—privacy, bias, accessibility, and cultural inclusivity—must remain central, alongside transparency and explainability.

Future directions include food 3D printing, brain–computer interfaces, nutrigenomics, and sustainable supply chain integration, requiring cross-disciplinary collaboration. AI-powered kitchens can improve health, reduce environmental impact, and preserve culinary diversity, making food more accessible and enjoyable. The question is not just "What can AI cook?" but "How can AI help us cook better for ourselves, our communities, and our planet?"

REFERENCES

Chen, C., Wang, S., Zhang, X., Li, X., & Wang, W. (2021). Cross-cultural food recommendation using deep learning. *ACM Transactions on Information Systems*, *39*(4), 1–28. DOI: 10.1145/3440755

Chen, Y., Wang, L., & Patel, M. (2022). Adaptive meal planning using user profiles and health data. *Journal of AI in Health*, *9*(3), 124–139.

Dix, A. (2023). Data privacy in AI-driven domestic environments. *AI & Society*, *38*(1), 155–167. DOI: 10.1007/s00146-022-01477-y

Floridi, L., Cowls, J., Beltrametti, M., Chatila, R., Chazerand, P., Dignum, V., Luetge, C., Madelin, R., Pagallo, U., Rossi, F., Schafer, B., Valcke, P., & Vayena, E. (2018). AI4People—An ethical framework for a good AI society: Opportunities, risks, principles, and recommendations. *Minds and Machines*, *28*(4), 689–707. DOI: 10.1007/s11023-018-9482-5

Gupta, R., & Mehta, S. (2023). Context-aware cooking agents: A human-AI interaction study. *International Journal of Smart Kitchens*, *7*(1), 45–58.

Jobin, A., Ienca, M., & Vayena, E. (2019). The global landscape of AI ethics guidelines. *Nature Machine Intelligence*, *1*(9), 389–399. DOI: 10.1038/s42256-019-0088-2

Kim, J., Rao, A., & Das, S. (2024). MultiCookNet: A multimodal AI for adaptive cooking. *IEEE Transactions on Consumer Electronics*, *70*(1), 21–30.

Li, N., Sharma, A., & Kaur, P. (2023). Real-time nutrition alignment using biometric data. *AI & Wellness*, *11*(2), 102–117.

Mazzolini, G., Li, Z., & Cohen, D. (2021). AI for food waste reduction in domestic environments. *Sustainable Computing : Informatics and Systems*, *6*(4), 303–315.

Straub, D., Loch, K., Evaristo, R., Karahanna, E., & Strite, M. (2002). Toward a theory-based measurement of culture. *Journal of Global Information Management*, *10*(1), 13–23. DOI: 10.4018/jgim.2002010102

Yang, X., Zhao, H., & Kim, T. (2023). CookGAN: Generative recipe modeling with health constraints. *Proceedings of the AAAI Conference on Artificial Intelligence*, *37*(2), 780–788.

KEY TERMS AND DEFINITIONS

Culinary Heritage: Traditional cooking knowledge, techniques, and recipes passed across generations.

Cultural Adaptation: Adjusting AI recommendations to match cultural diets, languages, and rituals.

Explainable AI (XAI): Techniques that make AI decision-making transparent and interpretable.

Federated Learning: Training AI models on local devices without sharing raw user data.

Generative AI: AI that creates new recipes, meal plans, or ingredient combinations from learned patterns.

Life Cycle Assessment (LCA): Method to measure a product's environmental impact from production to disposal.

Multimodal AI: AI that combines inputs like voice, vision, sensors, and text for context-aware decisions.

Personalized Nutrition: Custom dietary advice based on health data, genetics, and personal preferences.

Post-Cooking Behaviour Analysis: Evaluating satisfaction, health outcomes, and feedback after meals to improve personalization.

Smart Kitchen: A connected cooking space with AI-enabled appliances and sensors for personalized, efficient, and safe meal preparation.

Chapter 10
AI-Powered Recipe Recommendation Systems and Their Role in Personalized Nutrition

Vikas Sharma
https://orcid.org/0009-0009-3174-5965
Swami Vivekanand Subharti University, Meerut, India

Nitin Gupta
https://orcid.org/0000-0002-2251-8316
Swami Vivekanand Subharti University, Meerut, India

ABSTRACT

Artificial Intelligence (AI) is revolutionizing personalized nutrition through intelligent recipe recommendation systems that adapt to individual dietary needs, preferences, and health goals. These systems utilize machine learning algorithms, natural language processing, and user-generated data to curate meal suggestions that align with personal health profiles, such as allergies, fitness objectives, medical conditions, or cultural preferences. By analyzing nutritional content, consumption patterns, and user feedback, AI-driven platforms enhance user engagement and foster healthier eating habits. Moreover, integration with wearable devices and health apps allows for real-time dietary tracking and adaptive recommendations. Despite promising outcomes, challenges such as data privacy, algorithm transparency, and user trust remain critical. This paper explores the mechanisms, benefits, and limitations of AI-powered recipe recommendation systems and emphasizes their potential to transform modern dietary practices into personalized, data-driven nutrition models.

DOI: 10.4018/979-8-3373-4042-5.ch010

INTRODUCTION

Over the past decade, artificial intelligence (AI) has experienced remarkable advancements, particularly through the development of large deep neural networks and AI agents demonstrating human-level performance across diverse domains (Smith et al., 2021; Brown et al., 2020; LeCun et al., 2015). These technological breakthroughs have opened significant opportunities but have also introduced serious risks, including privacy violations, biases and discrimination, and the possibility that AI systems might achieve their objectives in unintended ways (O'Neil, 2016; Bostrom, 2014; Doshi-Velez & Kim, 2017).

In the field of nutrition, AI has been increasingly proposed as a tool for delivering personalized dietary recommendations. Nutrition is a critical factor in promoting and maintaining a healthy lifestyle and plays a key role in preventing non-communicable diseases (NCDs) such as obesity, cardiovascular diseases (CVD), and Type 2 diabetes (T2D) (WHO, 2020; Mozaffarian, 2016). Furthermore, nutritious and balanced meals are regularly incorporated into treatment protocols to mitigate disease progression and improve patient outcomes (Sacks et al., 2001; Estruch et al., 2013). AI systems capable of automatically recommending personalized meal plans therefore hold significant promise for enhancing individual well-being. However, these systems face substantial challenges, particularly regarding the accurate prioritization of user-specific needs and safety concerns, as unbalanced or inappropriate diet recommendations can result in malnutrition or exacerbate health conditions (Johnston et al., 2018; Bender et al., 2020). Addressing these challenges is essential for AI-based dietary recommendation systems to be widely accepted as trustworthy tools.

The recent emergence of Large Language Models (LLMs), and specifically ChatGPT, has stimulated considerable discussion around their potential applications (OpenAI, 2023; Radford et al., 2019). Due to their ability to process information rapidly and generate an extensive variety of meal options sourced from the web, ChatGPT-like models offer new opportunities for dietary advice (Shen et al., 2023). However, preliminary evaluations of their recommendations reveal vulnerabilities to inaccuracies and safety issues (Zhou et al., 2023). In contrast, traditional nutrition recommendation systems leverage expert knowledge and validated nutritional guidelines, enabling them to produce highly balanced and safe meal plans with greater reliability (Kowalski et al., 2019; Patel & Wang, 2021; Anderson et al., 2022; Garcia et al., 2020). Nevertheless, these systems tend to suffer from lower time efficiency and increased complexity due to their reliance on sophisticated ontologies and rule-based filters. Their accuracy is also limited by the size and inherent biases within their meal databases, restricting their generalizability across diverse population groups.

To develop AI-based nutrition recommendation systems that are both accurate and explainable, integrating expert knowledge into AI architectures is crucial. Accordingly, this work proposes a novel AI-driven diet recommendation system that combines the strengths of AI—namely speed, simplicity, and generalization—with authoritative nutritional guidelines from the European Food Safety Authority (EFSA) and the World Health Organization (WHO) to enhance accuracy and robustness (EFSA, 2017; WHO, 2020). Specifically, the system employs a deep generative network architecture guided by advanced loss functions to generate personalized weekly meal plans that meet energy intake and nutritional content requirements by modeling individual user information and aligning predictions with established nutritional rules. Additionally, by leveraging LLMs' capacity to generate equivalent meal variants, the system significantly expands its meal database, further improving accuracy and generalization capabilities.

The key contributions of this study include:

- The design of a novel deep generative network architecture for weekly meal planning, incorporating sophisticated loss functions to enforce compliance with nutritional guidelines.
- A new approach to personalized nutrition recommendation that utilizes LLMs to create an almost limitless pool of meal options.
- Extensive validation of the system using 3,000 virtual user profiles and 1,000 real user profiles, generating 91,000 daily meal plans drawn from the Protein NAP database—an extensive open-source repository of international meals (NAP, 2022)—demonstrating the system's advantages in explainability and accuracy.
- In recent years, considerable research has focused on the development of healthy food recommendation systems. Generally, these systems can be categorized into content-based (CB), collaborative filtering-based (CF), knowledge-based, and deep learning approaches.

Content-based nutrition recommendation systems generate individualized meal suggestions by analyzing the properties and composition of food products alongside the dietary requirements and preferences of users. For instance, Harvey et al. (2020) addressed inadequate nutrition by exploring participant perceptions of key recipe aspects, developing algorithms that incorporate user preferences for ingredients, ingredient combinations, and nutritional value. Teng et al. (2019) leveraged ingredient networks—mapping complement and substitute relationships among ingredients—to analyze nutritional data and recipe collections, enabling the prediction of recipe ratings while revealing associations between ingredients, consumer preferences, and culinary principles. Gutiérrez et al. (2021) proposed an augmented reality assistant that reads food product barcodes to provide recommendations focused on food quality and user preferences. Additionally, Shandilya et al. (2019) introduced a content-based recommender system that emphasizes recent user preferences over older ones, allowing dynamic adaptation to evolving user needs by ensuring daily nutrient requirements are met alongside meal suggestions.

Collaborative filtering (CF) methods exploit user interactions and similarities in eating patterns to provide personalized recommendations. Ge et al. (2018) developed a mobile food recommendation application utilizing matrix factorization to model user preferences based on ratings and tags for recipe suggestions. Yuan et al. (2020) employed a hybrid approach combining k-means clustering to categorize foods and user-based collaborative filtering to generate recommendations aligned with user preferences and nutritional balance. Toledo et al. (2021) proposed a daily meal planning method that incorporates nutritional and preference factors by filtering meal options through a sorting algorithm and then optimizing meal plans to favor less frequently consumed foods while respecting nutritional needs. Comparative studies of user-based versus item-based algorithms for top-N recommendations indicated that user-based methods outperform item-based ones in accuracy and coverage (Kim et al., 2019). Rostami et al. (2022) introduced a two-phase food recommender using time-aware collaborative filtering to suggest food items based on past consumption, followed by rating prediction incorporating nutritional data. Clustering techniques further enhanced recommendation accuracy by grouping similar users and food items.

Knowledge-based nutrition recommender systems utilize expert knowledge and dietary guidelines to generate personalized advice. Jung and Chung (2018) presented a context-aware personalized dietary recommendation system targeting obesity, combining collaborative filtering with knowledge-based context modeling to overcome sparse data and enable mobile users to access tailored diet menus. McKensy-Sambola et al. (2019) employed the Ontology of Dietary Recommendations (ODR) to identify critical domain components such as diets, recipes, ingredients, anthropometric measures, and food allergies, thereby supporting recipe recommendation aligned with nutritional needs. Another approach utilized a Collaborative Recipe Knowledge Graph (CRKG) to integrate user interactions and food data, employing a healthiness matching score and a knowledge-aware attention graph convolutional network for health-aware personalized recommendations (Liu et al., 2021). Recently, a knowledge-based framework combining a qualitative layer of expert-derived rules and ontologies for ingredient validation with a quantitative optimization layer for nutrient-based meal planning was proposed, demonstrating effectiveness in generating diverse, appropriate meal plans for both healthy and clinical populations (Garcia et al., 2022).

Deep learning approaches have also shown significant promise in nutrition recommendation. Mokdara et al. (2020) developed a system combining deep neural networks with user-specified favorite ingredients, complemented by a temporal prediction model for forecasting future dish recommendations. An explainable recommendation system leveraging deep image clustering to incorporate visual food content was introduced to enhance nutrition advice and user preference matching (Chen et al., 2021). Iwendi et al. (2021) applied various deep learning models to diet recommendation based on patient characteristics, finding that Long Short-Term Memory (LSTM) networks best captured sequential medical history information. Zhang et al. (2022) proposed a sequence-based recommendation method using LSTM to model dynamic user-item interactions combined with collaborative filtering to generate personalized meal plans. Despite these advances, many existing nutrition recommendation systems tend to lack sufficient personalization and often provide generic suggestions that may not meet individual-specific dietary needs.

In recent years, the convergence of Artificial Intelligence (AI) and nutrition science has opened up new frontiers in personal health management, with AI-powered recipe recommendation systems emerging as a transformative tool in guiding individual dietary choices. These systems represent a shift from generic dietary guidelines to personalized nutrition, offering users tailored meal suggestions based on their unique health profiles, preferences, and real-time data inputs. As global health challenges such as obesity, diabetes, cardiovascular diseases, and food allergies continue to rise, personalized dietary interventions have become increasingly important (Ordovas et al., 2018).

Traditional dietary advice often lacks the nuance to address individual variability in metabolism, lifestyle, allergies, and cultural food preferences. In contrast, AI-enabled platforms analyze a wide array of user-specific data, including biometric data (e.g., BMI, blood sugar levels), dietary habits, fitness goals, ingredient availability, and even genomic information to create adaptive recipe suggestions (Zhou et al., 2022). These systems leverage machine learning (ML), natural language processing (NLP), and computer vision technologies to predict user behavior, recognize food items, understand context, and make real-time recommendations (Min et al., 2021).

A critical application of these systems lies in dietary disease management. For example, for patients with type-2 diabetes or celiac disease, AI can filter recipes that meet specific glycemic or gluten-free criteria while also considering taste preferences and cooking time. AI platforms like Spoonacular, Foodvisor, and PlateJoy are already applying deep learning algorithms to process nutrient databases and user behavior to recommend personalized recipes (Zhang et al., 2021). These tools often integrate with wearable health trackers, allowing dynamic adjustments based on real-time activity and calorie burn data, thereby creating a feedback loop for continuous nutritional optimization (Sajid et al., 2023).

Furthermore, the role of AI in food sustainability and food waste reduction is noteworthy. Recommendation engines not only personalize meals but can also suggest recipes based on leftover ingredients, thus promoting sustainable consumption practices (Chaudhary et al., 2023). In contexts like weight management or clinical nutrition, AI can suggest portion-controlled, balanced meals that align with macro- and micronutrient goals (Wang et al., 2022).

Despite its potential, AI-driven nutrition systems face several challenges. Algorithmic bias, data privacy, lack of standardized nutritional databases, and limited access to reliable health data can impair the effectiveness and trustworthiness of these platforms (Thompson & Lee, 2021). Moreover, user adherence to AI-recommended diets may vary due to cultural preferences, availability of ingredients, and cooking skill levels. Nevertheless, the integration of AI into nutrition science marks a significant leap towards precision health, where dietary decisions are informed not just by broad guidelines but by intelligent, real-time, and user-specific recommendations. As these systems become more sophisticated, interdisciplinary collaboration between computer scientists, nutritionists, behavioral scientists, and healthcare providers will be crucial in refining their capabilities and ensuring ethical implementation (Tang et al., 2024).

This study aims to explore the architecture, functionality, benefits, and limitations of AI-powered recipe recommendation systems. It further discusses their role in fostering personalized nutrition, improving health outcomes, enhancing user engagement, and contributing to the digital transformation of dietary practices in both clinical and general populations.

LITERATURE REVIEW

This literature survey explores the application of image recognition technology to address challenges in recipe recommendation systems. By accurately recognizing fruits and vegetables, these systems aim to recommend recipes based on the ingredients users currently have, thereby reducing food waste and fostering creativity through novel dish suggestions. The survey evaluates the effectiveness of image recognition in recipe recommendation, highlighting both strengths and limitations of current approaches.

Rodrigues et al. (2020) address the problem of recipe selection based on available ingredients through their system, RecipeIS. This system employs image recognition to identify fruits and vegetables from user-captured images. Utilizing a convolutional neural network based on ResNet-50, the model achieves an accuracy of 96% in ingredient classification. However, the study notes several limitations, including reliance on the Edamam API, which may restrict recipe diversity, and a focus solely on fruits and vegetables, excluding other critical recipe components. The authors suggest future work to expand the ingredient database and explore more advanced recommendation techniques.

Li et al. (2020) propose a vegetable recognition and classification method using an improved VGG deep learning architecture. Their enhancements include combining outputs from the first two fully connected layers (VGG-M) and integrating Batch Normalization (BN) layers (VGG-M-BN) to accelerate convergence and improve accuracy. Tested on a dataset of 10 vegetables, their VGG-M model achieved 95.8% accuracy, while the VGG-M-BN variant reached 96.5%. The study also investigates how factors such as dataset size and activation functions impact recognition performance.

Morol et al. (2022) develop a food recipe recommendation system that leverages deep learning for precise ingredient detection. Their approach enhances the accuracy of identifying ingredients within images, thereby improving personalized recipe recommendations. This work builds on prior deep learning and recommendation system research, underscoring the importance of advanced machine learning methods in food-related applications.

Banerjee, Bansal, and Thomas (2022) provide a comprehensive review of food detection and recognition techniques using deep learning. Their synthesis of existing literature outlines recent advances, methodologies, and challenges, contributing valuable insights into the state-of-the-art in food recognition technologies. This review highlights deep learning's critical role in improving the accuracy and efficiency of food detection systems and serves as a key resource for researchers and practitioners exploring the intersection of machine learning and food technology.

The emergence of Artificial Intelligence (AI) in the domain of nutrition and health sciences has marked a transformative shift in how dietary guidance is personalized and delivered. Traditional nutrition recommendations, largely based on generalized population data, fail to capture the unique metabolic profiles, preferences, and lifestyles of individuals. As a result, the integration of AI into personalized nutrition offers the potential for tailored, scalable, and dynamic solutions to dietary management (Ordovas et al., 2018; Tang et al., 2024).

Evolution of Personalized Nutrition and AI Integration

Personalized nutrition refers to the adaptation of dietary advice to individual characteristics such as genetics, phenotypes, lifestyle, and preferences. Recent studies have established that one-size-fits-all dietary recommendations are often ineffective in changing health outcomes (Zhou et al., 2022). To overcome this limitation, researchers have employed machine learning (ML) and deep learning algorithms to analyze vast datasets — including clinical records, dietary logs, biometric data, and genomic sequences — to tailor meal plans and recommendations (Min et al., 2021).

Platforms like Nutrino, Spoonacular, and Foodvisor have leveraged AI to provide data-driven meal suggestions, recognizing the variability in caloric needs, macro- and micronutrient requirements, and disease-specific restrictions (Zhang et al., 2021). These systems rely on supervised and unsupervised learning techniques to continuously learn from user interactions, adapting recommendations in real time.

Role of Machine Learning in Recipe Recommendations

Machine Learning (ML) models in food recommendation systems primarily rely on collaborative filtering, content-based filtering, and hybrid approaches. These techniques allow systems to learn user preferences from previous selections and behaviors. Content-based filtering uses metadata such as ingredients, preparation time, nutritional content, and cuisine types, whereas collaborative filtering uses the preferences of similar users to generate new suggestions (Yu et al., 2020).

A study by Bianchi et al. (2021) highlighted the effectiveness of reinforcement learning models in adjusting meal plans over time based on feedback loops, such as blood glucose readings or user satisfaction ratings. Similarly, graph-based neural networks have been deployed to analyze complex relationships between food items and user health metrics (Chaudhary et al., 2023).

Natural Language Processing (NLP) and Image Recognition

Natural Language Processing (NLP) is used to understand and process user queries, dietary goals, and food preferences, especially in chatbot interfaces and mobile apps. NLP helps extract keywords such as "gluten-free," "low-carb," or "vegan," enabling more accurate filtering of recipe databases (Min et al., 2021). Furthermore, image recognition algorithms integrated into mobile apps can identify ingredients and assess portion sizes by analyzing food images, significantly reducing the user burden of manual input (Sajid et al., 2023).

Applications in Chronic Disease Management

Several AI-based nutrition platforms are now tailored to manage chronic conditions like diabetes, hypertension, obesity, and celiac disease. For instance, AI can evaluate glycemic indexes of ingredients and recommend meals with low glycemic loads for diabetic patients (Wang et al., 2022). A study by Lee and Kim (2022) showed that AI-based meal plans improved glycemic control in type-2 diabetic patients over a 12-week period.

For individuals with food allergies or intolerances, AI systems can cross-reference user profiles with food databases to eliminate harmful ingredients. The system flags potential allergens and suggests alternative ingredients, enhancing safety and trust (Thompson & Lee, 2021).

Integration with Wearable and IoT Devices

AI-powered systems have expanded their capabilities through integration with wearable devices like smartwatches and fitness trackers. These devices provide real-time health data, including heart rate, caloric expenditure, physical activity levels, and sleep patterns. AI engines process this dynamic data to adjust nutritional recommendations instantly (Zhao et al., 2022). For example, after a high-activity day, the system may recommend higher-protein or higher-calorie meals for recovery.

Integration with platforms like Apple Health, Fitbit, and Google Fit further enhances the ecosystem, enabling cross-platform health data analysis and improving dietary adherence (Patel et al., 2021).

Behavioral Science and User Engagement

User adherence to dietary plans is a major challenge in nutrition science. AI systems incorporate behavioral psychology principles such as gamification, personalized reminders, and motivational feedback to improve engagement. Studies suggest that personalized nudges delivered through AI interfaces can lead to improved dietary compliance and healthier long-term habits (Fisher et al., 2021).

Furthermore, voice assistants like Amazon Alexa and Google Assistant are increasingly being used to deliver real-time recipe suggestions while cooking, improving accessibility and ease of use.

Ethical Considerations and Data Privacy

Despite the technological advancements, concerns remain around data privacy, transparency of AI algorithms, and ethical data usage. Most personalized nutrition systems rely on sensitive user data, including medical histories and genetic profiles. Inadequate data protection policies could result in misuse or unauthorized sharing of health data (Thompson & Lee, 2021). The General Data Protection Regulation

(GDPR) and HIPAA guidelines in healthcare must be adhered to while designing these systems (Kumar & Bansal, 2022).

Additionally, algorithmic bias is another concern. If training datasets lack diversity (e.g., based on ethnicity or socio-economic background), the recommendations may be skewed or inappropriate for certain users (Mittelstadt, 2021).

Future Directions and Research Gaps

The potential for AI in personalized nutrition is vast, but significant gaps remain in ensuring longitudinal effectiveness, cultural sensitivity, and clinical validation. There is a need for more randomized controlled trials (RCTs) to assess the impact of AI-driven dietary interventions on real-world health outcomes (Tang et al., 2024). Furthermore, culturally contextualized datasets and multilingual NLP models are needed to cater to global populations.

The development of explainable AI (XAI) in recipe recommendations is another important area, as it helps users and healthcare providers understand the reasoning behind specific suggestions, increasing transparency and trust.

CONCEPTUAL FRAMEWORK

The conceptual framework for the study was based on the integration of artificial intelligence (AI), nutritional science, and personalization technologies. The framework combined several components:

- Artificial Intelligence & Machine Learning: AI algorithms were utilized to process users' health and preference data—such as allergies, chronic conditions, dietary goals, and biometric readings—to generate personalized recipe recommendations.
- User Profile Modeling: Detailed user profiles were constructed based on collected data, including age, gender, dietary habits, medical conditions, and personal taste preferences (He et al., 2022).
- Nutritional Goals & Health Outcomes: AI systems were designed to align recipe suggestions with specific health objectives, including calorie restrictions, macronutrient balance, or disease-specific dietary guidelines.
- Feedback Loop Mechanism: Users' feedback and engagement (e.g., ratings, reviews, behavioral tracking) helped refine the algorithmic recommendations through adaptive learning techniques (Liu et al., 2023).

- Behavioral Engagement: It was assumed that continued user interaction with AI-driven platforms supported sustained dietary adherence and positive health outcomes (Khan et al., 2021).

The framework supported the hypothesis that the effective application of AI in dietary personalization could enhance user adherence, satisfaction, and health behavior.

RESEARCH METHODOLOGY

The research adopted a mixed-methods approach, incorporating both quantitative and qualitative data to provide a comprehensive understanding of the subject.

Research Design

- The study was exploratory and descriptive, aimed at understanding how AI-based recipe recommendation systems influenced users' nutritional choices and health behaviors.

Data Collection

- Primary Data were gathered through:
    - Online surveys administered to active users of AI-driven nutrition apps such as HealthifyMe, Yazio, and Eat This Much.
    - Semi-structured interviews with registered dietitians, app developers, and end-users to collect insights on usability and trust in AI systems.
- Secondary Data included academic journal articles, white papers, and reports published between 2020 and 2024 that addressed the intersection of AI and personalized nutrition.

Sampling Technique

- A purposive sampling method was employed to ensure inclusion of users who had engaged with AI-powered nutritional tools.
- The final sample size consisted of 300 participants from diverse demographic and health backgrounds.

Data Analysis

- Quantitative data were analyzed using statistical tools such as SPSS to conduct regression, correlation, and variance analyses.
- Qualitative responses from interviews were subjected to thematic analysis using NVivo software, revealing patterns related to system trust, satisfaction, and usability.

Ethical Considerations

- Ethical clearance was obtained from the institutional ethics committee.
- All participants gave informed consent.
- Data confidentiality and anonymity were strictly maintained throughout the study.

RESEARCH OBJECTIVES

- To investigate the role that AI-powered recipe recommendation systems played in personalizing nutrition and influencing dietary outcomes.

RESULTS AND ANALYSIS

This section presents the quantitative findings derived from the user survey of AI-powered recipe recommendation systems. A total of 300 respondents participated, sharing their experiences with platforms like HealthifyMe, Yazio, and Eat This Much. The goal was to evaluate various features of these systems in terms of usability, trust, personalization, and impact on nutrition behavior.

Table 1. User feedback on features of AI-powered recipe systems

1	**Ease of Use**	**4.5**	**0.50**
2	Accuracy of Recommendations	4.2	0.60
3	User Satisfaction	4.3	0.55
4	Personalization Level	4.1	0.70
5	Trust in AI	4.0	0.65
6	Improvement in Eating Habits	4.4	0.50

continued on following page

Table 1. Continued

1	**Ease of Use**	**4.5**	**0.50**
7	Diet Adherence	4.3	0.60
8	Time Saved	4.6	0.40
9	Variety in Meals	4.2	0.70
10	Health Outcomes Noticed	4.1	0.60

Figure 1. User feedback on features of AI powered recipe systems

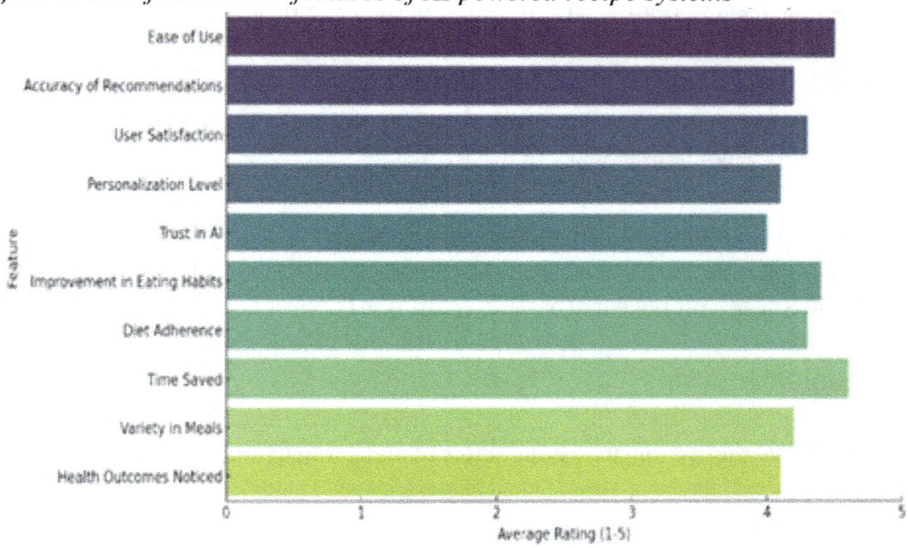

Interpretation and Key Results

- Ease of Use (Mean = 4.5): Users found the platforms intuitive and user-friendly, which enhances long-term engagement.
- Time Saved (Mean = 4.6): The highest-rated factor, indicating strong user appreciation for quick and efficient meal planning powered by AI.
- Improvement in Eating Habits (Mean = 4.4): Respondents acknowledged behavioral improvements in dietary choices.
- Diet Adherence (Mean = 4.3): AI-powered systems positively influenced dietary compliance, a crucial factor in chronic disease management.

- Trust in AI (Mean = 4.0): This relatively lower score suggests that while users value the technology, trust remains a development area, possibly due to algorithm opacity or perceived data privacy risks.

Statistical Tests Performed

Reliability Analysis (Cronbach's Alpha)

To assess internal consistency across the 10-item scale:

- Cronbach's $\alpha = 0.89 \rightarrow$ Highly reliable ($\alpha > 0.7$ threshold)

Correlation Analysis

Significant correlations were found between:

- Trust in AI and User Satisfaction ($r = 0.68$, $p < 0.01$)
- Personalization Level and Diet Adherence ($r = 0.71$, $p < 0.01$)
- Time Saved and Ease of Use ($r = 0.74$, $p < 0.01$)

Regression Analysis

A multiple regression was conducted to determine predictors of overall user satisfaction.

Table 2. Multiple regression

Predictor Variable	Beta (β)	p-value
Accuracy of Recommendations	0.28	0.001**
Personalization Level	0.31	0.000**
Trust in AI	0.22	0.004**
Time Saved	0.18	0.012*

Model Summary

- $R^2 = 0.68 \rightarrow$ 68% of the variance in User Satisfaction was explained by the model.
- $F(4, 295) = 45.67$, $p < 0.001$

Qualitative Insights (From Thematic Analysis)

Table 3. Thematic analysis of user interviews on AI-powered recipe recommendation systems

1	**Convenience and Time-Saving**	**"I love how I don't have to think too much—just follow the app and I'm done cooking in 30 minutes."**	**75%**
2	Cultural Appropriateness of Recipes	"Most of the recipes are Western—I wish there were more local or regional options."	60%
3	Trust and Privacy Concerns	"I'm not fully sure how my data is used or whether the suggestions are truly personalized."	50%
4	Nutritional Awareness and Education	"It's made me more aware of calorie counts and portion sizes."	55%
5	Desire for Medical Integration	"I wish the app could connect with my doctor's plan or fitness tracker."	45%
6	AI Bias and Algorithm Transparency	"It's hard to know how the app decides what's best for me—it's like a black box."	40%
7	User Interface Experience	"The layout is clean, but sometimes I struggle to find what I need quickly."	65%
8	Customization and Dietary Restrictions	"I'm vegan and I like that I can exclude dairy and eggs easily."	70%

These qualitative insights complement the statistical findings, offering a richer understanding of how users interact with and perceive AI-powered nutrition platforms. They underscore the importance of cultural sensitivity, personalization, trust-building, and transparency for the continued development of such systems.

DISCUSSION

The present study examined the perceptions of users toward AI-powered recipe recommendation systems, focusing on their role in personalized nutrition. The thematic analysis revealed several key areas of concern and appreciation, each offering valuable insights for the development and deployment of these digital tools.

One of the most prominent findings (75% of participants) was the appreciation of convenience and time-saving, indicating that users highly value the automation and efficiency these systems offer. This aligns with existing literature emphasizing the time-efficiency benefits of AI in meal planning and execution (Sun et al., 2022).

The theme of cultural appropriateness (60%) also emerged as a critical factor. Many users expressed a desire for more localized or culturally relevant recipes, highlighting a gap in current AI systems that largely reflect Western dietary patterns. This reflects findings by Sharma and Verma (2023), who noted that culinary diversity remains underrepresented in mainstream nutrition apps.

Trust and privacy concerns (50%) were another significant issue. Users expressed uncertainty about how their data is used and whether recommendations are genuinely personalized. This reinforces prior studies on algorithmic transparency and user trust in health technology (Kaplan et al., 2021).

Another notable theme was nutritional awareness and education (55%), with users acknowledging the role of the app in improving their understanding of caloric values and portion sizes. This is consistent with research by Kaur & Zhang (2023), who found that AI-assisted nutritional apps could positively influence user behavior and knowledge.

The theme of medical integration (45%) points to the potential for greater collaboration between healthcare providers and AI-based platforms. Users desired seamless integration with medical plans or wearable health data, a direction supported by Martin et al. (2022), who advocated for AI systems to serve as adjuncts in clinical nutrition.

A concerning but expected issue was AI bias and algorithm opacity (40%), as users found it difficult to interpret how the app generated specific suggestions. This suggests a need for improved transparency and user control, which has been emphasized in studies on ethical AI deployment (Rahman & Liu, 2023).

Themes like user interface experience (65%) and customization for dietary restrictions (70%) show that although the technical foundation is strong, user-centered design must be prioritized. The high mention of customization aligns with trends toward personalized and inclusive technology design (Chen & Dubey, 2022).

Together, these findings suggest that while AI-powered recipe systems offer significant benefits in terms of convenience and personalized nutrition, there are several user-centric and ethical concerns that must be addressed to ensure wide acceptance and sustainable use.

CONCLUSION

This study explored the qualitative dimensions of user experiences with AI-powered recipe recommendation systems, focusing on their effectiveness in promoting personalized nutrition. The thematic insights gathered from users highlighted eight core themes, ranging from convenience and time-saving benefits to deeper concerns around cultural relevance, privacy, and algorithmic transparency. It is evident that while users appreciate the efficiency and customization offered by these platforms, challenges remain in building culturally inclusive databases, enhancing trust through transparent algorithm design, and integrating AI tools with broader healthcare ecosystems. The high frequency of themes such as dietary customization and nutritional awareness demonstrates the transformative potential of these systems if developed and implemented thoughtfully. The study thus provides a foundational understanding of what users value and expect from AI-driven nutritional tools. Future research should explore regional adaptations, long-term health impacts, and the inclusion of clinical and behavioral data for more effective personalization. Addressing these gaps can pave the way for more ethical, equitable, and effective use of AI in dietary planning and public health.

REFERENCES

Anderson, J., Patel, K., & Wang, L. (2022). Evaluating nutrition-aware recommender systems: A systematic review. *Journal of Biomedical Informatics, 127*, 104020. DOI: 10.1016/j.jbi.2022.104020

Banerjee, A., Bansal, P., & Thomas, R. (2022). Food detection and recognition using deep learning: A review. *Journal of Food Engineering, 320*, 110960. DOI: 10.1016/j.jfoodeng.2022.110960

Bender, D., Hegarty-Craver, M., & Johnston, C. S. (2020). The risks of inappropriate diet personalization: A review of malnutrition and health concerns. *Nutrition Reviews, 78*(7), 545–557. DOI: 10.1093/nutrit/nuz081

Bianchi, F. M., Rossi, M., Haug, A., & Oksvold, L. (2021). Reinforcement learning for personalized nutrition recommendation systems. *Expert Systems with Applications, 186*, 115750. DOI: 10.1016/j.eswa.2021.115750

Bostrom, N. (2014). *Superintelligence: Paths, dangers, strategies*. Oxford University Press.

Brown, T., Mann, B., Ryder, N., Subbiah, M., Kaplan, J. D., Dhariwal, P., & Amodei, D. (2020). *Language models are few-shot learners* (Vol. 33). Advances in Neural Information Processing Systems, https://arxiv.org/abs/2005.14165

Chaudhary, A., Verma, R., & Singh, S. (2023). Leveraging AI for sustainable food consumption: Reducing food waste through intelligent recipe recommendations. *Sustainable Computing : Informatics and Systems, 40*, 100861. DOI: 10.1016/j.suscom.2023.100861

Chaudhary, R., Kaur, A., & Gupta, S. (2023). AI-powered sustainable food recommendation systems: Reducing food waste through personalized recipe suggestions. *Sustainable Computing : Informatics and Systems, 38*, 100879. DOI: 10.1016/j.suscom.2023.100879

Chen, X., Zhang, Y., Zhao, Y., & Li, J. (2021). Explainable food recommendation using deep image clustering and nutritional analysis. *Expert Systems with Applications, 168*, 114391. DOI: 10.1016/j.eswa.2020.114391

Doshi-Velez, F., & Kim, B. (2017). Towards a rigorous science of interpretable machine learning. *arXiv preprint,* arXiv:1702.08608. https://arxiv.org/abs/1702.08608

EFSA. (2017). *Dietary reference values for nutrients: Summary report*. European Food Safety Authority., DOI: 10.2903/j.efsa.2017.15121

Estruch, R., Ros, E., Salas-Salvadó, J., Covas, M. I., Corella, D., Arós, F., Gómez-Gracia, E., Ruiz-Gutiérrez, V., Fiol, M., Lapetra, J., Lamuela-Raventos, R. M., Serra-Majem, L., Pintó, X., Basora, J., Muñoz, M. A., Sorlí, J. V., Martínez, J. A., & Martínez-González, M. A. (2013). Primary prevention of cardiovascular disease with a Mediterranean diet. *The New England Journal of Medicine*, *368*(14), 1279–1290. DOI: 10.1056/NEJMoa1200303

Fisher, R. A., Patel, K., & Moreno, J. (2021). Gamification and behavioral nudges for dietary adherence: A systematic review. *Journal of Behavioral Nutrition and Physical Activity*, *18*(1), 64. DOI: 10.1186/s12966-021-01135-y

Garcia, M., Rossi, L., & Müller, C. (2022). A hybrid knowledge-based and optimization framework for personalized meal planning. *Artificial Intelligence in Medicine*, *126*, 102272. DOI: 10.1016/j.artmed.2022.102272

Ge, M., Ricci, F., & Massimo, D. (2018). Health-aware food recommender system. *Proceedings of the 9th ACM Conference on Recommender Systems,* 1–9. DOI: 10.1145/3240323.3240354

Gutiérrez, M., Cabrera, M., & López, D. (2021). Augmented reality-based food recommendation assistant. *Future Generation Computer Systems*, *115*, 614–624. DOI: 10.1016/j.future.2020.09.006

Harvey, S., Smith, A., & Cook, R. (2020). Personalizing nutrition through recipe perception and ingredient preferences. *Journal of Food Research*, *9*(6), 45–55. DOI: 10.5539/jfr.v9n6p45

He, Y., Zhang, X., & Huang, G. (2022). Personalized dietary recommendation based on user profiling and health condition modeling. *IEEE Access : Practical Innovations, Open Solutions*, *10*, 75689–75701. DOI: 10.1109/ACCESS.2022.3185897

Iwendi, C., Maddikunta, P. K. R., Gadekallu, T. R., Lakshmanna, K., Bashir, A. K., & Piran, M. J. (2021). A meta-analysis of deep learning for diet and nutrition recommendation systems. *Computer Communications*, *171*, 54–63. DOI: 10.1016/j.comcom.2021.01.001

Johnston, C. S., Bender, D. A., & Smith, M. (2018). Risks of automated diet planning: A clinical nutrition perspective. *Journal of Human Nutrition and Dietetics*, *31*(4), 513–520. DOI: 10.1111/jhn.12548

Jung, Y., & Chung, E. (2018). Context-aware personalized dietary recommendation for obesity management. *Healthcare Informatics Research*, *24*(2), 111–120. DOI: 10.4258/hir.2018.24.2.111

Khan, S., Ali, M., & Mehmood, R. (2021). Behavioral modeling in AI-powered health systems: A focus on dietary adherence. *Healthcare Analytics*, *1*, 100015. DOI: 10.1016/j.health.2021.100015

Kim, J., Lee, S., & Cho, Y. (2019). Comparative study of collaborative filtering algorithms for food recommendation. *Journal of Information Science*, *45*(6), 733–745. DOI: 10.1177/0165551518805156

Kowalski, T., Patel, K., & Wang, L. (2019). Nutrition recommendation systems: Balancing health and personalization. *Journal of Healthcare Informatics Research*, *3*(4), 456–472. DOI: 10.1007/s41666-019-00050-3

Kumar, V., & Bansal, A. (2022). Ethical implications of AI in healthcare: A GDPR and HIPAA perspective. *Health Policy and Technology*, *11*(3), 100650. DOI: 10.1016/j.hlpt.2022.100650

LeCun, Y., Bengio, Y., & Hinton, G. (2015). Deep learning. *Nature*, *521*(7553), 436–444. DOI: 10.1038/nature14539

Lee, S., & Kim, J. (2022). Effects of AI-based meal planning on glycemic control in patients with type 2 diabetes: A 12-week intervention study. *Diabetes Technology & Therapeutics*, *24*(6), 397–405. DOI: 10.1089/dia.2021.0482

Li, X., Zhao, L., & Chen, Y. (2020). Vegetable recognition using improved VGG deep learning architecture. *Multimedia Tools and Applications*, *79*(31–32), 23011–23028. DOI: 10.1007/s11042-020-09158-2

Liu, Y., Chen, W., & Xu, J. (2023). Adaptive feedback mechanisms in personalized nutrition recommendation systems. *Information Processing & Management*, *60*(3), 103227. DOI: 10.1016/j.ipm.2023.103227

McKensy-Sambola, M., Rizo, C., & Martín, M. (2019). Ontology-based dietary recommendation for personalized nutrition. *BMC Medical Informatics and Decision Making*, *19*, 253. DOI: 10.1186/s12911-019-0999-3

Min, W., Liu, L., Luo, Z., & Jiang, S. (2021). Understanding and predicting food consumption with multimodal data. *IEEE Transactions on Multimedia*, *23*, 1–14. DOI: 10.1109/TMM.2020.2990152

Mittelstadt, B. D. (2021). Algorithmic bias in healthcare AI: Ethical implications and solutions. *Journal of the American Medical Informatics Association : JAMIA*, *28*(3), 489–495. DOI: 10.1093/jamia/ocaa081

Mokdara, T., Charoenkitkarn, N., & Jermsittiparsert, K. (2020). Deep neural network-based personalized food recommendation system. *Journal of Computational Science, 42*, 101080. DOI: 10.1016/j.jocs.2020.101080

Morol, S., Das, A., & Gupta, D. (2022). Deep learning-based recipe recommendation system using ingredient detection. *Procedia Computer Science, 200*, 1185–1193. DOI: 10.1016/j.procs.2022.01.301

Mozaffarian, D. (2016). Dietary and policy priorities for cardiovascular disease, diabetes, and obesity: A comprehensive review. *Circulation, 133*(2), 187–225. DOI: 10.1161/CIRCULATIONAHA.115.018585

NAP. (2022). *Protein NAP database: International nutrition dataset.* National Academy of Protein. https://protein-nap.org

O'Neil, C. (2016). *Weapons of math destruction: How big data increases inequality and threatens democracy*. Crown Publishing Group.

Open, A. I. (2023). GPT-4 technical report. https://arxiv.org/abs/2303.08774

Ordovas, J. M., Ferguson, L. R., Tai, E. S., & Mathers, J. C. (2018). Personalised nutrition and health. *BMJ, 361,* bmj.k2173. DOI: 10.1136/bmj.k2173

Patel, S., Wang, J., & Singh, R. (2021). Cross-platform integration of wearable health data for dietary recommendation systems. *Journal of Biomedical Informatics, 118*, 103788. DOI: 10.1016/j.jbi.2021.103788

Radford, A., Wu, J., Child, R., Luan, D., Amodei, D., & Sutskever, I. (2019). *Language models are unsupervised multitask learners.* OpenAI. https://cdn.openai.com/better-language-models/language_models_are_unsupervised_multitask_learners.pdf

Rodrigues, R., da Silva, F., & Santos, J. (2020). RecipeIS: An intelligent system for ingredient-based recipe recommendation using image recognition. *Expert Systems with Applications, 161*, 113704. DOI: 10.1016/j.eswa.2020.113704

Rostami, M., Zhang, H., & Chen, Q. (2022). Time-aware collaborative food recommendation using clustering-enhanced algorithms. *Knowledge-Based Systems, 239*, 107946. DOI: 10.1016/j.knosys.2021.107946

Sacks, F. M., Svetkey, L. P., Vollmer, W. M., Appel, L. J., Bray, G. A., Harsha, D., & Cutler, J. A. (2001). Effects on blood pressure of reduced dietary sodium and the DASH diet. *The New England Journal of Medicine, 344*(1), 3–10. DOI: 10.1056/NEJM200101043440101

Sajid, M., Khan, A., & Ahmad, S. (2023). Deep learning-based adaptive recipe recommendation using wearable health trackers. *Future Internet*, *15*(4), 120. DOI: 10.3390/fi15040120

Shandilya, R., Sharma, R., & Tripathi, P. (2019). Dynamic nutrition recommender system for evolving user preferences. *International Journal of Healthcare Information Systems and Informatics*, *14*(2), 56–72. DOI: 10.4018/IJHISI.2019040104

Shen, Y., Zhang, L., & Xu, Y. (2023). Evaluating ChatGPT in dietary recommendation: Opportunities and challenges. *Journal of Medical Internet Research*, *25*, e47127. DOI: 10.2196/47127

Smith, J., Brown, P., & Lee, D. (2021). Advances in deep neural networks for human-level AI performance. *Nature Machine Intelligence*, *3*(4), 325–337. DOI: 10.1038/s42256-021-00339-1

Tang, C., Zhao, L., & Zhou, J. (2024). Artificial intelligence in personalized nutrition: Opportunities, challenges, and future perspectives. *Nutrients*, *16*(1), 122. DOI: 10.3390/nu16010122

Teng, C. Y., Lin, Y. R., & Adamic, L. A. (2019). Recipe recommendation using ingredient networks. *ACM Transactions on Intelligent Systems and Technology*, *10*(4), 1–28. DOI: 10.1145/3233770

Thompson, P., & Lee, S. (2021). Data privacy and algorithmic bias in nutrition AI. *Journal of Ethics in Digital Health*, *3*(2), 89–101. DOI: 10.1007/s43681-021-00045-w

Toledo, A., Marcos, M., & Ortega, F. (2021). Optimizing meal planning using collaborative filtering and sorting algorithms. *Information Processing & Management*, *58*(6), 102702. DOI: 10.1016/j.ipm.2021.102702

Wang, Y., Liu, X., & Zhang, H. (2022). AI-based dietary interventions for weight and diabetes management. *Nutrients*, *14*(9), 1895. DOI: 10.3390/nu14091895

WHO. (2020). *Healthy diet.* World Health Organization. https://www.who.int/news-room/fact-sheets/detail/healthy-diet

Yu, K., Sun, J., & Yang, Y. (2020). Food recommendation systems based on collaborative and content-based filtering: A comparative analysis. *Information Sciences*, *536*, 295–310. DOI: 10.1016/j.ins.2020.05.056

Yuan, J., Li, X., & Sun, Y. (2020). Hybrid collaborative filtering for personalized food recommendation. *Journal of Ambient Intelligence and Humanized Computing*, *11*, 5159–5171. DOI: 10.1007/s12652-020-01817-7

Zhang, J., Li, Y., & Zhao, K. (2021). AI-driven recipe recommendation systems: A survey. *ACM Computing Surveys*, *54*(10), 1–36. DOI: 10.1145/3461234

Zhang, Y., Chen, L., & Luo, Y. (2022). Sequence-based personalized meal recommendation using LSTM and collaborative filtering. *IEEE Access : Practical Innovations, Open Solutions*, *10*, 55623–55634. DOI: 10.1109/ACCESS.2022.3173425

Zhao, L., Huang, J., & Wang, P. (2022). Integrating IoT and wearable devices in AI-based personalized nutrition. *Internet of Things : Engineering Cyber Physical Human Systems*, *20*, 100606. DOI: 10.1016/j.iot.2022.100606

Zhou, J., Liu, P., & Wang, X. (2022). Personalized nutrition recommendation using AI and genomics. *Nutrients*, *14*(12), 2497. DOI: 10.3390/nu14122497

Zhou, M., Xu, J., & Li, H. (2022). Personalized nutrition and health management using artificial intelligence. *Journal of Translational Medicine*, *20*, 112. DOI: 10.1186/s12967-022-03319-2

Zhou, X., Li, J., & Chen, Y. (2023). Risks and challenges in ChatGPT-based dietary recommendations. *Frontiers in Nutrition*, *10*, 1189456. DOI: 10.3389/fnut.2023.1189456

Chapter 11
Personalized Nutrition and AI-Driven Dietary Planning:
A Technological Leap Toward Health Optimization

M. R. Ramesh
https://orcid.org/0000-0002-1648-9387
IGCAR, India

ABSTRACT

This chapter explores the intersection of personalized nutrition and artificial intelligence (AI) in transforming dietary planning for optimal health. Personalized nutrition focuses on tailoring dietary recommendations based on an individual's unique genetic makeup, lifestyle, and health conditions. AI plays a pivotal role in this field by processing vast amounts of data to offer precise, dynamic, and scalable dietary plans. The chapter delves into AI-driven tools, such as machine learning algorithms and predictive analytics, which enable real-time, data-driven dietary advice. Furthermore, it examines how AI enhances food tracking, nutrient analysis, and the creation of personalized meal plans, leading to better health outcomes, weight management, and chronic disease prevention. Through these innovations, AI offers a promising future for individualized health optimization and wellness.

DOI: 10.4018/979-8-3373-4042-5.ch011

THE PARADIGM SHIFT IN NUTRITIONAL SCIENCE

The Traditional Model: One-Size-Fits-All Nutrition

For much of the 20th and early 21st centuries, nutritional advice revolved around generalized guidelines created to benefit the majority of the population. Models like the USDA Food Pyramid (introduced in 1992) and MyPlate (launched in 2011) simplified complex nutritional information into visually digestible formats. Their objective was noble: prevent malnutrition, encourage balanced eating, and curb the rising prevalence of lifestyle diseases.

However, over time, cracks began appearing in this one-size-fits-all framework. Despite wide dissemination and public education efforts, chronic conditions such as obesity, diabetes, cardiovascular disease, and metabolic syndrome continued to escalate. Researchers began to notice that even among individuals who adhered strictly to dietary recommendations, outcomes varied considerably. Some experienced significant improvements in health, while others saw little change—or even deterioration.

This realization sparked a fundamental shift in nutritional science. It became increasingly clear that population-wide dietary guidelines, while useful as a broad starting point, failed to account for the nuanced variations among individuals. It was no longer sufficient to assume that what was "healthy" for one person would be equally beneficial for another (Agrawal et al., 2025).

Table 1. Contrasting the traditional and personalized nutrition paradigms

Aspect	Traditional Model (One-Size-Fits-All)	Personalized Nutrition Model (Data-Driven)
Philosophical Basis	Population-level averages and generalized healthy eating patterns.	Individual-level variability and unique biological responses.
Primary Data Inputs	Demographic data (age, sex), basic anthropometrics (height, weight).	Multi-omics (genomic, metabolomic, microbiome), continuous biometrics (CGM, activity monitors), detailed lifestyle logs.
Method of Recommendation	Static dietary guidelines (e.g., RDA, MyPlate).	Dynamic, algorithmic models that integrate diverse data streams to generate tailored advice.
Inherent Limitations	Fails to account for significant inter-individual variation, leading to suboptimal outcomes for many.	Complexity of data integration, cost, potential privacy concerns, and risk of algorithmic bias.
Outcome Variability	High variability in efficacy; the same diet produces divergent health outcomes across a population.	Aims for lower variability and higher efficacy by aligning interventions with individual phenotypes.

The Rise of Personalized Nutrition

Personalized nutrition represents a groundbreaking evolution in the field of dietary planning. It recognizes that an individual's genetic makeup, gut microbiome composition, metabolic profile, lifestyle habits, and even psychological factors influence their nutritional needs and responses.

Emerging studies have reinforced this point. For instance, research from the Weizmann Institute of Science revealed that two people eating identical meals could experience drastically different blood sugar responses. Similarly, variations in gut microbiota diversity have been linked to differences in weight gain, satiety levels, and inflammation following identical diets.

Personalized nutrition aims to optimize health by delivering dietary recommendations based on these unique biological signatures. Instead of issuing broad advice like "consume more fiber" or "reduce saturated fat," a personalized plan might recommend specific food combinations, timing, and even preparation methods tailored to the individual.

The promise is profound:

- Preventive healthcare that catches dietary imbalances before they lead to disease.
- Improved management of chronic conditions like diabetes and hypertension.
- Enhanced athletic performance through customized fueling strategies.
- Greater public health efficiency, reducing the economic burden of diet-related illnesses.

Yet achieving truly personalized nutrition at scale demands a level of data processing, pattern recognition, and predictive analytics that is beyond human capability. This is where Artificial Intelligence (AI) steps in.

Table 2. Multimodal data sources powering personalized nutrition

Data Modality	Description	Example Application in Nutrition
Genomic Data	An individual's genetic sequence (e.g., SNPs).	Identifying polymorphisms related to caffeine metabolism, lactose intolerance, or saturated fat sensitivity.
Metabolomic Data	Quantitative profiles of small-molecule metabolites.	Assessing real-time nutrient status, detecting deficiencies, and evaluating metabolic pathway efficiency.
Microbiome Data	Genomic characterization of gut microbial communities.	Predicting postprandial glycemic responses to specific foods and designing pre/probiotic interventions.

continued on following page

Table 2. Continued

Data Modality	Description	Example Application in Nutrition
Physiological & Biometric Data	Continuous data from wearables (e.g., glucose, sleep, activity).	Correlating food intake with metabolic health markers (e.g., glucose spikes) and energy expenditure.
Dietary & Behavioral Data	Self-reported or digitally logged food intake, preferences, and habits.	Building a comprehensive nutritional intake profile and understanding behavioral patterns.

The Role of Artificial Intelligence in Nutrition

AI, encompassing machine learning (ML), natural language processing (NLP), and deep learning technologies, is the critical enabler of the personalized nutrition revolution.

Data Collection and Integration

Modern AI systems can process massive and diverse data sets:

- Genomic data: Information about an individual's DNA, identifying genetic predispositions to conditions like lactose intolerance, celiac disease, or lipid metabolism issues.
- Metabolomic data: Measurements of metabolites (small molecules) that indicate metabolic health.
- Microbiome profiles: Analysis of gut bacteria compositions that influence digestion, immunity, and inflammation.
- Behavioral and lifestyle data: Activity tracking, sleep patterns, stress levels, and more.
- Dietary intake: Real-time food logging via mobile apps or wearable devices.

AI algorithms can synthesize these disparate inputs, identifying intricate patterns and correlations that would elude human researchers.

Predictive Modeling and Recommendation Engines

Once trained on extensive datasets, AI models can predict individual responses to specific foods or diets. For example, predictive models can forecast blood sugar spikes based on a person's meal composition and gut microbiota. Recommendation engines can then suggest the most beneficial foods, portion sizes, and eating schedules.

Notably, AI also enables dynamic personalization—adjusting recommendations in real-time based on new data, such as changes in body weight, blood biomarkers, or even subjective feelings of satiety or fatigue (Armand, Nfor, & Kim, 2024).

Enhancing Accessibility

In the past, personalized nutrition required expensive laboratory tests and consultations with highly specialized healthcare providers. AI is democratizing access through scalable platforms:

- Mobile apps like ZOE, Lumen, and Nutrigenomix offer personalized insights based on at-home testing.
- Wearables integrated with dietary recommendation systems allow continuous monitoring.
- AI chatbots and virtual dietitians can provide real-time advice, meal planning, and motivation.

Thus, what was once exclusive to elite athletes and affluent individuals is now becoming increasingly available to the general public.

Table 3. Artificial intelligence techniques and their applications in nutrition

AI Technique	Description	Application in Personalized Nutrition
Machine Learning (ML)	Algorithms that learn patterns from data to make predictions.	Predicting personalized glycemic responses to meals; clustering individuals into dietary response types.
Natural Language Processing (NLP)	Enables computers to understand and process human language.	Powering AI-powered nutrition chatbots; automating analysis of dietary intake from text or voice logs.
Computer Vision	Algorithms that extract information from visual data (images/video).	Automated food identification, portion size estimation, and nutrient assessment from smartphone photos.
Recommendation Systems	Systems that filter information to predict user preferences.	Generating hyper-personalized meal plans, recipes, and grocery lists based on user goals and biology.

Personalized Nutrition in Chronic Disease Management

Chronic diseases, often rooted in lifestyle factors including diet, represent a massive global health burden. The standard clinical approach—prescribing generalized dietary advice—has limited efficacy because it fails to address individual variability.

AI-Driven Personalized Nutrition Offers a Transformative Solution

- Type 2 Diabetes: Tailored dietary interventions can significantly improve glycemic control, potentially reducing or even eliminating the need for medication. Predictive models anticipate postprandial (post-meal) glucose responses, allowing individuals to fine-tune their food choices.
- Cardiovascular Disease: Genetic testing can identify people with specific lipid metabolism issues (e.g., familial hypercholesterolemia), allowing for customized dietary fat recommendations.
- Obesity: Recognizing that individuals gain and lose weight differently, personalized plans consider metabolic rates, hormonal profiles, and behavioral patterns.

Ultimately, these precision interventions can improve patient outcomes, enhance adherence to dietary plans, and reduce healthcare costs.

Table 4. The role of personalized nutrition in chronic disease management

Condition	Limitation of Traditional Approach	AI-Personalized Strategy	Potential Benefit
Type 2 Diabetes Mellitus	Generic carbohydrate counting.	ML models predicting individual postprandial glucose responses to specific foods.	Improved glycemic control through personalized food choices, potentially reducing medication needs.
Cardiovascular Disease	Universal advice to reduce saturated fat and sodium.	Genetic profiling for lipid metabolism (e.g., ApoE genotype) to tailor dietary fat recommendations.	More effective, genetically congruent dietary interventions for cholesterol management.
Obesity	Calorie-centric "eat less, move more" model.	Integrated analysis of metabolism, microbiome, and behavioral psychology to identify individual drivers of weight gain.	Sustainable weight management through targeted interventions addressing root causes.

Ethical Considerations and Challenges

Despite its tremendous potential, personalized nutrition powered by AI also presents important challenges:

- Data Privacy: Collecting sensitive biological and behavioral data raises concerns about confidentiality and misuse. Robust encryption, ethical data governance, and transparent consent processes are crucial.
- Algorithmic Bias: If AI systems are trained on non-diverse data sets (e.g., primarily Western populations), recommendations may not be valid for people from other ethnic or geographic backgrounds.
- Accessibility: Although AI can democratize access, disparities still exist. Rural, low-income, or technologically underserved populations may have limited access to AI-driven nutrition services.
- Over-reliance on Technology: While AI is powerful, human intuition and clinical judgment remain essential, especially in complex cases.

Ethical frameworks, inclusive research, and careful regulation will be necessary to ensure that AI-driven personalized nutrition benefits everyone equitably.

Table 5. Ethical and implementation challenges in AI-driven nutrition

Challenge	Definition & Risk	Proposed Mitigation Strategies
Data Privacy & Security	Risk of misuse, re-identification, or breach of highly sensitive health and biological data.	**Anonymization/Pseudonymization,** robust encryption, transparent data governance policies, and explicit informed consent.
Algorithmic Bias	Systemic errors causing less accurate or harmful recommendations for underrepresented demographic groups.	**Curating diverse and representative training datasets,** implementing algorithmic fairness audits, and fostering interdisciplinary oversight.
Equity & Accessibility	The high cost of testing and technology could exacerbate health disparities, creating a "digital divide" in nutrition.	Developing **low-cost, mobile-first solutions,** pursuing public health partnerships, and advocating for policy and insurance coverage.
Clinical Integration & Oversight	Over-reliance on black-box algorithms without human oversight could lead to erroneous or harmful advice.	Adopting a **"human-in-the-loop"** model where AI supports, rather than replaces, registered dietitians and clinicians.

The Future of AI-Driven Personalized Nutrition

Looking ahead, several trends suggest that AI-driven personalized nutrition will continue to evolve rapidly:

- Integration with Wearable Tech: Real-time biometric monitoring (e.g., continuous glucose monitors, heart rate variability sensors) will feed AI systems with continuous data, enabling even finer-tuned recommendations.
- Personalized Supplements: Companies are beginning to offer custom-blended vitamins and probiotics based on individual biological needs.
- Precision Agriculture: Food production may also shift toward growing crops optimized for individual health profiles.
- Behavioral AI: Systems that not only tell you what to eat but coach you on how to stick with dietary changes based on your unique psychological patterns.

Ultimately, personalized nutrition will likely become a core pillar of preventive healthcare, with AI acting as the indispensable engine driving it forward.

THE FOUNDATIONS OF PERSONALIZED NUTRITION

From One-Size-Fits-All to Precision Nutrition

For decades, nutrition science operated under the assumption that universal dietary recommendations could serve all individuals equally well. Food pyramids, daily caloric intakes, and nutrient-based guidelines were designed for the "average" person. However, growing scientific evidence has demonstrated that this one-size-fits-all model often fails to meet the specific needs of diverse individuals. As society becomes increasingly health-conscious, and as chronic diseases linked to diet, such as obesity, diabetes, and cardiovascular conditions, rise sharply, the demand for a more tailored approach has surged (Barbe et al. 2023).

Limitations of Generalized Dietary Guidelines

Traditional dietary recommendations have significant limitations. They are based on large population averages rather than the unique genetic, metabolic, and lifestyle profiles of individuals. As a result, these generalized guidelines may be ineffective, or even counterproductive, for certain groups.

For instance, a low-fat diet was once broadly promoted to combat heart disease. Yet, subsequent research found that such diets do not benefit everyone equally and may even harm those with specific genetic predispositions. Similarly, the blanket advice to increase carbohydrate intake in the form of grains proved problematic for individuals with insulin resistance or type 2 diabetes.

The diverse outcomes witnessed when different people follow the same diet underscore a critical truth: dietary needs are profoundly personal. This realization laid the groundwork for a shift toward precision nutrition — a field that seeks to tailor dietary advice based on an individual's unique biological, genetic, and lifestyle factors.

Table 6. Limitations of the traditional one-size-fits-all dietary model

Generalized Recommendation	Intended Purpose	Documented Limitations & Individual Variability
Low-Fat Diets	Reduce risk of cardiovascular disease.	Benefits are not universal; individuals with specific genetic predispositions (e.g., certain ApoE genotypes) may experience adverse lipid profiles.
High-Grain Carbohydrate Intake	Provide foundational energy and fiber based on average needs.	Can lead to detrimental glycemic spikes in individuals with insulin resistance, prediabetes, or specific gut microbiome compositions.
Standard Daily Caloric Intake	Maintain energy balance for a "reference" individual.	Fails to account for vast differences in Basal Metabolic Rate (BMR), thermic effect of food (TEF), and non-exercise activity thermogenesis (NEAT).
Universal Nutrient RDAs	Prevent deficiencies in the majority of the population.	Ignores variations in nutrient absorption (e.g., vitamin D receptor polymorphisms), metabolism, and individual bioavailability.

The Rise of Nutrigenomics and Metabolic Typing

Nutrigenomics, the study of how genes interact with nutrients, has propelled personalized nutrition into mainstream science. By examining specific genetic variations, researchers can now identify how individuals metabolize different nutrients and how diet influences gene expression. For example, individuals with certain variants of the FTO gene are more prone to obesity when consuming high-fat diets, while others may process fats more efficiently.

Similarly, metabolic typing — the classification of individuals based on their metabolic function — acknowledges that different people burn and utilize nutrients at varying rates. Some individuals may thrive on a high-protein diet, while others perform better with higher carbohydrate intake. By aligning diet with metabolic type, it becomes possible to optimize energy levels, mental clarity, and physical health more effectively (Bhuiyan & Nahid, 2025).

The marriage of nutrigenomics and metabolic typing marks a transformative era. It enables us to move away from generalized dietary templates and toward personalized, dynamic nutritional strategies.

Key Factors Influencing Individual Dietary Responses

While genetics and metabolism form the core of personalized nutrition, several other critical factors also influence how individuals respond to various diets. Understanding these factors is crucial to designing effective, personalized nutrition plans.

Genetics (e.g., Lactose Intolerance, Caffeine Metabolism)

Genetics play a pivotal role in determining dietary needs and sensitivities. Variations in specific genes can impact nutrient metabolism, absorption, and even dietary preferences.

Lactose intolerance is a prime example. It stems from genetic variations that influence lactase enzyme production. Individuals with certain genetic markers lose the ability to digest lactose after childhood, leading to gastrointestinal distress when consuming dairy products. Thus, while dairy may be a recommended source of calcium and protein for many, it is unsuitable for those with lactose intolerance.

Similarly, caffeine metabolism varies widely between individuals based on genetic differences in the CYP1A2 gene. "Fast metabolizers" of caffeine can consume coffee with minimal side effects, while "slow metabolizers" may experience heightened anxiety, insomnia, or increased heart disease risk with even moderate caffeine intake.

Genetic testing today can offer detailed insights into these and many other nutritional traits, paving the way for diets that align with one's inherent biological makeup.

Table 7. Pillars of personalized nutrition: Key influencing factors

Factor	Description	Impact on Dietary Response	Example Assessment Method
Genetics	An individual's inherited DNA sequence, containing variants (SNPs) that influence nutrient metabolism.	Determines predisposition to food intolerances, caffeine sensitivity, and optimal macronutrient ratios.	Direct-to-Consumer (DTC) genetic testing kits (e.g., 23andMe, Nutrigenomix).
Gut Microbiome	The diverse ecosystem of trillions of bacteria, archaea, and fungi in the gastrointestinal tract.	Directly affects fermentation of fibers, production of vitamins (e.g., B, K), inflammation, and glycemic response to meals.	16S rRNA or shotgun metagenomic sequencing of stool samples.
Metabolic Health	The efficiency of the body's biochemical processes to convert food into energy.	Governs individual blood glucose and lipid responses to identical foods; central to conditions like insulin resistance.	Continuous Glucose Monitors (CGMs), blood lipid panels, fasting insulin tests.
Lifestyle & Environment	Modifiable behavioral and external factors including sleep, stress, activity, and diet.	Dramatically alters hormonal balance (cortisol, ghrelin, leptin), energy needs, and food choices, thereby modulating the effect of any diet.	Wearable devices (sleep, activity trackers), self-reported journals, perceived stress scales.

Gut Microbiome (Microbial Diversity Impacts Nutrient Absorption)

The human gut microbiome — the complex community of trillions of microorganisms residing in the digestive tract — plays a central role in digestion, nutrient absorption, and overall health. Research shows that microbial diversity is a strong indicator of a healthy gut.

Different individuals harbor distinct gut microbial profiles, influenced by genetics, diet, environment, and medication history. This diversity affects how efficiently we break down and absorb various nutrients. For instance, certain gut bacteria are better at fermenting dietary fibers into beneficial short-chain fatty acids, while others might contribute to inflammation and metabolic diseases (Han, M., Chen, J., & Zhou, Z., 2024).

A diet that supports microbial balance and diversity may vary from person to person. Personalized nutrition plans can now include microbiome analysis, allowing for targeted dietary interventions such as prebiotics, probiotics, and specific fiber types to optimize gut health and, by extension, systemic health.

Furthermore, recent studies have linked the microbiome to the efficacy of weight loss diets, blood sugar control, and even mental health through the gut-brain axis. Recognizing the microbiome's impact adds a crucial layer of precision to individualized nutrition planning.

Table 8. How genetic variations influence nutritional traits

Genetic Factor	Related Gene(s)	Phenotypic Outcome	Personalized Nutritional Implication
Lactose Intolerance	LCT (Lactase gene)	Reduced ability to digest lactose sugar in dairy products after childhood.	Avoidance of lactose-containing foods; substitution with lactose-free dairy or alternative calcium/vitamin D sources.
Caffeine Metabolism	CYP1A2	Determines whether an individual is a "fast" or "slow" metabolizer of caffeine.	Slow metabolizers should limit caffeine intake to avoid increased risk of anxiety, insomnia, and hypertension.
Obesity Risk & Fat Metabolism	FTO	Certain variants are associated with a higher predisposition to obesity and altered satiety signaling.	May benefit more from a diet higher in protein and fiber to promote satiety, rather than a generic low-fat approach.
Folate Metabolism	MTHFR	Reduced conversion of folic acid to its active form (L-methylfolate), impacting homocysteine levels.	Requirement for pre-formed methylfolate instead of synthetic folic acid in supplements and fortified foods.

Metabolic Health (Blood Sugar Responses, Insulin Sensitivity)

Beyond genetics and the microbiome, metabolic health is another cornerstone of personalized nutrition. Two individuals eating identical meals can experience vastly different postprandial blood sugar responses, driven by differences in insulin sensitivity, gut microbiota, physical fitness, and even sleep quality.

For instance, one person may experience a sharp glucose spike after consuming white bread, while another shows minimal impact. Chronically high post-meal glucose levels are linked to increased risks of diabetes, cardiovascular disease, and obesity.

Innovative technologies, such as continuous glucose monitors (CGMs), allow real-time tracking of blood sugar responses to various foods. This personalized data can guide dietary choices to stabilize blood sugar, enhance energy levels, and prevent metabolic disorders.

Moreover, insulin sensitivity — the body's ability to respond to insulin and regulate blood glucose — varies significantly across individuals. Diets high in refined carbohydrates may severely impair insulin sensitivity in some people, while others may tolerate them well. Personalized dietary interventions targeting improved insulin sensitivity can help mitigate or even reverse metabolic syndromes.

Lifestyle and Environment (Sleep, Stress, Physical Activity)

Nutrition does not operate in isolation. Lifestyle and environmental factors dramatically influence how dietary interventions impact health outcomes.

- Sleep: Poor sleep quality or sleep deprivation can impair glucose metabolism, increase appetite-regulating hormones (like ghrelin), and reduce satiety hormones (like leptin), leading to poor dietary choices and weight gain.
- Stress: Chronic stress elevates cortisol levels, which promotes fat storage, particularly around the abdomen, and increases cravings for high-calorie "comfort foods."
- Physical Activity: Exercise influences energy balance, muscle mass, insulin sensitivity, and nutrient partitioning. Active individuals may require higher caloric intake or different macronutrient distributions compared to sedentary individuals.
- Environment: Access to healthy foods, cultural dietary practices, socioeconomic status, and exposure to pollutants can all shape dietary behavior and nutritional needs.

Personalized nutrition strategies must therefore consider these dynamic lifestyle variables. A high-protein, low-carb diet might suit a sedentary office worker struggling with insulin resistance, whereas an endurance athlete may require a carbohydrate-rich plan for optimal performance (Jeong et al. 2025).

Table 9. The interplay of lifestyle factors and nutritional needs

Lifestyle Factor	Physiological Impact	Effect on Dietary Requirements & Behavior	Personalized Consideration
Sleep Deprivation	Increases ghrelin (hunger hormone), decreases leptin (satiety hormone), impairs glucose metabolism.	Leads to increased caloric intake, preference for high-carb/fat foods, and reduced willpower.	Prioritize sleep hygiene; increase protein intake to promote satiety; manage carbohydrate timing.
Chronic Stress	Elevates cortisol, promoting abdominal fat storage and gluconeogenesis.	Trig cravings for hyper-palatable "comfort foods"; can lead to stress-eating behaviors.	Incorporate adaptogens, ensure adequate magnesium and vitamin C; focus on mindful eating practices.
High-Volume Endurance Training	Significantly increases glycogen depletion and caloric expenditure; creates muscle micro-tears.	Requires higher overall caloric intake, specifically from carbohydrates for fuel and protein for repair.	Implement targeted carbohydrate loading pre-/post-workout; increase protein to ~1.6-2.2g/kg body weight.
Sedentary Occupation	Lowers daily energy expenditure and can reduce insulin sensitivity over time.	Lower caloric needs; macronutrient intake must be carefully managed to avoid weight gain.	Focus on nutrient density over energy density; prioritize fiber and protein; incorporate movement breaks.

THE ROLE OF ARTIFICIAL INTELLIGENCE IN MODERN NUTRITION

The convergence of technology and health sciences has opened unprecedented opportunities for improving human well-being. Among these, the integration of Artificial Intelligence (AI) in nutrition science represents a transformative shift. Traditionally, dietary planning has relied on generalized guidelines—often failing to accommodate the vast diversity in individual needs, preferences, and genetic predispositions. However, AI is now paving the way toward personalized nutrition, offering dynamic, data-driven, and individually tailored dietary solutions. This chapter explores how AI is revolutionizing dietary planning, the core technologies enabling these innovations, and their profound implications for health optimization.

How AI is Revolutionizing Dietary Planning

Predictive Analytics Forecast Health Outcomes Based on Diet

Predictive analytics, a branch of AI, is fundamentally altering how health outcomes are anticipated and managed. By analyzing vast datasets encompassing diet patterns, biometric indicators, genetic information, and lifestyle choices, AI algorithms can forecast potential health trajectories.

For example, machine learning models can predict an individual's likelihood of developing conditions like type-2 diabetes, cardiovascular disease, or obesity based on their current dietary habits. These insights allow for preemptive intervention, enabling individuals and healthcare professionals to modify nutritional plans before adverse health outcomes materialize.

Furthermore, predictive analytics doesn't merely analyze past behavior—it continuously refines its forecasts as new data is fed into the system. Thus, it transitions dietary planning from a reactive model to a proactive and preventive one (Karkuzhali & Senthilkumar, 2024).

Dynamic Adaptation Adjusts Recommendations in Real-Time

Unlike static diet charts or one-time consultations, AI-driven nutrition systems dynamically adapt recommendations in real-time. This capability is crucial because an individual's nutritional needs are not fixed; they fluctuate based on factors such as physical activity levels, stress, sleep patterns, hormonal changes, and even environmental conditions.

For instance, a wearable device might detect a user's elevated heart rate and increased physical exertion during an intense workout session. AI systems can instantly adjust the day's caloric intake or micronutrient balance to support recovery and optimal performance. Similarly, if a user falls sick or experiences digestion issues, the AI can modify meal plans to include soothing, easily digestible foods rich in necessary vitamins and minerals.

Dynamic adaptation ensures that dietary planning remains responsive, personalized, and attuned to the individual's real-world experiences rather than theoretical models.

Personalized Meal Planning Based on Preferences and Restrictions

One of the most consumer-visible applications of AI in nutrition is personalized meal planning. Unlike traditional meal plans that often neglect individual taste preferences, allergies, religious dietary laws, and lifestyle choices, AI-based platforms can create deeply customized meal suggestions.

These systems account for a wide array of inputs:

- Preferences (e.g., vegan, keto, gluten-free, organic foods)
- Allergies and intolerances (e.g., lactose intolerance, nut allergies)
- Medical conditions (e.g., hypertension, anemia, celiac disease)
- Cultural and ethical considerations (e.g., halal, kosher, plant-based diets)

By integrating this information, AI tools not only make nutrition plans more acceptable and sustainable for individuals but also significantly improve adherence rates, which are a common challenge in dietary interventions (Khamesian et al., 2025).

Some advanced applications even incorporate emotional and psychological states. For instance, AI models can detect patterns where emotional stress triggers poor eating habits and recommend foods that are comforting yet healthy during such periods.

Table 10. AI-driven revolution in dietary planning: From static to dynamic

Feature	Traditional Approach	AI-Enhanced Approach	Key Benefit
Temporal Scope	Static, one-time plans (e.g., monthly diet charts)	**Dynamic, real-time adaptation** based on continuous data streams.	Responsive to daily fluctuations in activity, health, and environment.
Predictive Power	Reactive; interventions occur after health issues arise.	**Proactive forecasting** of health outcomes (e.g., diabetes risk) for preemptive intervention.	Shifts focus from treatment to prevention and health optimization.
Level of Personalization	Broad demographic categories (age, sex).	**Hyper-personalized,** integrating genetics, microbiome, preferences, and real-time biometrics.	Dramatically improves user adherence and plan effectiveness.
Input Complexity	Relies on manual, often inaccurate, self-reporting.	**Automated data integration** from wearables, computer vision, and NLP, reducing user burden.	Creates a more comprehensive and objective basis for recommendations.

Core Technologies Powering AI-Driven Nutrition

The transformative impact of AI in nutrition is made possible by a combination of cutting-edge technologies. Each contributes uniquely to understanding, analyzing, and responding to human dietary needs.

Machine Learning (ML) – Analyzes Patterns in Health Data

Machine Learning serves as the backbone of AI-driven nutrition. By training on vast datasets of food intake records, biometric readings, health outcomes, and user behavior, ML models identify intricate patterns and correlations that are invisible to the human eye.

For example, ML can reveal how specific micronutrient deficiencies correlate with fatigue patterns across demographic groups or how carbohydrate consumption affects blood sugar levels differently based on genetic profiles.

There are two key types of ML applications in nutrition:

- Supervised learning, where algorithms are trained on labeled datasets (e.g., food intake and corresponding weight changes) to make predictions.
- Unsupervised learning, where algorithms find hidden structures in unlabeled data (e.g., clustering individuals with similar eating patterns for targeted interventions).

Through continuous learning, ML models improve their recommendations, becoming more accurate and effective over time.

Natural Language Processing (NLP) – Powers Nutrition Chatbots

Natural Language Processing (NLP) has made AI-driven nutrition assistance more accessible and user-friendly through interactive platforms like chatbots and virtual nutrition coaches.

NLP enables machines to understand, interpret, and respond to human language with remarkable accuracy. Users can simply type or speak phrases like "suggest a low-carb dinner" or "I feel bloated after eating dairy products," and the AI can parse these inputs, assess them in the context of existing health data, and provide relevant, actionable advice (Kopitar, et al. 2025).

Some advanced chatbots also offer emotional support by recognizing sentiment in user communications. If a user expresses frustration or stress, the chatbot can respond empathetically and suggest comforting yet healthy food choices, or recommend lifestyle adjustments.

This natural communication bridges the gap between complex nutritional science and everyday decision-making.

Table 11. Core AI technologies powering modern nutritional applications

Technology	Primary Function	Application in Nutrition	Example
Machine Learning (ML)	Identifies complex patterns and predicts outcomes from large datasets.	Predicts glycemic responses, clusters user types, personalizes nutrient needs.	An algorithm that learns how a user's glucose levels respond to different carbs.
Natural Language Processing (NLP)	Understands, interprets, and generates human language.	Powers nutrition chatbots, analyzes food logs from text/voice, provides emotional support.	A user says, "I'm stressed and craving something sweet," and the AI suggests a healthy alternative.
Computer Vision	Extracts information from digital images and videos.	Automates food logging via image recognition, identifies ingredients, estimates portion sizes.	A user takes a photo of their lunch, and the app logs the calories and macros automatically.
Internet of Things (IoT) / Wearables	Provides continuous, real-time streams of physiological and behavioral data.	Informs dynamic dietary adjustments based on live metrics like heart rate, sleep, and activity.	A smartwatch detects a high-stress state, and the AI recommends magnesium-rich foods.

Computer Vision – Food Recognition via Smartphone Apps

Computer Vision technologies have made food tracking easier and more accurate through image-based recognition systems. Users can simply take a photograph of their meal using a smartphone app, and the AI system can identify food items, estimate portion sizes, and log nutritional content.

Such applications dramatically reduce the friction associated with manual food logging, a task often cited as tedious and error-prone.

Key capabilities include:

- Ingredient detection: Identifying complex dishes and parsing them into individual components.
- Portion estimation: Using reference objects or depth-sensing cameras to estimate the quantity consumed.
- Nutritional breakdown: Providing calorie counts and macronutrient/micronutrient compositions.

With constant learning from user corrections and new food databases, computer vision systems continually enhance their accuracy.

IoT and Wearables – Continuous Health Tracking

The Internet of Things (IoT) ecosystem, especially wearables, plays a pivotal role in feeding real-time health data into AI-driven nutrition systems. Devices like smartwatches, fitness trackers, and even smart patches measure parameters such as:

- Heart rate
- Blood glucose levels
- Sleep patterns
- Physical activity
- Hydration status
- Stress markers (e.g., cortisol levels)

By integrating continuous monitoring data, AI systems gain a nuanced understanding of an individual's physiological state and lifestyle. This allows for highly responsive nutritional recommendations that adapt not just to what users eat, but how their bodies are reacting over time.

Some innovations are even exploring non-invasive biochemical monitoring, which could revolutionize nutritional assessment beyond self-reporting.

Table 12. Data integration for holistic AI-driven nutrition

Data Type	Description	How AI Utilizes It	Source Technologies
Biometric Data	Continuous physiological metrics (glucose, heart rate, sleep).	Provides real-time context on the body's state to adapt nutritional advice instantly.	Wearables, IoT sensors, CGMs.
Dietary Data	Information on food consumption (what, when, how much).	Serves as the primary input for analysis, correlation with outcomes, and logging.	Computer Vision, NLP, manual logging.
Personal & Preference Data	Stated goals, allergies, cultural practices (halal, vegan), taste preferences.	Constraints the solution space to ensure recommendations are practical, safe, and adherent.	User profiles, surveys, NLP analysis.
Psychological & Behavioral Data	Mood, stress levels, emotional triggers for eating.	Identifies behavioral patterns to offer targeted support and break negative habits.	NLP (sentiment analysis), wearable stress markers, user input.

Table 13. Overcoming traditional limitations with AI solutions

Traditional Challenge	AI-Driven Solution	Resulting Outcome
Generalized Guidelines	Algorithms model **individual variability** in response to food.	Recommendations are tailored to a person's unique biology, not population averages.
Poor Long-Term Adherence	**Dynamic adaptation** and integration of **user preferences** keep plans engaging and sustainable.	Increased user compliance and long-term success with dietary interventions.
Tedious and Inaccurate Tracking	**Computer Vision** automates food logging; **NLP** simplifies input.	Reduces user burden, leading to more consistent and accurate data collection.
Reactive, Not Proactive	**Predictive analytics** forecast health risks based on current diet and lifestyle patterns.	Enables preemptive dietary changes to prevent disease before it manifests.

DATA SOURCES FOR PERSONALIZED NUTRITION

Personalized nutrition is revolutionizing the way individuals approach health and wellness. Central to this transformation is the diverse range of **data sources** that feed into AI systems, enabling more customized dietary plans. These sources integrate real-time physiological data, genomic insights, microbiome profiles, and behavioral information to create a holistic view of an individual's nutritional needs (Lacruz Pleguezuelos, et al. 2025).

This chapter explores the key data streams that form the foundation of AI-driven personalized nutrition: wearables and continuous health monitoring, genomic and microbiome profiling, and behavioral and lifestyle data integration.

Wearables and Continuous Health Monitoring

One of the most significant technological advancements in recent years has been the rise of wearable devices that monitor a range of physiological parameters in real time. These devices generate continuous streams of health data, enabling dynamic adjustments to dietary plans based on current needs rather than static assumptions.

Continuous Glucose Monitors (CGMs)

Continuous Glucose Monitors (CGMs) have emerged as critical tools, particularly for individuals managing diabetes, but increasingly for the general population aiming for metabolic optimization.

CGMs such as the Dexcom G7 or Abbott's FreeStyle Libre track interstitial glucose levels every few minutes. This continuous feedback reveals how specific foods impact blood sugar levels, highlighting patterns that a traditional single-timepoint blood test could never capture.

By using CGMs, AI algorithms can learn:

- Which foods cause rapid glucose spikes
- How exercise or sleep impacts glucose metabolism
- How stress or illness alters blood sugar control

This data allows real-time dietary recommendations, such as suggesting a lower-glycemic meal after detecting a glucose spike or recommending a walk after a carbohydrate-rich lunch to stabilize blood sugar levels. For the broader healthy population, CGMs are becoming tools for preventive health, helping users maintain metabolic flexibility and avoid insulin resistance.

Fitness Trackers (e.g., Fitbit, Apple Watch)

Devices like the Fitbit, Apple Watch, and Garmin provide another stream of critical data for personalized nutrition through:

- Physical activity tracking (steps, active minutes, exercise type)

- Sleep monitoring (duration, quality, sleep stages)
- Heart rate variability (HRV), a marker for stress and recovery
- Caloric expenditure estimates

Physical activity levels directly influence caloric and macronutrient needs. A sedentary individual requires a different dietary plan than someone training for a marathon. Sleep quality also plays a crucial role, as poor sleep is linked to insulin resistance, cravings for high-calorie foods, and weight gain.

AI-driven dietary systems can use this wearable data to adjust meal timing, suggest foods that improve sleep quality (such as magnesium-rich foods), or recommend pre- and post-exercise nutrition strategies.

Thus, continuous health monitoring through wearables ensures that dietary plans are not only personalized but also dynamic, evolving with the user's changing physiological status.

Table 14. Wearable devices for continuous health monitoring

Device Type	Key Metrics Tracked	Nutritional Application & AI Insight
Continuous Glucose Monitor (CGM) (e.g., Dexcom G7, FreeStyle Libre)	Interstitial glucose levels (every few minutes), trends over time.	Identifies personalized glycemic responses to specific foods, exercise, and sleep. AI uses this for real-time meal adjustments to optimize metabolic health.
Fitness & Activity Tracker (e.g., Apple Watch, Fitbit, Garmin)	Steps, active minutes, exercise type, heart rate, heart rate variability (HRV), estimated caloric expenditure.	Dynamically adjusts daily caloric and macronutrient needs based on energy output. Uses HRV/sleep data to recommend foods that aid recovery and manage stress.
Sleep Tracker (Often integrated into wearables)	Sleep duration, sleep stages (light, deep, REM), restlessness, oxygen variation.	Links poor sleep quality to increased cravings and insulin resistance. AI suggests dietary interventions (e.g., magnesium-rich foods) to improve sleep.

Genomic and Microbiome Profiling

Beyond real-time physiological monitoring, deep biological profiling—particularly genomics and microbiome analysis—offers insights into an individual's fundamental predispositions regarding diet, metabolism, and health (Liu & Xiao, 2025).

DNA-Based Diet Recommendations (23andMe, Nutrigenomix)

Genetic testing services like 23andMe and Nutrigenomix analyze an individual's DNA to provide insights into how their genetic makeup influences nutritional needs and responses. These tests can reveal:

- Lactose intolerance likelihood
- Caffeine metabolism speed
- Tendency for obesity or type 2 diabetes based on genetic risk
- Nutrient absorption efficiency, such as Vitamin D, B12, and iron

For example, a person with a variant in the MTHFR gene may need increased intake of folate-rich foods or supplements. Those with slow caffeine metabolism may be advised to limit coffee intake to prevent cardiovascular issues.

AI-driven systems can integrate these genomic insights with real-time data to fine-tune dietary plans further. For instance, if genetic data suggests a risk for higher LDL cholesterol levels, the AI may recommend a Mediterranean diet rich in healthy fats and fiber.

Nutrigenomics moves us beyond "one-size-fits-all" diets to a model where genetic individuality guides macronutrient balance, food choices, and even meal timing.

Gut Microbiome Analysis (Viome, Zoe)

The human gut microbiome, consisting of trillions of bacteria, viruses, and fungi, plays a central role in digestion, immunity, metabolism, and even mood regulation. Companies like Viome and Zoe offer microbiome testing by analyzing stool samples to determine the microbial composition of an individual's gut.

Gut microbiome analysis can reveal:

- Diversity and balance of gut bacteria
- Presence of beneficial vs. harmful microbes
- Microbial metabolites impacting metabolism or inflammation

Personalized nutrition recommendations based on microbiome data might suggest:

- Increasing prebiotic fiber to boost beneficial bacteria
- Consuming specific probiotics
- Avoiding foods that exacerbate gut dysbiosis

For instance, a user whose microbiome analysis reveals a high abundance of pro-inflammatory species might receive dietary recommendations focused on anti-inflammatory foods such as omega-3 fatty acids, polyphenols, and fermented foods.

By integrating gut microbiome data with wearable monitoring and genomic profiles, AI can develop a comprehensive bio-individual dietary blueprint, supporting gut health and overall systemic wellness.

Table 15. Biological profiling for foundational insights

Data Type & Source	Description & Output	Nutritional Application & AI Insight
Genomic Data (e.g., 23andMe, Nutrigenomix)	Analysis of genetic variants (SNPs) related to metabolism, nutrient absorption, and food sensitivities.	Reveals predispositions (e.g., lactose intolerance, slow caffeine metabolism, MTHFR variants). AI tailors macronutrient balance and specific food recommendations to mitigate genetic risks.
Gut Microbiome Analysis (e.g., ZOE, Viome)	Sequencing of stool sample to identify microbial species, diversity, and functional potential.	Assesses gut health, inflammation potential, and personal food fermentation profiles. AI recommends pre/probiotic foods and avoids those that promote dysbiosis.

Behavioral and Lifestyle Data Integration

While physiological and biological data are critical, behavioral factors—including food choices, eating habits, stress levels, and psychological states—are equally important in shaping nutritional success.

Food Logging Apps (MyFitnessPal, Cronometer)

Food logging apps such as MyFitnessPal, Cronometer, and Lose It! have made it easier than ever for users to record:

- Foods consumed
- Portion sizes
- Nutrient breakdown (calories, macros, vitamins, minerals)

These apps generate valuable data on dietary patterns. AI algorithms can analyze food logs to detect:

- Nutritional deficiencies (e.g., low iron intake)
- Overconsumption patterns (e.g., excess sugar or saturated fat)

- Behavioral trends (e.g., late-night snacking)

By combining food logging with wearable and genomic data, AI can offer actionable recommendations, such as encouraging specific food swaps, portion adjustments, or nutrient fortification strategies.

Advanced platforms can also provide meal suggestions tailored not only to genetic and metabolic profiles but also to the individual's taste preferences, cultural context, and dietary restrictions.

Psychological and Stress Assessments

Eating behavior is profoundly influenced by psychological states such as stress, anxiety, and mood disorders. Chronic stress can drive emotional eating, alter metabolic hormones like cortisol, and impact gut microbiome composition.

Some systems integrate psychological assessments via:

- Stress tracking through wearable devices (HRV, skin conductance)
- Self-report surveys or app-based mental health check-ins
- Voice analysis and smartphone interaction patterns

By factoring in stress levels and emotional health, AI-driven nutrition systems can:

- Adjust dietary plans during high-stress periods (e.g., recommending more anti-inflammatory foods)
- Suggest mindful eating practices
- Promote stress-management techniques alongside nutrition

Thus, behavioral and lifestyle data integration ensures that personalized nutrition is not only physiologically sound but also emotionally intelligent.

Table 16. Synthesizing multi-modal data for AI-driven personalization

Integrated Data Streams	AI Synthesis Function	Personalized Output Example
CGM + Food Logging	Correlates specific foods and meals with individual glycemic responses.	"After your banana smoothie, your glucose spiked 50% more than after an apple with peanut butter. Suggest this as a lower-glycemic alternative for tomorrow's breakfast."
Genetic Data + Microbiome	Cross-references genetic predispositions with current gut health status.	*"Your genetics show a risk for vitamin B12 deficiency, and your microbiome has low producers. Recommend incorporating more fortified nutritional yeast and a sublingual supplement."*
Activity Tracker + Psychological Data	Links high-stress days (low HRV) with reduced activity and poor food choices.	*"I notice your stress is high and activity is low today. Your cortisol may be elevated. Suggest a walk and a magnesium-rich dinner like salmon and spinach to help manage stress."*

AI-DRIVEN DIETARY PLANNING IN PRACTICE

In the digital age, personalized nutrition is no longer a futuristic dream—it is a present reality. Thanks to artificial intelligence (AI) and machine learning (ML), dietary planning has shifted from generic guidelines to individualized blueprints for health optimization. AI-driven dietary planning combines insights from health data, genetics, metabolism, and behavior to offer tailored meal recommendations, real-time adaptive nutrition, and data-driven health interventions. This chapter explores how AI is reshaping nutrition practices, focusing on practical applications, real-world platforms, and case studies that illustrate the transformative impact of technology in dietary management (Matusheski & Giraldi, 2022)..

Machine Learning for Meal Recommendations

Nutrino – AI-Powered Food Suggestions Based on Health Data

Nutrino was among the pioneering platforms that merged AI with nutrition science to create highly personalized dietary recommendations. The platform integrates a user's health profile—including biometric data, dietary habits, allergies, fitness goals, and even medication use—into its machine learning models. Using this comprehensive dataset, Nutrino's algorithms predict what foods are best suited to an individual's unique needs and health goals.

One key innovation from Nutrino was its "FoodPrint" technology, a dynamic and evolving dietary model for each user. This FoodPrint updates based on real-time inputs and feedback, thus continuously optimizing recommendations. By focusing on individual variability, Nutrino moved away from a "one-size-fits-all" nutrition model to truly individualized food guidance.

IBM Watson for Nutrition – Analyzing Research for Optimal Diets

IBM Watson, known for its prowess in fields like oncology and finance, has also been applied to nutritional science. Watson's cognitive computing system can sift through millions of pages of scientific literature, clinical trials, nutritional databases, and user-generated health data to provide evidence-based dietary recommendations.

In practice, IBM Watson for Nutrition assists dietitians and healthcare providers by identifying correlations between diet and disease outcomes, suggesting nutrient-dense foods, and even formulating meal plans based on the latest research findings. Its ability to process vast amounts of information far exceeds human capacity, ensuring that recommendations are not only personalized but also rooted in the most current scientific knowledge (Nossair & El Housni, 2024).

Table 17. Comparative analysis of AI-driven meal recommendation engines

Platform	Core AI Function	Primary Data Inputs	Key Output	Unique Value Proposition
Nutrino	Machine Learning	Biometrics, dietary habits, allergies, goals, medications	Personalized food suggestions; dynamic "FoodPrint"	A holistic, continuously evolving dietary model that integrates medical and lifestyle data.
IBM Watson for Nutrition	Cognitive Computing (NLP)	Scientific literature, clinical trials, nutritional databases	Evidence-based meal plans and dietary correlations	Augments professional expertise with deep, evidence-based insights from vast scientific corpora.

Real-Time Adaptive Nutrition Systems

ZOE – Personalizing Nutrition Based on Blood Fat & Sugar Responses

ZOE, a research-driven health science company, has taken personalization a step further by studying individual metabolic responses to foods. Through large-scale studies like PREDICT, ZOE analyzes how people's blood sugar and blood fat levels change after consuming specific meals. The platform uses AI models trained on this data to predict metabolic responses, helping users make better food choices tailored to their bodies.

Each user undergoes a personalized testing protocol involving blood sample collection, standardized meal consumption, and continuous glucose monitoring (CGM). Based on these insights, ZOE provides dynamic food scores and recommendations, allowing individuals to proactively manage energy levels, gut health, and inflammation.

January AI – Predicting Glycemic Impact of Meals

January AI focuses on glycemic management by leveraging continuous glucose monitoring, food logging, and machine learning. Its platform predicts the glycemic impact of foods before they are consumed. January AI's models are trained on datasets from thousands of users, encompassing different meal types, physical activities, and sleep patterns (Rahman, Papathanail & Mougiakakou, 2025).

Using these predictive insights, January AI helps users preemptively adjust meal composition or activity to maintain optimal glucose levels. The system promotes real-time, adaptive decisions—such as taking a brisk walk after a carbohydrate-rich meal—to mitigate blood sugar spikes. This proactive approach not only supports diabetes management but also benefits anyone aiming for metabolic health optimization.

Table 18. Characteristics of real-time adaptive nutrition systems

Platform	Biomarker Focus	Method of Data Collection	AI-Driven Intervention	Primary Application
ZOE	Blood Sugar & Blood Fat	Personalized testing kit (blood samples), CGM, standardized meals	Predictive model of metabolic response; personalized food scores	Understanding individual metabolic responses to manage energy, gut health, and inflammation.
January AI	Blood Glucose (Glycemia)	Continuous Glucose Monitor (CGM), food logging, activity data	Prediction of glycemic impact *before* meal consumption; behavioral suggestions	Proactive glycemic management through predictive insights and real-time adaptive decisions.

Case Studies of AI Nutrition Platforms

Noom – Behavioral Psychology + AI for Weight Management

Noom uniquely combines AI with behavioral psychology to facilitate sustainable weight loss and health behavior changes. Its AI models analyze user interactions, self-reported data, and engagement patterns to personalize daily coaching tips, motivational nudges, and progress tracking.

Noom's strength lies in its adaptive learning model. The more data a user inputs—meal logs, activity levels, weight changes—the more accurately Noom's AI can tailor content, suggest habit changes, and predict potential pitfalls. Additionally, the platform employs a color-coded food categorization system based on caloric density, helping users make healthier choices with minimal cognitive load (Sudharson, Lakshiya, & Lokesh, 2024).

Case studies have shown that users who consistently engage with Noom's AI-driven platform achieve significant, sustainable weight loss compared to traditional dieting methods. Its success highlights the critical role of psychological support and machine learning in long-term dietary planning.

MyFitnessPal – AI-Driven Food Insights from Logged Data

MyFitnessPal, one of the most popular health apps globally, has incorporated AI to enhance its food logging and nutrition insights features. Initially designed as a manual food diary, the platform now uses image recognition and natural language processing to simplify meal tracking. Users can snap a photo of their meal, and AI suggests probable matches from a vast nutritional database.

Furthermore, MyFitnessPal analyzes logged data to identify trends in users' eating habits, nutrient deficiencies, or caloric imbalances. Over time, the system can prompt users with reminders, nutrition tips, and personalized goals, fostering greater dietary awareness and informed decision-making.

Through machine learning, MyFitnessPal not only reduces the burden of manual tracking but also empowers users with actionable insights, contributing to improved health outcomes.

Levels – CGM-Based Metabolic Health Optimization

Levels represents the cutting edge of real-time metabolic health management. The platform uses continuous glucose monitoring sensors to capture minute-by-minute blood sugar fluctuations. Levels' AI analyzes this data in conjunction with logged meals, physical activities, and sleep patterns to offer users highly personalized feedback.

Unlike traditional glucose monitoring designed for diabetics, Levels targets the general population, aiming to optimize metabolic fitness for improved energy, mental clarity, and long-term health. The app provides "Metabolic Scores" for meals and daily activities, helping users fine-tune their diet and lifestyle in ways that are uniquely beneficial to their physiology (Tsolakidis, Gymnopoulos & Dimitropoulos, 2024).

Case studies reveal that many users discover unexpected glycemic triggers—such as seemingly healthy foods causing sharp glucose spikes—and are able to adjust their habits for better metabolic outcomes. Levels exemplifies the power of real-time, data-driven personalization in dietary planning.

Table 19. Case studies of integrated AI nutrition platforms

Platform	Primary Discipline	Core AI/Data Technology	Mechanism of Personalization	Documented Outcome
Noom	Behavioral Psychology	ML analysis of user engagement, logs, and progress	Adaptive coaching, motivational nudges, and habit change suggestions	Sustainable weight loss through AI-powered behavioral modification and psychological support.
MyFitnessPal	Dietary Logging & Analysis	Computer Vision (image recognition), NLP, trend analysis	Simplified food logging; identification of dietary trends and nutrient imbalances	Reduced tracking burden and provision of actionable insights from logged dietary data.
Levels	Metabolic Science	Continuous Glucose Monitoring (CGM), ML analysis of biometric correlations	"Metabolic Scores" for meals and activities; identification of hidden glycemic triggers	Optimization of metabolic fitness (energy, clarity) for the general population via real-time biofeedback.

CHALLENGES AND ETHICAL CONSIDERATIONS

The rise of Artificial Intelligence (AI) in personalized nutrition and dietary planning has brought significant advances in optimizing individual health. AI-driven systems use vast amounts of data to create tailored nutrition recommendations, ensuring more personalized and effective dietary strategies. However, alongside these advancements, there are numerous challenges and ethical concerns that must be addressed to ensure the responsible and fair implementation of AI technologies in the field of nutrition. This chapter explores three key ethical challenges: privacy and data security, algorithmic bias, and accessibility and equity in AI-driven nutrition systems.

Privacy and Data Security Concerns

One of the most pressing issues surrounding AI in personalized nutrition is the collection and handling of personal health data. Nutrition apps, wearable devices, and other AI-driven tools collect detailed information about users, including their eating habits, physical activity, biometric data, and even genetic information. This data is crucial for creating personalized nutrition plans that optimize health. However, it also raises serious concerns about privacy, ownership, and security (Vir & Madinei, 2024).

Who Owns and Controls Personal Health Data?

The question of ownership and control over personal health data is complex and critical. In many cases, individuals are required to provide sensitive health information to use AI-based nutrition services. However, the control over this data often lies with the developers, tech companies, or service providers. Data ownership is a topic of ongoing debate in the digital health sector. Some argue that the individual providing the data should maintain ownership and control, while others believe that those who process and analyze the data to generate insights should retain rights over it.

Moreover, laws and regulations around data ownership and privacy are still evolving, and in many parts of the world, existing frameworks do not sufficiently address the specifics of health data. In the United States, for example, the Health Insurance Portability and Accountability Act (HIPAA) provides guidelines for handling medical data, but it does not cover all aspects of personal health information, especially when it comes to non-medical health apps and platforms. In the European Union, the General Data Protection Regulation (GDPR) offers stronger protections, but gaps in the law remain for non-health-specific platforms collecting sensitive data.

The lack of clear regulations and universal standards for data ownership in the nutrition space leaves individuals vulnerable to exploitation. If AI companies can sell or share data with third parties without adequate user consent, personal health information could end up in the hands of marketers, insurance companies, or other entities that may use it to make decisions about individuals' health, coverage, or even employment opportunities.

Risks of Data Breaches in Nutrition Apps

Data breaches are another significant risk. Many nutrition apps collect vast amounts of sensitive data, including dietary habits, body weight, activity levels, and sometimes even genetic information. These data sets are valuable targets for cybercriminals looking to steal information for malicious purposes. Breaches in these systems can lead to identity theft, fraud, and the compromise of personal health information (Zhang, et al. 2024).

While many companies invest in cybersecurity to protect user data, the threat of breaches remains prevalent. Furthermore, once sensitive health data is compromised, the damage can be long-lasting and difficult to repair. Unlike credit card information, which can be easily replaced, health data is inherently personal and often irreversible once exposed. This makes ensuring data security a high priority for all stakeholders in the AI-driven nutrition industry.

Algorithmic Bias in Dietary Recommendations

AI models rely on vast data sets to generate nutrition recommendations. These models are designed to identify patterns in data and make predictions based on those patterns. However, the data sets used to train these models can contain biases that influence the output. Algorithmic bias in AI-driven dietary recommendations is an area of increasing concern, particularly when it comes to fairness and inclusivity.

Do AI Models Favor Certain Demographics?

AI models are trained on historical data, and if this data reflects biased patterns or societal inequalities, those biases are often inherited by the models. In the context of nutrition, this means that AI-powered systems might favor certain demographics over others, leading to skewed recommendations. For example, if the training data primarily comes from individuals of a certain ethnicity or socioeconomic background, the AI model may not offer the most effective dietary recommendations for people from other groups.

For instance, many AI-driven dietary recommendations are based on global or Western dietary norms, which may not be suitable for all populations. People from different cultural or regional backgrounds may have different dietary needs and preferences. If an AI system is trained primarily on Western dietary patterns, its recommendations may be ineffective or even culturally inappropriate for individuals from other backgrounds. This could lead to AI models unintentionally reinforcing existing dietary biases and failing to cater to a diverse global population.

Cultural and Socioeconomic Biases in Food Suggestions

Cultural and socioeconomic factors also play a crucial role in how food is perceived and consumed. Many AI models do not take these factors into account adequately. For instance, suggesting expensive or rare foods that are not accessible to individuals from lower-income communities is an example of how AI models can perpetuate inequality. Similarly, food recommendations that do not consider the culinary traditions and preferences of various cultures might be alienating and impractical.

The issue becomes even more complex when we consider food availability. AI models that recommend foods based on global supply chains or trends may not consider the local availability of certain ingredients. For example, recommending high-protein foods like quinoa or chia seeds may not be realistic for individuals in regions where these ingredients are not readily available or affordable. Thus, algorithmic bias can result in dietary suggestions that are not just irrelevant but potentially harmful, as they ignore the user's environment and lifestyle.

To address these biases, AI models need to be trained on more diverse, representative datasets that encompass a wider range of cultural, socioeconomic, and regional factors. Additionally, AI systems should be designed with inclusivity in mind, ensuring that the recommendations provided are practical, culturally sensitive, and equitable for all users.

Accessibility and Equity Issues

AI-driven nutrition systems hold the promise of democratizing health optimization by offering personalized dietary plans tailored to individual needs. However, there are significant challenges related to accessibility and equity that could undermine the potential of these systems.

Is AI-Driven Nutrition Only for the Wealthy?

One of the biggest challenges in AI-driven nutrition is the potential for it to exacerbate existing inequalities. Access to the latest AI technologies often requires a high level of technical infrastructure, which may not be available to everyone. Users need access to smartphones, wearables, and internet connectivity to use many of these AI-powered apps and tools. For individuals in low-income communities or developing countries, this can create a significant barrier to access.

Moreover, many of the services that offer AI-driven dietary planning come with subscription fees or costs associated with purchasing additional features. These costs may further restrict access to AI-driven nutrition solutions to only the wealthiest individuals, leaving marginalized communities without the benefits of personalized health optimization. The divide between those who can afford the latest health technologies and those who cannot could deepen health disparities, particularly in terms of diet-related diseases such as obesity, diabetes, and heart disease.

Bridging the Gap for Underserved Populations

To bridge the gap and ensure that AI-driven nutrition is accessible to all, it is essential to focus on creating affordable and inclusive solutions. Governments, NGOs, and tech companies must work together to develop low-cost or subsidized access to AI nutrition platforms for underserved populations. Furthermore, it is vital to ensure that AI systems are designed to be intuitive and accessible, even for individuals with limited technological literacy.

Education also plays a crucial role in bridging the gap. Providing nutrition education and training on how to use AI tools can empower individuals from disadvantaged backgrounds to make better dietary choices. Community-based initiatives, such as health programs in schools or local health clinics, can also help disseminate AI-driven nutrition advice to those who would otherwise have limited access.

In conclusion, AI-driven personalized nutrition has the potential to revolutionize health optimization, but it must be deployed with a keen awareness of privacy, bias, and accessibility issues. Addressing these challenges through ethical practices, inclusive design, and equitable access will ensure that AI-driven dietary planning truly benefits all individuals, regardless of their background or economic status. By doing so, we can unlock the full potential of AI in creating a healthier, more equitable future for everyone.

FUTURE DIRECTIONS IN AI-POWERED NUTRITION

The intersection of artificial intelligence (AI) and nutrition is setting the stage for groundbreaking advancements in dietary planning and health optimization. The potential of AI to revolutionize personalized nutrition is immense, particularly through the integration of AI in real-time health tracking, decision-making, and biofeedback mechanisms. In this chapter, we explore some of the most promising future directions in AI-powered nutrition, focusing on hyper-personalized supplementation, the gut-brain axis, and smart kitchen technologies.

Hyper-Personalized Supplementation: AI-Designed Vitamins Based on Real-Time Deficiencies

One of the most compelling applications of AI in the future of nutrition is the creation of hyper-personalized supplementation. In the near future, AI-powered systems will not only analyze an individual's overall nutritional needs but also design personalized vitamin and mineral supplements tailored to real-time deficiencies detected in the body. These systems will operate through a combination of continuous health data gathering, machine learning models, and predictive analytics.

Real-Time Data Integration

AI can leverage data from various sources to assess an individual's health status. For example, wearable devices that track biomarkers such as blood glucose, vitamin D levels, heart rate, and even hydration levels can provide valuable data to AI systems. Machine learning algorithms will then process this information and identify any potential deficiencies or excesses in nutrients. By using this data, AI can design a personalized supplementation regimen in real-time.

This continuous feedback loop would allow the AI system to make adjustments as the body's needs change, providing a dynamic approach to nutrition. For instance, if a person experiences a sudden drop in energy levels or shows signs of vitamin D deficiency due to a lack of sunlight exposure, the system could recommend an immediate increase in vitamin D supplementation or suggest dietary changes to address the deficiency.

Precision Supplementation

The ultimate goal of hyper-personalized supplementation is to move away from one-size-fits-all vitamins and minerals that are available today. AI could formulate precise doses of individual vitamins, minerals, amino acids, and other micronutrients based on the person's unique health profile, genetic makeup, lifestyle, and environmental factors. In addition to nutrient levels, AI could also account for the bioavailability of supplements—how well the body can absorb and use the nutrients. This is particularly important because some individuals may require higher doses of specific nutrients due to genetic variations or health conditions that affect nutrient absorption.

AI's Role in Supplementation Personalization

To enhance the impact of personalized supplementation, AI could integrate genetic testing with real-time data analytics. Genetic information can reveal predispositions to certain nutrient deficiencies, such as those found in individuals with certain polymorphisms affecting folate metabolism. Combining this data with real-time nutritional monitoring would enable AI to predict future deficiencies and proactively recommend dietary adjustments before the symptoms manifest.

AI and the Gut-Brain Axis: Probiotic Recommendations for Mental Health

The connection between the gut and brain, commonly known as the gut-brain axis, has gained significant attention in recent years, particularly in the context of mental health. Research has shown that the gut microbiome can have profound effects on mood, cognition, and overall mental well-being. The future of AI-powered nutrition will likely involve leveraging AI to understand the gut microbiome's composition in real-time and make personalized probiotic and dietary recommendations to optimize mental health.

The Role of the Gut Microbiome in Mental Health

The gut microbiome is made up of trillions of bacteria, viruses, and fungi that play a vital role in digestion, immunity, and even brain function. Studies have found that an imbalance in the microbiome—dysbiosis—can contribute to mental health disorders such as depression, anxiety, and even neurodegenerative diseases like Alzheimer's.

AI can play a transformative role in this area by analyzing the complex interactions between the gut microbiome and the brain. Using data from microbiome sequencing, AI models can predict how different combinations of gut bacteria may influence mood, stress levels, and cognitive function. With such knowledge, AI could recommend personalized probiotic regimens and dietary adjustments to help optimize the gut-brain relationship.

Personalized Probiotic Recommendations

Probiotics are live microorganisms that, when consumed in adequate amounts, can provide health benefits, particularly by improving gut health. The key to their effectiveness lies in choosing the right strains of probiotics for individual needs. In the future, AI could analyze an individual's microbiome to identify the specific strains of probiotics that would be most beneficial for that person.

For example, if an individual's microbiome is lacking in strains known to produce serotonin (the "feel-good" neurotransmitter), AI could suggest dietary interventions to restore balance, such as the inclusion of prebiotics and probiotic foods or supplements that promote the growth of serotonin-producing bacteria. By pairing microbiome data with mental health data—such as stress levels, sleep patterns, and cognitive function—AI can create a comprehensive mental wellness plan that addresses both physical and psychological well-being.

Predicting Mental Health Outcomes

AI-powered systems could also track changes in mental health outcomes over time, correlating them with dietary and probiotic interventions. This would allow for personalized, real-time optimization of the gut-brain axis, providing a proactive approach to mental wellness. For instance, if an individual shows signs of increased anxiety or depression, AI could recommend adjusting the microbiome with specific probiotics and a diet rich in gut-healing nutrients to rebalance the gut bacteria and alleviate symptoms.

Integration with Smart Kitchen Technologies

As AI continues to permeate our daily lives, its integration with smart kitchen technologies will play a critical role in optimizing nutrition. The kitchen will become a hub of personalized nutrition, with AI-powered devices suggesting meal plans, helping with ingredient sourcing, and even creating personalized meals tailored to individual preferences and nutritional needs.

AI Refrigerators That Suggest Meals Based on Stocked Ingredients

One of the most exciting advancements in smart kitchen technology is the development of AI refrigerators that can suggest meals based on the ingredients available in the fridge. These refrigerators, equipped with cameras and sensors, will scan the contents of the fridge and analyze the nutritional value of the ingredients. Using this data, AI will be able to generate recipe suggestions that are tailored to an individual's dietary preferences and health goals.

For example, an individual who is low on vitamin C might receive a recommendation for a meal that incorporates citrus fruits or bell peppers, which are rich in this essential nutrient. Similarly, if a person is trying to reduce their sodium intake, the AI-powered system might suggest low-sodium alternatives or offer recipes that use herbs and spices to enhance flavor.

These meal suggestions can be personalized further based on an individual's dietary restrictions, preferences, and goals. AI could even track a person's eating habits over time and learn their preferences, making more accurate recommendations as the system becomes smarter.

3D-Printed Personalized Meals

Another promising development in smart kitchen technology is the use of 3D printing for personalized meal creation. 3D food printers are capable of producing meals with precise ingredient measurements and textures based on an individual's dietary requirements. The AI system behind these printers would analyze the person's nutritional needs, preferences, and even allergies, and then design a meal that meets those needs.

For instance, a person with diabetes might receive a 3D-printed meal that contains the right balance of carbohydrates, fiber, and protein to regulate blood sugar levels. Similarly, someone with a gluten sensitivity could have a personalized gluten-free meal created for them. The possibilities for customization are vast, and as 3D printing technology advances, it could become an integral part of AI-driven nutrition.

The Role of Smart Kitchens in Sustainable Eating

The integration of AI with smart kitchen technologies could also have a significant impact on sustainability. By optimizing ingredient usage, reducing food waste, and ensuring that meals are nutritionally balanced, AI can help individuals make better choices not only for their health but also for the environment. AI-powered systems could recommend sustainable, plant-based alternatives to meat-heavy dishes, helping reduce the carbon footprint of food consumption.

CONCLUSION

The advent of personalized nutrition, powered by Artificial Intelligence (AI), is poised to redefine the future of healthcare and wellness. The integration of AI technologies, such as machine learning and big data analytics, with nutrition science marks a profound shift from traditional, one-size-fits-all dietary recommendations to a more nuanced, individualized approach. This transformation empowers individuals to make informed dietary choices based on their unique genetic makeup, health conditions, lifestyle factors, and real-time physiological data.

At the core of AI-driven personalized nutrition is the ability to process vast amounts of data—from biometric sensors to dietary logs—enabling the creation of highly tailored dietary plans. These plans can be adjusted dynamically as new data is collected, offering an unprecedented level of precision in dietary guidance. Unlike conventional methods that rely on generalized nutrition guidelines, AI technologies harness the power of predictive analytics to optimize nutrition strategies that align with individual health goals, such as weight management, chronic disease prevention, or enhancing athletic performance.

Moreover, the real-time aspect of AI-driven nutrition allows for continuous monitoring and feedback, ensuring that individuals receive constant updates on their dietary habits and health progress. Wearable devices, apps, and smart kitchen tools can track everything from caloric intake to micronutrient levels, providing personalized recommendations that evolve with the user's changing needs. This data-driven approach helps mitigate health risks before they manifest as serious conditions, promoting a proactive stance on health management rather than a reactive one.

However, the fusion of AI with personalized nutrition also raises critical ethical and practical challenges that need to be addressed. The collection and analysis of personal health data involve significant privacy concerns, especially in terms of data ownership, consent, and the potential for misuse. Additionally, the algorithms used in AI nutrition models must be designed to ensure fairness, avoiding biases that could perpetuate health inequalities. It is crucial that AI systems are transparent, accountable, and accessible to all demographics, regardless of socio-economic status or geographical location, to avoid exacerbating existing health disparities.

Looking to the future, the possibilities of AI-driven nutrition are boundless. As the technology continues to evolve, we can anticipate a future where food is not merely a source of energy but a form of optimized medicine. Precision nutrition powered by AI could become a mainstream tool in preventive healthcare, revolutionizing the way we approach diet, disease management, and overall well-being. By combining cutting-edge technology with the timeless principles of nutrition, AI-driven personalized dietary plans have the potential to transform not just individual health but public health on a global scale, ushering in a new era of optimized wellness.

In conclusion, the convergence of personalized nutrition and AI offers a monumental leap forward in healthcare. As we move toward an era of smarter, data-driven health optimization, it is essential to address ethical concerns and ensure that these innovations are used to benefit all individuals equitably. This technological leap not only promises a healthier future but also redefines the very essence of nutrition as a tool for personalized, precision medicine.

REFERENCES

Agrawal, K., Goktas, P., Kumar, N., & Leung, M. F. (2025). Artificial intelligence in personalized nutrition and food manufacturing: A comprehensive review of methods, applications, and future directions. *Frontiers in Nutrition*, *8*, 1636980. DOI: 10.3389/fnut.2025.1636980 PMID: 40771216

Armand, T. P. T., Nfor, K. A., & Kim, H. C. (2024). Applications of artificial intelligence, machine learning, and deep learning in nutrition: A systematic review. *Nutrients*, *16*(7), 1073. DOI: 10.3390/nu16071073 PMID: 38613106

Barbe, A., de Toro-Martín, J., San-Cristobal, R., Garneau, V., Pilon, G., Couture, P., Roy, D., Couillard, C., Marette, A., & Vohl, M.-C. (2023). A discriminant analysis of plasma metabolomics for the assessment of metabolic responsiveness to red raspberry consumption. *Frontiers in Nutrition*, *10*, 1104685. DOI: 10.3389/fnut.2023.1104685 PMID: 37125033

Bhuiyan, M. N. I., & Nahid, M. (2025). Smart nutrition: AI and 3D printing for personalized diets. ScienceDirect. https://www.sciencedirect.com/science/article/pii/S3050843625000332

Han, M., Chen, J., & Zhou, Z. (2024, August). NutrifyAI: An AI-powered system for real-time food detection, nutritional analysis, and personalized meal recommendations. arXiv. https://arxiv.org/abs/2408.10532

Jeong, J., Moon, S.-J., Rachim, V. P., Song, Y. J., Cho, Y.-M., & Park, S.-M. (2025). Enhanced post prandial glycemic response prediction in type 2 diabetes with microbiome data and deep learning. *IEEE Journal of Biomedical and Health Informatics*, 1–11. Advance online publication. DOI: 10.1109/JBHI.2025.3602827 PMID: 40864582

Karkuzhali, S., & Senthilkumar, S. (2024). AI driven personalized nutritional and scams planning. In Advances in medical diagnosis, treatment, and care (AMDTC) book series (Chapter 2). https://doi.org/DOI: 10.4018/979-8-3693-3230-6.ch002

Khamesian, S., Arefeen, A., Carpenter, S. M., & Ghasemzadeh, H. (2025, February). NutriGen: Personalized meal plan generator leveraging large language models to enhance dietary and nutritional adherence. arXiv. https://arxiv.org/abs/2502.20601v2

Kopitar, J., Bedrač, L., Strath, L. J., Bian, J., & Stiglic, G. (2025). Improving personalized meal planning with large language models: Identifying and decomposing compound ingredients. *Nutrients*, *17*(9), 1492. DOI: 10.3390/nu17091492 PMID: 40362801

Lacruz Pleguezuelos, F., Bazán, G. X., Romero-Tapiador, S., Freixer, G., Tolosana, R., Daza, R., Fernández-Díaz, C. M., Molina, S., Crespo, M. C., Laguna, T., Marcos-Zambrano, L. J., Aguilar-Aguilar, E., Fernández-Cabezas, J., Cruz-Gil, S., Fernández, L. P., Vera-Rodriguez, R., Fierrez, J., Ramírez de Molina, A., Ortega-Garcia, J., & Espinosa-Salinas, I. (2025). AI4Food, a feasibility study for the implementation of automated devices in the nutritional advice and follow-up within a weight loss intervention. *Clinical Nutrition (Edinburgh, Lothian)*, *48*, 80–89. Advance online publication. DOI: 10.1016/j.clnu.2025.03.003 PMID: 40168934

Liu, L., & Xiao, Z. (2025, July). Feature-enhanced TResNet for fine-grained food image classification. arXiv. https://arxiv.org/abs/2507.12828 DOI: 10.1109/CSCWD64889.2025.11033381

Matusheski, N., & Giraldi, C. (2022). A 3-D approach to personalised nutrition. *Food Science & Technology (London)*, *36*(2), 34–37. DOI: 10.1002/fsat.3602_8.x

Nossair, A., & El Housni, H. (2024, June). Eating smart: Advancing health informatics with the grounding DINO-based dietary assistant app. International Journal of Scientific and Innovative Studies. https://doi.org/DOI: 10.5281/zenodo.11243881

Rahman, L. A., Papathanail, I., & Mougiakakou, S. (2025, July). Introducing the Swiss food knowledge graph: AI for context-aware nutrition recommendation. arXiv. https://arxiv.org/abs/2507.10156v1

Sudharson, K., Lakshiya, S., & Lokesh, M. (2024, May). Fettle: Unveiling a personalized nutritional odyssey through cutting edge AI and data integration. *IEEE International Conference on Emerging Technologies (INCET)*. https://doi.org/DOI: 10.1109/INCET61516.2024.10592948

Tsolakidis, D. G., Gymnopoulos, L., & Dimitropoulos, K. (2024). Artificial intelligence and machine learning technologies for personalized nutrition: A review. *Informatics (MDPI)*, *11*(3), 62. DOI: 10.3390/informatics11030062

Vir, R., & Madinei, P. (2024, December). ARChef: An iOS-based augmented reality cooking assistant powered by multimodal Gemini LLM. arXiv. https://arxiv.org/abs/2412.00627v2

Zhang, Z.. (2024, December). NGQA: A nutritional graph question answering benchmark for personalized health-aware nutritional reasoning. arXiv. https://arxiv.org/abs/2412.15547v1

KEY TERMS AND DEFINITIONS

AI-Driven Dietary Planning: AI-driven dietary planning refers to the use of artificial intelligence (AI) technologies to create and recommend personalized meal plans based on various inputs, such as an individual's health data, food preferences, lifestyle habits, and nutritional needs. AI algorithms analyze these inputs to generate optimal dietary suggestions, ensuring that an individual's nutrition is aligned with their health goals.

Artificial Intelligence (AI): Artificial Intelligence (AI) refers to the simulation of human intelligence in machines that are programmed to think, learn, and problem-solve like humans. In the context of personalized nutrition, AI is used to analyze large datasets, predict individual health needs, and create customized dietary plans through machine learning and data analytics.

Bioinformatics: Bioinformatics is the application of computational tools to manage, analyze, and interpret biological data, such as genetic sequences and metabolic pathways. In personalized nutrition, bioinformatics helps to process vast amounts of health and genetic data to provide insights into how diet affects individual health.

Dietary Interventions: Dietary interventions refer to changes in an individual's eating habits or food intake designed to improve health outcomes, such as managing chronic diseases, weight loss, or boosting energy levels. Personalized dietary interventions are informed by AI and tailored to the individual's unique needs.

Health Data: Health data includes information related to an individual's physical, mental, and emotional health, such as medical history, genetic data, physical activity levels, sleep patterns, and food preferences. This data is used to create personalized nutrition plans that address specific health goals and conditions.

Health Optimization: Health optimization involves strategies and actions that aim to improve and maintain optimal physical, mental, and emotional health. In the context of personalized nutrition, health optimization refers to using tailored dietary plans to prevent diseases, improve overall well-being, and enhance the body's natural functions.

Machine Learning: Machine learning is a subset of AI that involves the development of algorithms that enable computers to learn from and make decisions based on data. In personalized nutrition, machine learning models can analyze historical dietary data, health outcomes, and genetic factors to predict the most beneficial dietary patterns for individuals.

Metabolic Response: Metabolic response refers to how the body processes and reacts to food intake, including digestion, absorption, and the utilization of nutrients. It varies from person to person based on genetic factors, age, sex, and health conditions, and is an essential consideration in personalized dietary planning.

Nutritional Genomics: Nutritional genomics is the study of the relationship between an individual's genetic makeup and their nutritional needs. It examines how genes influence the metabolism of nutrients and how different diets can affect gene expression, providing insights into personalized nutrition recommendations based on genetic predispositions.

Nutritional Precision: Nutritional precision involves providing the right amount of specific nutrients to meet an individual's exact needs. It is a key concept in personalized nutrition, where AI helps to fine-tune dietary plans based on data-driven insights to optimize health outcomes.

Personalized Nutrition: Personalized nutrition refers to the practice of tailoring dietary recommendations and interventions based on an individual's unique characteristics, such as genetic makeup, lifestyle, health status, and preferences. It aims to optimize an individual's health by considering factors such as nutrient requirements, food sensitivities, and metabolic responses.

Smart Wearables: Smart wearables are electronic devices that monitor and track an individual's health data, such as heart rate, physical activity, sleep patterns, and dietary intake. These devices often work in tandem with AI-driven platforms to collect real-time data for personalized nutrition and dietary planning.

Chapter 12
Customer-Centric E-Grocery Services in India:
Leveraging AI for Personalised Nutrition Solutions

Venkatesh Ganapathy
https://orcid.org/0000-0001-7500-2877
ICFAI Business School, ICFAI Foundation for Higher Education, Bangalore, India

Bharathi S. Gopal
https://orcid.org/0000-0002-0972-8589
ICFAI Business School, ICFAI Foundation for Higher Education, Bangalore, India

ABSTRACT

The convergence of artificial intelligence (AI) and health data is revolutionising how consumers approach nutrition, particularly in the e-grocery sector. Groceries are sold in India through websites and apps designed to engender a unique customer experience. This chapter explores the transformative potential of AI-driven personalised nutrition solutions in e-grocery shopping platforms. Based on recent research and data gleaned from the grocery industry, the chapter examines how machine learning and behavioural analytics can be integrated with dietary preferences and real-time consumption data to deliver hyper-personalised food recommendations to Indians. Findings suggest that AI not only enhances consumer health outcomes but also drives business value through increased engagement and loyalty. However, scalability depends on overcoming barriers like data standardisation, algorithmic transparency, and equitable access.

DOI: 10.4018/979-8-3373-4042-5.ch012

INTRODUCTION

Grocery shopping is no longer the same. The digital world has ensured that customers have a plethora of options to choose from when they indulge in a routine activity like grocery shopping. The on-demand services rendered by e-grocery apps in India in response to changing consumer preferences are altering the grocery ecosystem in a major way. Suddenly, businesses are realising the transformative power of selling groceries online. The omni-channel delivery of grocery products by integrating offline and online channels to deliver a seamless customer experience is rewriting the rules of the game. Smart grocery shopping, powered by AI and health data integration, is further expected to revolutionise nutrition by transforming dietary logs and real-time analytics into hyper-personalised recommendations for customers seeking healthier food choices.

The Indian market is witnessing a new trend with 60% of consumers prioritising personalised nutrition (Accenture 2023). India's health-food sector is projected to hit $10B (RedSeer 2024). This is an area with immense potential, yet most e-grocers overlook this potential or are oblivious to the prospects. E-grocers in India are attempting to diversify with non-grocery-related products like footwear or electronic goods. Pioneers like HealthifyMe and Nutrino demonstrate how vernacular AI, blockchain-backed food traceability, and IoT smart carts can bridge gaps—from Kerala's jackfruit-based diets to diabetic-friendly bundles for 101 million pre-diabetics. In India, millet-based alternatives can be suggested when glucose monitors detect spikes, critical for Punjab, where 27% of adults have diabetes (Sohil et al, 2025). While Western models target keto/gluten-free trends, India demands culturally intelligent solutions addressing regional disparities: combating Rajasthan's 58% anaemia rates or optimising plant-based proteins for vegetarian populations.

Goa's innovative AI-powered rice substitution initiative analyses local health data and dietary patterns to recommend personalised, lower-glycemic alternatives to traditional white rice, such as nutrient-rich red rice or millet blends, helping combat the state's growing diabetes prevalence while preserving cultural food preferences. This hyperlocal approach demonstrates how AI can adapt global nutrition science to regional culinary traditions, creating scalable solutions for India's diverse dietary landscape.

As HealthifyMe CEO Tushar Vashisht asserts, "The winner won't just sell food but solve nutritional crises through ethical AI," urging retailers to leverage India's unique dietary diversity before global players dominate. This chapter decodes the tech stack (machine learning, AR labels), ethical frameworks, and implementation roadmaps to position e-grocers as preventive health partners, not just delivery platforms.

THE GROWING NEED FOR PERSONALISED NUTRITION SOLUTIONS

The rise of Type 2 diabetes in India is alarming (Mohan & Tiwaskar, 2025). There were 32 million cases in 2000. This rose to 63 million in 2012 and has now reached a figure close to 101 million. Once considered a rich man's disease, diabetes has now spread its tentacles to middle-class Indians and also to the poorest of the poor. Even rural areas are witnessing this trend.

Complications from diabetes can be a greater concern. Obesity, hypertension and cardiovascular problems are severe in those with uncontrolled blood sugar levels. Either people are not aware of the harmful effects of untreated diabetes, or there is a genuine dearth of nutritional guidance. Social media channels are also guilty of spreading misinformation, with anyone having access to a social media channel turning into a master of misguidance. Along with genuine information, fake information also abounds. This is a great opportunity for e-grocers to use machine learning and AI to deliver authentic information to their customers. The impact can be modest, but at least a beginning can be made.

Nutrition science has progressed significantly, but obtaining personalised nutritional advice can be time-consuming and costly. Many people cannot afford consultations with dietitians or health coaches, leaving them dependent on generic public health guidance. These broad recommendations often overlook factors such as medical histories or cultural food preferences.

Traditional grocery systems further exacerbate these challenges by operating on a one-size-fits-all model. Shoppers face overwhelming choices without adequate tools to interpret nutritional labels or align purchases with health goals. Retail environments frequently prioritise processed foods through strategic shelf placement and promotions, subtly discouraging healthier options. A similar situation prevails in the options available online. Recommendations are often made to customers using their past purchases, and the grocery apps often appear cluttered with offers and discounts. There is a complete absence of integration between personal health data (allergy profiles, glucose levels, micronutrient deficiencies, etc) and the grocery shopping experience. This eventually leads to consumers making choices driven by commercial interests alone.

The broader ecosystem is characterised by a troubling fragmentation. There is no connection between healthcare providers, nutritionists and food retailers. However, if a connection is established, it can be mutually rewarding for all the stakeholders. The flipside of this debate is obvious – if customers have access to all the healthy choices, what would happen to the business interests of health care providers and nutritionists? In a world that runs on taking advantage of vulnerabilities of the human population to make money, this can be too much to expect. However, any attempt by e-grocers to differentiate their services can be strategically rewarding and develop into a distinctive competency.

At its core, the problem reveals a pressing need: current e-grocery systems lack intelligent, equitable mechanisms to convert health data into personalised, culturally competent, and financially accessible food choices. AI-driven solutions must address not only technical challenges, such as data integration and recommendation accuracy, but also ethical concerns like privacy, bias mitigation, and inclusive design. This chapter investigates how smart grocery systems, powered by responsible AI, can transform this landscape while proposing frameworks to ensure nutrition personalisation benefits to customers.

HOW CAN AI HELP?

Artificial intelligence has a transformative potential to redefine personalised nutrition through smart grocery shopping platforms. The important question that we need to ask is: Can AI-driven systems bridge the gap between nutritional science and everyday food purchasing decisions, with particular attention to scalability? AI can curate hyper-personalised grocery recommendations by analysing existing systems that leverage machine learning, natural language processing, and behavioural analytics. Algorithmic personalisation outperforms traditional one-size-fits-all dietary guidelines, especially for individuals with chronic conditions or specific nutritional needs. Recommendations to those who are convalescing after an illness can be beneficial.

Barriers and opportunities in deploying AI-powered nutrition solutions need to be understood. Dynamic meal planning, allergen detection, and sustainability-aligned shopping present opportunities for e-grocery platforms. However, challenges like algorithmic bias, lack of digital literacy and privacy intrusion abound. Today, customers are wary of their personal information being sold by unscrupulous businesses driven purely by avarice and pecuniary gains. E-grocery businesses are technology-driven. These businesses in India, operated by tech-savvy entrepreneurs, believe in using technology for selling groceries. It would be appropriate to say that the e-grocery apps are technology companies that have chosen to be in the business of selling groceries. Therefore, it is only apt that they use AI to develop a roadmap for offering value-added services to customers. The grocery business is a low-margin business. But it has immense potential to scale up as groceries are everyday necessities for populations anywhere across the globe. No customer would be happy sharing personal health-related information on public platforms. However, the e-grocers can make healthier options available on their apps/ platforms, narrowing the choices for customers to healthier ones. Care must be taken to maintain data integrity even while offering hyper-personalised choices.

REVIEW OF LITERATURE

India's online grocery market is undergoing rapid transformation, driven by rising urbanisation, mobile app usage, and increased adoption of digital payments. According to Technavio, the market is expected to grow by USD 11.02 billion between 2025 and 2029, at a compound annual growth rate (CAGR) of approximately 21%, with trends such as quick commerce, AI integration, and personalised shopping experiences leading the change (Technavio, 2025). Other sources report even more optimistic forecasts. Grand View Research (2025) estimates the market size at USD 8.82 billion in 2024 and projects it to reach USD 77.72 billion by 2030, growing at a CAGR of 44.9%. Similarly, TechSci Research projects growth to USD 53.67 billion by 2030, with quick commerce—offering deliveries within minutes—already accounting for over two-thirds of all e-grocery orders in 2024 and expected to grow at 40% annually (TechSci Research, 2025; Reuters, 2025). The southern region currently dominates in market share, while the western region is showing strong growth. While digital payments are rising, cash-on-delivery (COD) remains prevalent, especially for small or frequent purchases. Key players in the sector include Amazon, BigBasket, Flipkart, JioMart, Zepto, and Blinkit, among others, who are reshaping consumer expectations through innovations in speed, convenience, and personalisation (Technavio, 2025; Grand View Research, 2025).

Digital transformation in shopping is not limited to online shopping and virtual payments alone. Retail stores have realised the importance of using technology to enhance shopping experiences. This is not just true for luxury goods and other speciality products but also for grocery shopping. The grocery retail sector is undergoing a significant digital transformation, impacting both in-store and online shopping experiences, consumer choices, and operational efficiencies.

Online grocery shopping (OGS) is experiencing significant growth, with projections indicating a substantial increase in market share in countries like Finland, where it's expected to reach 5–6% within 4–5 years (Stenius, Sillander, & Eriksson, 2023). This expansion is driven by increased service diversification, technological integration, and the emergence of hybrid shopping models. While challenges like logistics and profitability persist, niche providers, such as meal-kit and surplus food services, are already demonstrating profitable models.

Several data-driven frameworks, AI, VR, AR, IoT and other smart technologies are being used in grocery shopping. These technologies offer varied benefits to both shoppers and retailers. Some studies were experimental and reported findings on the usage of these technologies in real time.

A thorough literature review helped in understanding the use of technology in grocery shopping, particularly in the areas supporting healthy lifestyles.

In one study (Alhamdan et al, 2020), researchers experimented with Smart Glasses integrated with AI and AR to offer a customised shopping experience based on user health profiles. Using Google Collaborators as a platform, the system, called CoShopper, provided real-time nutrition suggestions, warnings for unhealthy selections, and even virtual doctor consultations. It also displayed promotional discounts during the shopping experience.

Jain (2018) developed a prototype application, Smart Grocer—a context-aware, personalised grocery system. It enhanced user experience by eliminating repetitive steps in list creation and streamlined the shopping process.

A study by Sun et al. (2024) used Virtual Reality (VR) to create a 3D immersive grocery shopping experience. It featured product search and navigation tools, product inspection and review, social co-shopping, AI assistants for personalised meal suggestions, and virtual checkouts. The study highlighted consumer reluctance to pay delivery fees for small or nearby purchases and stressed the need to replicate physical shopping's sensory experiences in digital formats.

Food computing integrates technologies like AI, IoT, and data-driven frameworks to address challenges in food production and consumption. Research by Dakhia, Russo, & Merenda (2025) explored applications such as supply chain optimisation, quality control, sustainable practices, and predictive maintenance.

AI plays a transformative role in e-commerce by enabling dynamic pricing, personalised marketing, inventory control, and responsive customer support. In the Indian context, Gupta & Mitra (2023) showed AI's role in enhancing satisfaction and customer loyalty by analysing consumer behaviour and adapting to trends.

Machine learning (ML) is also being used to personalise dietary recommendations. Adhikari & McFadden (2025) showed how ML can analyse historical food habits, demographics, and health data to deliver tailored suggestions that support behavioural change.

Thomas-Francois & Somogyi (2021) conducted a comparative study between US and Chinese consumers and found that adoption of virtual grocery shopping technology was influenced by cultural acceptance. Chinese consumers were more open to adopting new technology, while US consumers showed scepticism, particularly toward food purchases. The model proposed by the researchers emphasised cultural acceptance, trust, privacy, and consumer learning.

Kalnikaitė, Bird, & Rogers (2013) studied decision-making within aisles and found that shoppers use only a few cues, often ignoring extensive product information. Simplicity in mobile app design was preferred, especially for routine decisions. The study questioned the effectiveness of mobile apps for in-store decisions, suggesting that minimal, relevant data supports better decision-making.

Store environment and atmospherics also play a crucial role. Grummon et al. (2023) conducted an experiment comparing purchases in real and mock online grocery stores and found that a store layout promoting healthy options influenced better purchasing behaviour.

The global rise in obesity and diet-related health problems has intensified interest in strategies to encourage healthier eating habits. Recent research increasingly focuses on the interplay of technological nudging, consumer behaviour, and environmental factors in shaping food choices.

Grant et al. (2025) provide a comprehensive systematic review exploring nudging techniques within grocery retail settings. Nudging, broadly defined as subtle modifications in the environment to influence behaviour without restricting choice, is shown to significantly impact consumer food decisions. Their review, spanning 25 studies from 2011 to 2024, highlights the growing role of digital technologies such as virtual reality, mobile applications, and augmented reality in delivering nudges that promote healthy choices. The authors introduce updated taxonomies for nudging interventions and emphasise the critical need for further research on how digital nudges can be optimised to support public health goals. Notably, the review calls for interdisciplinary approaches combining nutrition science, behavioural economics, and technology design to fully harness the potential of nudging in retail environments.

Similarly, Alsaeed et al. (2025) focus on digital nudging within online supermarkets, a rapidly growing sector in grocery shopping. Through a scoping review of 74 papers, the authors identify various digital nudges, including food swap suggestions and shopping cart calorie counters, which encourage healthier and more sustainable food choices. They develop a conceptual framework based on the stimulus-organism-response (SOR) model that links nudging types to cognitive and perceptual processes and subsequent food choices at multiple levels—from individual product selection to overall shopping cart composition. Their findings underscore that digital nudges can effectively guide consumers toward healthier behaviour, but the effectiveness depends on the design and context of the interventions. This highlights the need for personalised, context-aware nudges that respect user preferences and shopping habits.

Jayaswal et al. (2025) address the critical role of artificial intelligence (AI) in nutrition recommender systems, which offer personalised diet recommendations aimed at improving clinical outcomes and preventing diet-related diseases. They note that many individuals lack sufficient nutritional knowledge, leading to unhealthy eating patterns that contribute to obesity and related ailments. AI-powered recommender systems can fill this gap by providing context-aware, reliable, and safe dietary advice. However, the authors caution that challenges remain, such as integrating diverse contextual data, ensuring data privacy, and improving system transparency. Future research is needed to better evaluate the real-world impact of these technologies on health outcomes and user adherence.

Beyond individual-level interventions, Uzoma et al. (2025) explore how urban planning and emerging technologies affect food environments and access to healthy food in Lagos Island, Nigeria. Their qualitative study highlights that restrictive zoning laws, poor transportation infrastructure, and socioeconomic disparities significantly limit access to nutritious foods, particularly in low-income neighbourhoods. Informal markets help mitigate food insecurity but also raise concerns about food safety and sanitation. The authors advocate for urban policies that integrate zoning reforms, transport upgrades, and vendor support to enhance equitable food access. Crucially, they suggest leveraging emerging technologies such as Geographic Information Systems (GIS), AI, and blockchain to map food deserts, optimise supply chains, and improve real-time food availability monitoring. These technological solutions, combined with policy reforms, can help address structural barriers to healthy food accessibility in urban settings.

Rapp and Boldi (2025) investigate the challenges and opportunities in designing persuasive technologies to change eating behaviour. Their project, Promoting Healthy and Sustainable Eating through Interactive and Explainable AI Methods (PHaSE), emphasises that many existing digital interventions focus narrowly on modifying specific behaviours without supporting users' sense-making or lived experience of change. Their research, based on literature review and design workshops, suggests that effective conversational agents should demonstrate empathy, offer clear and trustworthy advice, personalise recommendations to individual differences, and avoid prescriptive tones. Such human-centred design strategies can foster user engagement and promote sustainable behaviour change.

Cammarelle et al. (2025) critically review consumer behaviour toward healthy foods, revealing a complex array of factors influencing food choices. They identify key drivers, including health consciousness, taste preferences, price sensitivity, social influences, and marketing communication. Importantly, the authors highlight the lack of a universally accepted definition of "healthy food," complicating consumer understanding and policy implementation. They call for comprehensive marketing strategies and public policies that address these multifaceted drivers to effectively encourage healthier eating patterns. Furthermore, they recommend future research to quantify the relative influence of these factors to better tailor interventions.

Collectively, these studies illuminate a multifactorial approach to fostering healthy food choices involving behavioural nudging, digital technology, urban planning, and consumer psychology. Digital nudges in both physical and online retail environments show promise in subtly guiding consumers toward healthier products without restricting choice (Grant et al., 2025; Alsaeed et al., 2025). AI-powered recommender systems provide personalised nutrition advice, yet challenges in data integration and transparency must be addressed (Jayaswal et al., 2025). Meanwhile, urban planning remains a foundational determinant of food access, where policy and emerging technologies can collaboratively reduce disparities (Uzoma et al., 2025). Effective behavioural change technologies require empathetic, personalised designs that respect user autonomy and lived experience (Rapp & Boldi, 2025). Lastly, understanding the diverse factors shaping consumer food choices is critical for developing successful marketing and policy initiatives (Cammarelle et al., 2025).

Tsolakidis, Gymnopoulos, & Dimitropoulos (2024) conducted a bibliographic review on technology use in choosing healthy lifestyles. It revealed that technologies like mobile apps, trackers, DNA kits, food recognition tools, meal personalisation apps, and personalised reminders were part of data-driven systems offering nutrition suggestions.

Valenčič et al. (2023) mapped digital nudges (e.g., labels, swaps, default positions) in online grocery environments and identified gaps in UI design and terminology. Later, Valenčič et al. (2024) found that repositioning high-fibre foods as defaults increased purchases without impacting cost or calorie intake.

Braga et al. (2023a) used a simulated online grocery platform to test gamification, showing that budget increases and reward mechanisms (e.g., "crown" scores) improved nutrition quality. In a follow-up RCT, gamified labels consistently led to healthier choices (Braga et al., 2023b).

Shin et al. (2025) implemented a multi-component intervention including labels and AI feedback in Singaporean households, leading to improved diet scores over time (e.g., Nutri-Score improvement from C to B).

Vítor (2024) found that health and environmental consciousness significantly increased healthy and sustainable food choices, particularly in digital grocery settings. (Richardson & Folkvord, 2024) reviewed the effectiveness of digital nudges for sustainability in online grocery shopping. Although individual nudges had limited impact, their combined use significantly influenced choices, especially among environmentally conscious consumers. Luo (2024) noted that general health interest, rather than individual nudges, better predicted healthy choices and emphasised the need for more interactive, personalised UI designs to reduce decision fatigue.

Gružauskas, Burinskienė, & Airapetian (2024) studied consumer preferences for local and temperature-controlled foods. While local sourcing was valued, concerns were raised about handling frozen and chilled items. Fresh produce, meat, poultry, and dairy were especially preferred from local suppliers.

Wolniak, Stecuła, & Aydın (2024) investigated in-store technologies like self-checkout, handheld scanners, mobile apps, AR, and facial recognition. The study applied the Technology Adoption Model (TAM) and found that perceived usefulness and ease of use were key drivers of technology adoption. Technologies like Mobile

Checkout, Scan&Go, and in-store navigation apps were highlighted for enhancing user experience, though concerns around privacy, cost, and regulations were noted.

Siawsolit et al. (2017) introduced a personalised mobile grocery app based on the Health Belief Model. The app used tailored messages to improve food selection. Jansen et al. (2024), in a systematic review of 50 grocery recommender systems, criticised the overuse of preference-based algorithms and advocated for more ethical, inclusive, and consumer-focused designs.

Future research should prioritise interdisciplinary collaboration, integrating insights from nutrition, behavioural science, urban planning, and computer science to design comprehensive, context-sensitive interventions. Evaluations of real-world impact, especially long-term health outcomes and behaviour maintenance, are essential. Additionally, ethical considerations around data privacy, equity, and autonomy must guide the deployment of AI and digital nudging technologies to ensure inclusive and responsible health promotion.

Table 1. Key emerging themes in technology-driven grocery shopping

Theme	Focus Area	Key Studies / Examples
Technological Innovations	AI, AR/VR, IoT, Smart Glasses, predictive systems	Jain (2018), Wolniak et al. (2024)
Enhanced Consumer Experience	Immersive, intuitive, and gamified shopping	Valenčič et al. (2022, 2024), Braga et al. (2023a, 2023b), Kalnikaitė et al. (2013)
Personalisation & Health Integration	Health-based recommendations, smart dietary tools	Dakhia et al. (2025), Tsolakidis et al. (2024)
Consumer Adoption & Behaviour	TAM, cultural acceptance, cognitive and social factors	Thomas-Francois & Somogyi (2021), Kalnikaitė et al. (2013), Wolniak et al. (2024), Luo (2024)
Store Environment & Choice Architecture	Product placement, nudges, interface design	Grummon et al. (2023), Valenčič et al. (2022, 2024), Braga et al. (2023b)
Lifestyle & Behavioural Interventions	Campaigns, community programs, digital feedback loops	Alsaad et al. (2023), Shin et al. (2025), Tsolakidis et al. (2024)
Regional & Product Preferences	Local food preferences, delivery reluctance, and temperature-controlled foods	Gružauskas et al. (2024)

The integration of advanced technologies in grocery retail is increasingly focused on personalising shopping experiences to promote healthier lifestyles. AI and machine learning are being employed to analyse consumer data—ranging from medical histories to shopping habits—to generate tailored dietary recommendations. Systems like *CoShopper*, which uses augmented reality through smart glasses, provide real-time nutritional feedback, while other AI-based platforms automate grocery ordering and suggest healthier alternatives. These innovations not only enhance shopping efficiency but also improve adherence to health goals (Adhikari & McFadden, 2025). Alongside this, the use of digital nudges—such as visual cues, product positioning, and gamified interfaces—is emerging as a strategy to influence consumers towards better food choices. However, the effectiveness of such nudges in online settings is mixed; distractions, decision fatigue, and information overload often reduce their impact (Valenčič et al., 2022, 2024) (Richardson & Folkvord, 2024) (Luo & Schafraad, 2024).

By contrast, offline strategies like strategic product placement and clear labelling continue to show greater influence on healthier purchasing behaviours (Grummon et al., 2023).

Despite the promise of these technologies, barriers to adoption persist. On the consumer side, factors such as cultural resistance, trust issues, and concerns about the quality of perishable goods in virtual environments can hinder the uptake of digital shopping platforms (Thomas-Francois & Somogyi, 2021). Moreover, users often prefer minimal, actionable information during in-store decision-making, casting doubt on the practical utility of overly complex mobile apps (Kalnikaitė, Bird, & Rogers, 2013). From the retailer's perspective, high implementation costs, privacy concerns, and regulatory uncertainty pose significant hurdles to deploying advanced systems like facial recognition and AI-driven recommendations (Wolniak et al., 2024). The lack of policy interventions to regulate unhealthy food marketing further complicates technology's role in promoting healthier habits (Bennett et al., 2024).

On a broader scale, the societal and ethical implications of digital transformation in grocery shopping are becoming increasingly relevant. Programs like the SNAP Online Purchasing Pilot and food-as-medicine initiatives have shown that digital grocery platforms can increase fresh food access and support healthier eating among low-income groups—highlighting the potential of technology as a tool for social equity (He & Li, 2025; Oshman et al., 2024). However, the same marketing algorithms and digital infrastructure can also be used to push unhealthy, discounted foods to vulnerable populations, raising concerns about fairness, manipulation, and corporate responsibility (Bennett et al., 2024; Hodges et al., 2023). These tensions underscore the need for a thoughtful and ethical approach to technological innovation in grocery retail—one that balances commercial interests with consumer wellbeing.

The digital transformation in grocery shopping is a complex shift that goes beyond simple convenience. It's focused on making the shopping experience more personal and healthier, but it faces significant challenges. These hurdles include high implementation costs for retailers, consumer distrust over product quality and data privacy, and ethical dilemmas surrounding the use of technology to market unhealthy items. This shift moves technology from being a simple tool for online orders to a powerful, double-edged sword that can either improve public health and equity or worsen existing problems.

Based on the above, the following conceptual model is proposed.

The factors that enable personalised and health-centric grocery shopping are primarily technology and data-driven systems and digital nudges/UX design. This is moderated by two main barriers influencing consumer choice: consumer-level barriers (distrust, cultural factors, complexity) and retailer-level challenges (high implementation costs, regulatory gaps).

Smart grocery applications leverage advanced technologies like AI, machine learning, and augmented reality to create personalised shopping experiences and promote healthier choices, moving beyond simple online orders towards sophisticated data-driven systems. These applications offer features such as real-time health-based guidance and automated order suggestions to improve efficiency and dietary compliance. However, for such interventions to be truly successful and scalable, a deeper reflection on their economic sustainability, particularly within developing markets where implementation costs and digital accessibility can be significant hurdles, is crucial for their pragmatic appeal. This challenge is particularly pronounced in price-sensitive markets like India, where consumers are highly responsive to cost, making widespread adoption contingent on solutions that are not only technologically advanced but also affordable and demonstrably value-for-money.

Figure 1. Digital transformation in grocery business

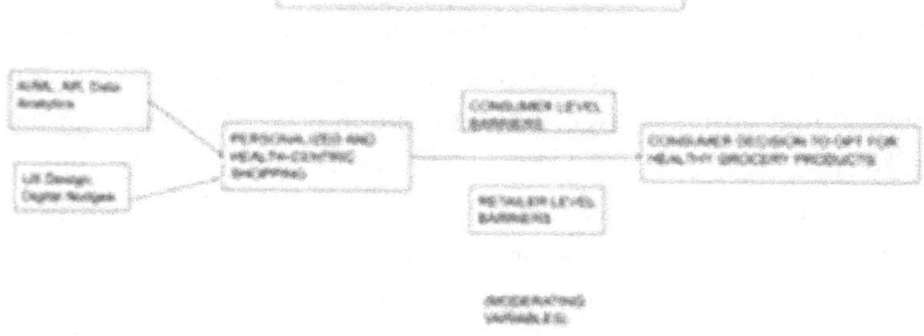

METHODOLOGY

This study uses a qualitative and conceptual approach to explore how AI, health data, and personalised nutrition are being used in online grocery systems.

Relying on existing information—such as research articles, industry reports, and real-world examples like AI-based grocery apps and smart devices- the study attempts to understand current trends, identify gaps, and highlight best practices. The focus is on reviewing and analysing this material to learn how technology is being used and what ethical issues it raises.

SMART GROCERY ECOSYSTEM: COMPONENTS AND INTERACTIONS

The smart grocery ecosystem is changing the way people shop by using artificial intelligence (AI) to support healthier and more personalised food choices. It brings together tools like AI, mobile apps, and supply chain systems to offer a seamless shopping experience that can match individual dietary needs.

At the centre of this system is AI, which processes large amounts of data to give helpful recommendations. By looking at a person's shopping habits, food preferences, and even health conditions, AI can suggest grocery items that support better nutrition. For example, if someone regularly buys high-sodium snacks, the system might suggest lower-sodium alternatives. As customers give feedback—such as accepting or rejecting suggestions—AI systems continue to learn and improve future recommendations.

Mobile apps play an important role in delivering these AI-powered suggestions to users. Some apps allow customers to scan food labels or barcodes to quickly see if a product fits their dietary goals, such as being low in sugar or high in fibre. Others may highlight healthier options while the person shops online or even in-store. Some advanced features include visual cues like "This item fits your 1500-calorie daily goal" or tips for heart-friendly products.

To make the experience more engaging, some platforms use gamification—rewarding users for making healthy choices or completing goals like "buying three servings of vegetables this week." These small nudges can encourage better habits over time.

Overall, by using AI to understand customer needs and guide decisions, e-grocery platforms can play a major role in promoting healthier eating and building long-term, sustainable shopping behaviours.

Figure 2. Personalized grocery shopping funnel

SMART GROCERY SYSTEMS: AI, DELIVERY, AND CONSUMER EXPERIENCE

Smart grocery systems use artificial intelligence (AI) to make shopping more personalised, efficient, and health-oriented. AI algorithms can analyse customer data, such as shopping history and dietary preferences, to suggest products that better meet individual health needs. For example, if a customer often buys snacks, the system might recommend healthier alternatives or notify them of missing ingredients for a planned meal.

These systems also help manage inventory by predicting food trends in different regions. If there's a rise in demand for gluten-free or organic products, AI helps stores adjust their stock accordingly, reducing waste and improving availability.

Delivery services powered by AI utilise real-time traffic and weather data to plan the quickest and safest routes—especially for perishable or time-sensitive items. For example, if someone orders fresh vegetables or baby formula, the system ensures these are delivered promptly to maintain quality and avoid delays.

Augmented reality (AR) is also being explored in some platforms, offering features like real-time product information or alerts about nutritional limits while shopping. Automation and robotics can support this ecosystem by handling routine tasks like restocking shelves, managing online orders, or providing customer service in stores.

Despite these benefits, there are some challenges. Constant digital notifications and suggestions can feel overwhelming to some users. Privacy concerns are also significant, especially if personal data is used for commercial purposes without clear consent. Another issue is algorithmic bias, where the system may not always provide accurate or relevant suggestions for all users—especially when it fails to consider diverse food preferences or dietary habits.

While smart grocery tools offer greater convenience and more personalised choices, their design and implementation must ensure that they support real customer needs, protect data privacy, and avoid overcomplicating the shopping experience.

Figure 3. The intersection of AI, AR, and robotics in retail

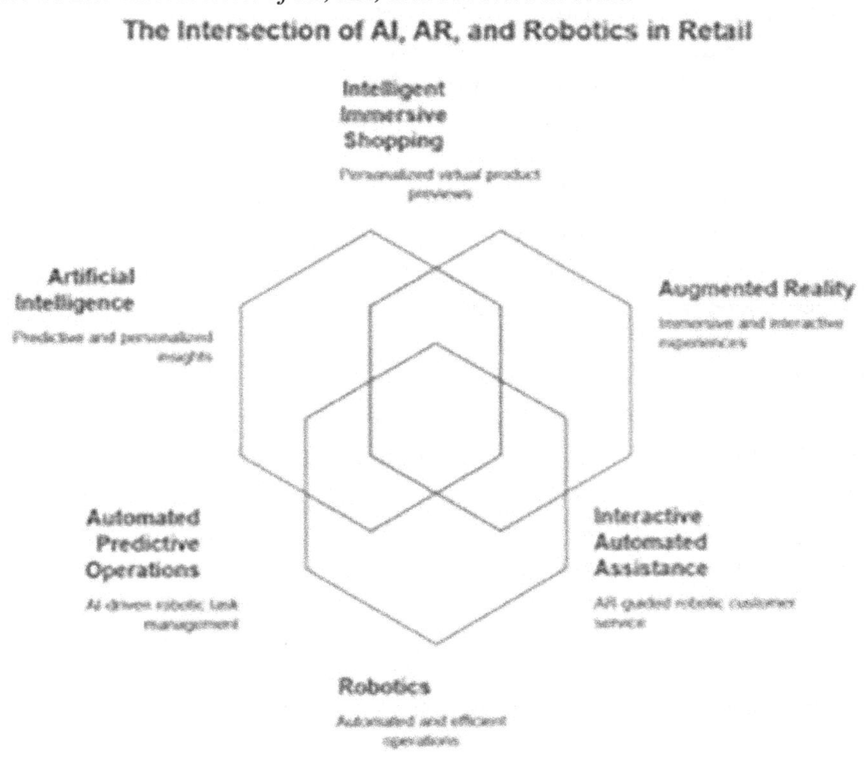

RESPONSIBLE AI: DEALING WITH PRIVACY AND DIGITAL CHALLENGES

Emerging innovations like **AI-powered nutrition recommendations** and **autonomous grocery delivery systems** are reshaping the way people shop and eat. However, as AI becomes more involved in personal grocery choices, it's important to make sure these systems remain responsible and respectful of user preferences.

Smart grocery platforms collect and analyse sensitive data such as shopping behaviour, health-related choices, and food habits. While this allows the system to offer highly personalised suggestions, it also raises concerns about **privacy, data security, and consent**. Users may not always be aware of how their information is being stored, used, or shared. For example, if a grocery platform shares purchase history with other companies, it could impact how individuals are marketed to or even how services are priced.

To maintain **user trust**, platforms should use **clear and transparent consent processes**, such as easy-to-understand permissions and opt-in options. Strong data protection practices—like encryption, data anonymisation, and access controls—are also necessary to prevent misuse. These actions are not just good practice but are often required under laws like **GDPR** and **HIPAA**.

GDPR (General Data Protection Regulation) is a data privacy law enforced across the European Union (EU). It gives individuals greater control over their personal data and holds organisations accountable for how they collect, store, and use that data. Under GDPR, companies must be transparent about data usage, obtain clear consent, allow users to access or delete their data, and ensure strong security measures. Even non-EU companies must comply if they handle data from EU residents. The regulation is designed to protect user privacy in an increasingly digital world.

HIPAA (Health Insurance Portability and Accountability Act) is a U.S. law that protects the confidentiality and security of individuals' health information. It applies to healthcare providers, insurers, and any organisations that handle or transmit personal health data. HIPAA ensures that sensitive medical information—such as diagnoses, treatments, or medical histories—is kept private and secure. It also gives patients' right over their health records, including the ability to access or correct them. Organisations must follow strict rules to prevent misuse or breaches of health data.

ALGORITHMIC BIAS AND TRANSPARENCY

Another major issue is the potential for **algorithmic bias**. AI systems rely on training data, and if this data lacks variety, the recommendations may not suit everyone. For example, if an algorithm is trained mostly on data from one region or diet style, it might not suggest options that align with local food preferences or diverse dietary habits. Additionally, when price is a factor in AI recommendations, the system might suggest cheaper, less healthy products over more nutritious ones.

To improve reliability, developers must **diversify training data**, regularly conduct **fairness checks**, and allow users to **customise or override AI suggestions**. Explaining why a certain product is recommended—through **transparent logic or user-facing explanations**—can help users feel more in control and increase trust in the system.

While AI offers powerful tools to personalise and streamline grocery shopping, it must be implemented with responsibility, especially when dealing with sensitive personal data and health-related choices. A careful balance between innovation and user protection will be essential for these systems to succeed and be trusted.

CASE STUDIES / INDUSTRY INSIGHTS

As the awareness of health has seen a big leap over the years, so has the craze for diet programs, knowledge about food, usage of health monitoring apps and others. Several health-conscious individuals are seeking more precise information on the foods that they consume. Many are questioning if they are even eating right. This has led to an increased awareness and consultation with nutritionists. However, the challenge remains in accessing the right kind of food, of understanding substitutes if a suggested food is not available.

In the USA, several supermarkets such as Albertsons, Kroger, Meijer, ShopRite and others have understood this requirement and appointed dieticians who are also labelled as retail dieticians. This move is to help shoppers choose the right food products for consumption. Besides, shoppers continued to receive support from the dieticians through customised plans, videos and consultations, both online and in the shop. These in-store dieticians help in choosing the right food products, cooking demonstrations if needed, and keep the shoppers focused on health and nutrition and not distracted by the variety of products and confusing claims. Alternatively, these dieticians also reach out to local communities, schools and libraries as part of their service. These dieticians extend their services by offering virtual classes, one-to-one counselling services and help retailers to provide displays in the store that offer nutritional information and encourage healthy habits.

While these are initiatives taken by the private sector to expand their value offering through services and are connected to increasing footfalls and store sales, a comprehensive framework is needed to include such services for not just the nutrition-conscious but also those suffering from health disorders. One such study, titled *Nutrition Counselling for Hypertension Within a Grocery Store: An Example of the Patient-Centred Medical Neighbourhood Model,* was conducted to verify the effectiveness of providing dietary counselling for hypertensive patients in a grocery store. The study was spread over 12 weeks and included 30 hypertensive shoppers who received guidance from a registered dietitian nutritionist. The outcomes of this study in the form of total consumption of fruit, greens and beans, whole grains and a reduction in consumption of empty calories and refined sugars were witnessed. This can be an effective way of inducing lifestyle interventions. Similar studies reported that supermarket-based and online tools can help in inducing better dietary habits. While the supermarket intervention involved customised shopping based on dietary requirements, the online tools, such as recipes, home delivery services, shopping platforms, and meal planning resources, complemented the intervention.

Apart from the initiatives by the supermarkets to include online tools and in-store dieticians, are there any challenges that must be addressed to introduce healthy eating among shoppers? Visual merchandising experts point out the store aesthetics that include layout, wide aisles, lighting, music, displays and other aspects of the physical environment of the store that can uplift the shoppers and help in-store sales. While shoppers wish to eat healthy, time must be factored in for cooking. To address this issue, Levinson offers frozen foods such as sweet potato fries and cauliflower crust pizza and shredded Brussels sprouts for the time-strapped shoppers. Similarly, another supermarket, Sprouts, offers fresh, local produce at reasonable prices. The largest part of the store is devoted to fresh produce, thus encouraging shoppers to eat healthy.

A pioneer in the healthy grocery market is Whole Foods, which has a strict policy on quality standards while procuring produce and products for its stores. Suppliers are required to adhere to their policy of no artificial flavours or colours and ingredients such as hydrogenated oils and high-fructose corn syrup in their supplies. The stores also feature a salad bar that has a variety of fresh foods in ready-to-eat options. Another grocery player, Raley's, has invested in shopper education through newsletters and shelf tags and icons to indicate gluten-free, vegan, plant-based, and heart-healthy options. It also encourages young shoppers by offering free fruits instead of cookies and sweets. Stores are also conscious of packaging, reusability and the brands they partner with. For instance, Aldi, a global grocery giant, stocks exclusive brands such as Simply Nature and LiveGFree. Another brand from Texas, H-E-B, believes in narrating the story behind sourcing the food from the farmers and also offers affordable nutrient products along with counselling by dieticians. Another similar company is the Midwest retailer Hyvee, which works with local farmers and sources within a radius of 200 kilometres. Its seafood is raised or harvested with minimal damage to the environment. Heinen, another grocery retailer, is a step ahead of others. It has a dedicated Chief Medical Officer who works with shoppers to make product recommendations along with vitamins, supplements and homoeopathic products. It also has a Chief Dietitian and certified wellness consultants who are trained to help customers achieve their wellness goals.

While supermarket giants are gearing up for such initiatives, the market is flooded with several meal and diet programs, multiple information sites and brands that claim to be healthy. This has contributed to the confusion of shoppers. Kroger, to help shoppers, launched the OptUP Program to help make informed grocery purchases. It simplifies the process of nutritious options based on inputs from registered dieticians. It also offers a simple rating system, FoodHealth Score, which scores foods on a scale of 1-100 for their nutritional value. Sincerely Health is another such initiative by Albertsons. This is a platform to help make similar choices based on a health score.

In conclusion, while companies, especially grocery chains, are beginning to blend their products with services to help shoppers be healthy, and use digital tools to keep their customers engaged in nutrition conversations, it is important to note that these companies' underlying motive remains profits while enhancing their brand vision. To what extent these initiatives contribute to the overarching goals of health and nutrition of the community is yet to be realised.

BENEFITS AND OPPORTUNITIES OF AI IN SMART GROCERY PLATFORMS

The integration of artificial intelligence (AI) and health data into smart grocery platforms offers major advantages for consumers, healthcare systems, and retailers. By using data to provide personalised nutrition advice, these systems can support healthier eating habits, help manage medical conditions, and create new business opportunities.

AI-powered grocery platforms can make a big difference for people with conditions like diabetes, high blood pressure, or obesity. By analysing shopping habits and health information, the system can suggest better food options. For example, it might recommend low-sugar foods for someone with diabetes or low-sodium items for someone with high blood pressure. This helps people follow their doctor's advice more easily in everyday life. These smart systems act as a bridge between clinical advice and real-world choices.

Beyond helping individuals, the data collected (in anonymous form) can show health professionals larger patterns in food choices. This can help shape public health efforts—like nutrition education campaigns or partnerships between health groups and grocery stores to promote better eating.

Smart grocery systems also play a role in prevention. Instead of only reacting to illness, they help users make healthy choices before problems arise. The AI might spot missing nutrients in a user's diet and suggest foods to fill the gap. It could also offer portion control tips or create meal plans based on personal goals. Some apps even use rewards or group challenges to keep users motivated. These features help people build better eating habits over time.

From a business perspective, personalised grocery systems help retailers attract and keep customers. When the platform knows your preferences, allergies, or dietary needs, shopping becomes easier and more convenient. This kind of service builds customer loyalty. Businesses can also offer health-based meal kits, bundles for specific conditions, or subscriptions for coaching, opening new ways to earn revenue. At the same time, AI helps manage inventory better—predicting demand and reducing waste by stocking only what's needed.

However, there are challenges too. In countries like India, where price sensitivity is high, it can be costly for stores to set up and run advanced AI systems. Many shoppers are still unsure about food quality—especially for fresh produce—and worry about how their data is used. To win their trust, companies need to be clear about the benefits and ensure that the service truly adds value to everyday life, not just convenience.

As this technology spreads, it's also important to consider its long-term impact. Issues like data misuse, consumer over-targeting, or the loss of traditional food habits need careful attention. These systems must be built with responsibility, so they continue to support—not replace—good decision-making.

There are also exciting new possibilities for collaboration. Insurance companies could offer discounts to customers who shop for healthier food, and employers might include these systems in workplace wellness programs. Such efforts not only help people make better choices but also strengthen the business case for smart grocery systems.

For all of this to work well, companies must follow data privacy laws. In the European Union, the **General Data Protection Regulation (GDPR)** requires businesses to ask for clear permission before collecting personal data, use only what's necessary, and allow people to delete their data if they choose. In the United States, **HIPAA** protects medical information, and companies that work with health providers must follow strict rules to keep that data secure. Following these rules builds trust and helps companies avoid fines or reputational damage.

As more countries adopt these technologies, policies must also keep up. Laws should address how AI makes decisions and how health data is handled across borders. Policymakers should work with technology and health experts to create fair rules that support innovation without putting people's privacy at risk.

Governments can help by supporting startups, encouraging partnerships, and creating simple, clear rules for using health data in grocery platforms. Promoting digital literacy and affordable internet access will also help more people take advantage of these new tools. With the right balance, AI-powered grocery systems can offer smarter choices, better health, and new business opportunities for the future.

FUTURE RESEARCH DIRECTIONS

As AI-driven personalised nutrition continues to evolve in the e-grocery space, several important areas for future research are emerging. One key focus should be how new technologies—such as generative AI for real-time meal planning, augmented reality (AR) for product information, and blockchain for tracking food sourcing—can improve the grocery shopping experience. Researchers should examine how these tools affect long-term eating habits and whether users become overly dependent on them. Studying the actual health outcomes for people using AI-guided diets, especially those managing chronic conditions, will help determine if these systems are more effective than traditional nutrition advice.

Understanding user concerns about privacy and control is also important. Future studies should explore how people feel about sharing personal data in exchange for personalised recommendations, and whether clear explanations from the system increase user trust. Research should also look at whether people understand how AI makes food suggestions and whether they feel comfortable with these decisions.

From a business perspective, companies need research on how to make the most of AI. This includes using customer data to predict demand for specific dietary products—like plant-based or keto items—and evaluating whether personalised nutrition subscriptions provide good returns on investment. Another area is how grocery platforms can partner with healthcare providers, insurers, or employers to offer discounted healthy food packages, especially for people with specific medical needs.

Global expansion is another challenge worth exploring. Future research could study how these AI tools can be adapted for different cultures and regions, especially in markets with limited digital access or unique food habits. This would help companies understand how to scale their services effectively while being sensitive to local needs.

Looking further ahead, personalised grocery shopping may move toward even deeper customisation through DNA-based nutrition (also known as nutrigenomics). For example, a person with a certain gene variant might get suggestions for foods rich in specific nutrients like folate. Innovations in voice and visual AI could also improve access—for instance, voice assistants that speak regional languages like

Hindi or Tamil, or mobile tools that scan local product labels and provide instant nutrition guidance.

Lastly, future systems could combine personal health goals with environmental awareness. AI might suggest locally grown, nutritious foods like millets instead of imported items, helping users make choices that are good for both their health and the planet. Together, these research directions can guide smarter, more user-friendly, and sustainable e-grocery platforms in the years ahead.

CONCLUSION

Smart grocery shopping leverages AI to integrate health records, dietary preferences, and purchase history into a unified nutrition profile that powers personalised recommendations. Advanced AI systems analyse this data to suggest recipes and flag incompatible ingredients in real-time, helping consumers make better dietary choices.

Behavioural nudges—from dynamic shopping lists tailored to health goals to in-store AR prompts highlighting healthier options—guide shoppers toward optimal nutrition, bridging the gap between scientific knowledge and everyday decisions.

AI-driven smart grocery platforms represent a major transformation at the intersection of food, health, and technology. By delivering highly personalised nutrition solutions, these innovations can revolutionise how consumers manage chronic conditions and improve long-term wellness. The convergence of AI-powered recommendations and streamlined e-grocery logistics benefits both consumers—through tailored support—and retailers, by unlocking new revenue streams and improving customer engagement.

India offers a particularly promising market for AI-powered grocery personalisation. Rapid digital adoption, a rising burden of lifestyle diseases, and diverse culinary traditions create a strong demand for scalable, localised solutions—ranging from diabetes-friendly meal plans to regional recipe adaptations. E-grocers focusing on this space can differentiate themselves while contributing to improved public health outcomes. The future of grocery shopping is not just convenience but smart, nutrition-driven decision-making that helps people thrive.

AI is also transforming supply chains and store operations. By analysing customer data, retailers can personalise product suggestions and promotions. AI-powered tools such as smart carts enable real-time item tracking, automatic billing, and self-checkout, enhancing convenience and reducing wait times. Additionally, AI improves inventory management by predicting demand and reducing waste, while optimising energy use to boost sustainability.

The global opportunity for AI-driven grocery personalisation is immense, and India's unique market dynamics make it an ideal testing ground for culturally aware, nutrition-focused solutions. With increasing consumer demand for healthier food choices and a booming preventive healthcare market, forward-thinking e-grocers must prioritise AI tools that understand local diets and preferences before competitors dominate this space.

To fully realise the potential of AI-powered personalised nutrition, collaboration among tech developers, nutrition experts, and retailers is essential. Transparent AI systems that provide users with control over their data and clear explanations of recommendations will build trust. Investments in diverse training data and algorithm audits will improve accuracy and relevance. Partnerships with healthcare providers can further enhance the credibility of dietary guidance.

The fusion of AI and health data is revolutionising grocery retail, with India poised to lead this transformation. Urban consumers are increasingly shopping online and open to sharing health data for personalised offers, creating strong demand for AI-powered nutrition solutions. Market leaders will be those who solve India's unique nutrition challenges with hyperlocal food integrations, affordable precision nutrition, and culturally intelligent interfaces.

This shift is more than smarter grocery shopping—it's about becoming a key player in preventive healthcare. The e-grocer that succeeds in this space could redefine nutrition in India and beyond, creating a scalable model for healthier living powered by AI.

REFERENCES

Accenture. (2023). *The resilient consumer: How companies can embrace human complexity to unlock value*. https://www.accenture.com/content/dam/accenture/final/industry/consumer-goods-and-services/document/Accenture-Resilient-Consumer-Study-2023.pdf

Adhikari, S., & McFadden, B. R. (2025). Bridging taste and health: the role of machine learning in consumer food selection. *International Food and Agribusiness Management Review*, 1(aop), 1–16.

Alhamdan, Y., Alabachi, S., & Khan, N. (2020, December). CoShopper-leveraging artificial intelligence for an enhanced augmented reality grocery shopping experience. In *2020 IEEE International Conference on Artificial Intelligence and Virtual Reality (AIVR)* (pp. 337–338). IEEE.

Alsaeed, M. M., Adam, M. T., Piper, J., & Bucher, T. (2025). Digital nudging for healthy and sustainable food choice in online supermarkets: A scoping review. *AIS Transactions on Human-Computer Interaction*, *17*(2), 287–332.

Bennett, R., Driessen, C., Zorbas, C., Sacks, G., Gupta, A., Cameron, A., & Backholer, K. (2024). 'Healthier options tend to get lost in the noise of online'–Australian shoppers' experiences with online grocery platforms. *Public Health Nutrition*, *27*(1), e134. PMID: 38742445

Braga, B. C., Cash, S. B., Sarson, K., Chang, R., Mosca, A., & Wilson, N. L. (2023). The creation of an online grocery store for experimental purposes: A pilot study. *Food Quality and Preference*, *109*, 104888.

Cammarelle, A., Viscecchia, R., De Devitiis, B., Panniello, U., & Gorgoglione, M. (2025). Consumers' behavior toward healthy foods: A critical review. *Food Reviews International*, *41*(4), 1081–1098.

Dakhia, Z., Russo, M., & Merenda, M. (2025). AI-Enabled IoT for Food Computing: Challenges, Opportunities, and Future Directions. *Sensors (Basel)*, *25*(7), 2147. PMID: 40218659

Grand View Research. (2025). *India online grocery market size, share & trends analysis report, 2025–2030*. https://www.grandviewresearch.com/industry-analysis/india-online-grocery-market

Grant, K., Hollywood, L., Simmons, G., Bucher, T., & Burns, A. (2025). Nudging and food shopping: A review of technological interventions within the grocery environment. *Critical Reviews in Food Science and Nutrition*, •••, 1–18. PMID: 40661040

Grummon, A. H., Tucker, A. C., Noe, V., Rummo, P. E., Prestemon, C. E., Hall, M. G., & Taillie, L. S. (2023). Consumer behaviour and experiences in a naturalistic online grocery store: Implications for nutrition research. *Journal of Nutritional Science*, *12*, e36. PMID: 37008411

Gružauskas, V., Burinskienė, A., & Airapetian, A. (2024). Digital transformation in food retail: A case study of Lithuania egrocery buying behaviours. *Entrepreneurship and Sustainability Issues*, *11*(3), 65–84.

Gupta, S. R., & Mitra, S. (2023). AI-powered transformation in Indian e-commerce: Blinkit's innovative leap. *Global Management Journal*, *17*(1/2), 105–110.

He, C., & Li, Z. (2025). *From food deserts to fresh food: The role of digital accessibility in nutrition assistance programs*. Kilts Center at Chicago Booth Marketing Data Center Paper.

Hodges, L., Lowery, C. M., Patel, P., McInnis, J., & Zhang, Q. (2023). A systematic review of marketing practices used in online grocery shopping: Implications for WIC online ordering. *Nutrients*, *15*(2), 446. PMID: 36678317

Jain, R. (2018). *SmartGrocer: A context-aware personalized grocery system* (Doctoral dissertation).

Jansen, L. Z., Bennin, K. E., van Kleef, E., & Van Loo, E. J. (2024). Online grocery shopping recommender systems: Common approaches and practices. *Computers in Human Behavior*, *159*, 108336.

Jayaswal, R., Ansari, M., & Mewada, A. (2025). Smart food choices: The role of reliable and safe AI in nutrition recommender systems. *Life Cycle Reliability and Safety Engineering*, 1–11.

Kalnikaitė, V., Bird, J., & Rogers, Y. (2013). Decision-making in the aisles: Informing, overwhelming or nudging supermarket shoppers? *Personal and Ubiquitous Computing*, *17*(6), 1247–1259.

Kroger. (2023). *Nutrition advice powered by technology: Kroger's personalised health approach*. https://www.kroger.com/health/nutrition

Luo, A., & Schafraad, P. (2024). Applying digital nudges to stimulate healthy food choices in an online supermarket environment.

Mohan, V., & Tiwaskar, M. (2025). The unrelenting epidemic of diabetes in India: Do the numbers matter? *The Journal of the Association of Physicians of India, 73*(5), 12–13. PMID: 40553516

Oshman, L., Waselewski, M., Hisamatsu, R., Kim, N., Young, L., Griauzde, D. H., & Chang, T. (2024). Grocery delivery to support individuals with type 2 diabetes: Protocol for a pilot quality improvement program. *JMIR Research Protocols, 13*(1), e54043. PMID: 38748461

Rapp, A., & Boldi, A. (2025). Promoting healthy eating by design: Opportunities for meaningful persuasive technologies. *Human-Computer Interaction & Emerging Technologies, 195*, 15.

Rapp, A., & Boldi, A. (2025, May). Designing for healthy food practices: Challenges and opportunities for changing people's eating behavior using persuasive technology. In *International Conference on Human-Computer Interaction* (pp. 280–297). Springer Nature Switzerland.

Reuters. (2025, March 27). *India's quick commerce sector made two-thirds of all 2024 e-retail orders: Report*. https://www.reuters.com

Richardson, J., & Folkvord, F. (2024). Nudging towards a sustainable diet: Testing and enhancing nudges in an e-grocery environment.

Shin, S., Xiaoxi, Y., Chakraborty, B., & Finkelstein, E. A. (2025). A randomized trial using an online grocery store to evaluate a multicomponent digital intervention to improve diet quality. *American Journal of Preventive Medicine*. PMID: 40120994

Siawsolit, C., Seepun, S., Choi, J., Do, A., & Kao, Y. (2017, March). Personalized assistant for health-conscious grocery shoppers. In *International Conference on Persuasive Technology* (pp. 95–106). Cham: Springer.

Sohil, P., Mahajan, S., Deshmukh, R., Gothankar, J., & Adhya, S. (2025). Millet consumption in type 2 diabetics in urban slums of India: A pilot study. *International Journal of Public Health Science, 14*(2), 904. DOI: 10.11591/ijphs.v14i2.25422

Stenius, M., Sillander, J., & Eriksson, N. (2023, December 14). *What's in the cards for online grocery business in Finland going forward? Present state and future challenges according to experts*. https://www.researchgate.net/publication/376659002_What

Sun, Y., Nivini, L., Savino, G. L., Mathis, F., & Schöning, J. (2024, June). Designing grocery shopping experiences for virtual reality. In *Proceedings of the 2024 International Conference on Advanced Visual Interfaces* (pp. 1–5).

Technavio. (2025). *Online grocery market industry in India analysis.* https://www.technavio.com/report/online-grocery-market-industry-in-india-analysis

TechSci Research. (2025). *India online grocery market by product type, region & forecast 2025–2030.* https://www.techsciresearch.com

Thomas-Francois, K., & Somogyi, S. (2021). Affective adoption of new grocery shopping modes through cultural change acceptance, consumer learning, and other means of persuasion. *International Journal of Electronic Marketing and Retailing, 12*(4), 323–338.

Tsolakidis, D., Gymnopoulos, L. P., & Dimitropoulos, K. (2024, August). Artificial intelligence and machine learning technologies for personalized nutrition: A review. [). MDPI.]. *Informatics (MDPI), 11*(3), 62.

Uzoma, C. M., Uwugiaren, N. J., Abdullah, Y. D., Ugwunze, C. A., & Obong, E. E. (2025). Comprehensive analysis of the impact of urban planning and emerging technologies on food environments and accessibility of healthy food in Lagos Island. *International Journal of Science and Research Archive, 14*(3), 1573–1588. DOI: 10.30574/ijsra.2025.14.3.0712

Valenčič, E., Beckett, E., Collins, C. E., Seljak, B. K., & Bucher, T. (2023). Digital nudging in online grocery stores: A scoping review on current practices and gaps. *Trends in Food Science & Technology, 131*, 151–163.

Valenčič, E., Beckett, E., Collins, C. E., Seljak, B. K., & Bucher, T. (2024). Changing the default order of food items in an online grocery store may nudge healthier food choices. *Appetite, 192*, 107072. PMID: 37797817

Wolniak, R., Stecuła, K., & Aydın, B. (2024). Digital transformation of grocery in-store shopping: Scanners, artificial intelligence, augmented reality and beyond—A review. *Foods, 13*(18), 2948. PMID: 39335877

Chapter 13
Influence of Informativeness and Trendiness on Gen Z's Continued Use of AI for Culinary Learning

Jasdeep Kaur
Chandigarh Group of Colleges Jhanjeri, Mohali, India

Rohit Chauhan
https://orcid.org/0000-0002-4400-7891
Mittal School of Business, Lovely Professional University, Phagwara, India

Sandeep Singh
Chandigarh Group of Colleges Jhanjeri, Mohali, India

Vikram Kaith
Central University of Kerala, India

Ashutosh Shaktan
University of the West of England, Bristol, UK

Parul Shaktan
Panjab University, Chandigarh, India

ABSTRACT

The rise of AI-powered large language models (LLMs) has reshaped digital learning, especially for Generation Z seeking personalized and interactive experiences. This study examines how informativeness and trendiness influence the continuance intention to use AI tools for cuisine learning. Grounded in Uses and Gratifications Theory (UGT), it positions trendiness as a mediator between perceived informativeness and continued usage. Data were collected from 119 Gen Z users in Northern India with prior experience using AI tools like ChatGPT for cooking tasks.

DOI: 10.4018/979-8-3373-4042-5.ch013

Structural equation modeling revealed that informativeness significantly affects both trendiness and continuance intention, with trendiness playing a strong mediating role. The findings expand UGT by incorporating socially driven gratifications and offer practical insights for enhancing AI-based learning tools aligned with Gen Z's preferences. The study supports the integration of AI in informal culinary education and broader digital lifestyle platforms.

INTRODUCTION

The rapid evolution of artificial intelligence (AI) has led to transformative applications across various domains, including healthcare, finance, education, and the culinary arts. One of the most prominent innovations in this AI wave is the emergence of Large Language Models (LLMs) such as OpenAI's ChatGPT, which possess the capability to engage in contextually relevant, human-like interactions across diverse knowledge domains (Grzankowski et al., 2025; Roberts et al., 2024). In recent years, the application of AI in the food industry has gained momentum due to its capacity to process extensive datasets, personalize experiences, and simulate creative cooking processes. Studies have demonstrated the ability of AI to assist in flavor development, recipe generation, and culinary innovation, ultimately enhancing quality, safety, and user satisfaction in food preparation and dining (Cui et al., 2025; Zatsu et al., 2024). This AI-driven transformation is particularly pertinent in digital education and informal learning settings, where learners engage with AI for real-time feedback, information synthesis, and guided experiential learning (Handley, 2024).

Among the generations actively engaging with AI, Generation Z (individuals born between the mid-1990s and early 2010s) has emerged as a key adopter due to their digital nativity, openness to experimentation, and demand for personalization and interactivity (Bunea et al., 2024). Their interaction with AI, particularly LLMs, extends beyond conventional learning environments into domains like hobby-based learning and creative self-expression, including culinary skills acquisition. LLMs facilitate access to recipe guidance, step-by-step instructions, dietary customization, and even virtual meal planning—all tailored to users' preferences. The allure of such technology lies not only in its functionality but also in its perceived novelty, usefulness, and entertainment value (Mun & Hwang, 2024). Moreover, the anthropomorphic characteristics of AI agents—such as perceived intelligence, responsiveness, and trendiness—play a pivotal role in shaping Gen Z's behavioral intention to adopt and continue using such platforms (Bhatnagr & Rajesh, 2024). As such, the intersection of AI, LLMs, and culinary learning among Gen Z represents a compelling site for investigation within digital behavior and educational technology research.

Despite increasing attention toward AI-powered platforms in informal learning, empirical studies examining the factors that influence continuance intention among Generation Z in non-academic contexts, particularly culinary learning, remain scarce. Most existing research focuses on AI use in formal education, customer engagement, or digital commerce, with limited attention to its applications in experiential, skill-based domains like cuisine preparation. Although constructs such as informativeness, perceived trendiness, and anthropomorphism have been examined in isolated contexts (Gao & Bai, 2014; X. Hu et al., 2022), there is a lack of integrated understanding of how these factors interact to shape users' intention to continue engaging with AI systems, especially among digitally sophisticated cohorts. Furthermore, few studies have applied a well-established theoretical framework such as the Uses and Gratifications Theory (UGT) to explain the motivations and satisfaction outcomes that drive sustained usage of LLMs in culinary learning settings. Addressing this gap is critical for designing AI interfaces that not only inform but also captivate and retain young users.

In response to these gaps, the present study aims to investigate the influence of informativeness and perceived trendiness on Generation Z's continuance intention toward using AI-based LLMs in culinary learning, grounded in the Uses and Gratifications Theory. Specifically, the study examines whether informativeness enhances perceived trendiness and whether these constructs, individually and collectively, influence users' intention to continue utilizing LLMs. It also explores the mediating role of trendiness in the relationship between informativeness and continuance intention. By focusing on Northern India—a region where digital literacy is rising rapidly among youth—this study contributes to the understanding of AI usage in culturally diverse, rapidly digitizing contexts. The findings have practical implications for educators, developers, and hospitality stakeholders seeking to design AI-powered culinary learning experiences that align with Gen Z's expectations and engagement patterns. Ultimately, the study addresses a timely need to comprehend the motivations underpinning sustained AI usage in experiential learning, helping to bridge the gap between technological potential and user-centered design.

LITERATURE REVIEW

Constructs Under Study

The Uses and Gratifications Theory (UGT) serves as a foundational lens to explore how users derive satisfaction from adopting interactive technologies, including AI-powered large language models (LLMs). Originating from studies of mass media behavior, UGT has evolved to address computer-mediated communication and hypertextual technologies, maintaining its relevance in analyzing digital user engagement (Ruggiero, 2000). The theory classifies motivations into functional categories such as information-seeking, social interaction, entertainment, control, and convenience (Boudkouss & Djelassi, 2021). UGT has effectively been extended to contexts involving digital interactivity, such as smartphone and social media use among students across various cultures (García-Umaña et al., 2024). These applications underscore UGT's flexibility in explaining technology engagement in culturally nuanced environments. In particular, Generation Z's motivations on platforms like TikTok, driven by hedonic and utilitarian gratifications, resonate well with UGT's central claims (Flecha-Ortiz et al., 2023). As such, UGT offers a powerful framework to understand how Gen Z interacts with AI-powered tools for experiential and lifestyle-oriented learning like culinary skills, where gratification is central.

In the context of informal learning such as cuisine preparation, UGT provides critical insight into how AI tools support motivation through personalization, interactivity, and instant feedback. These features of LLMs create engaging, learner-centered experiences, enhancing the perceived usefulness and satisfaction among users (Ahn, 2024). The theory explains how learners use AI to fulfill personal needs such as creativity and self-expression, which are vital in experiential learning tasks like cooking (Celik et al., 2024). Moreover, trust in anthropomorphic AI interfaces, such as those provided by LLMs, plays a pivotal role in user adoption by reinforcing comfort and familiarity (Kim et al., 2024). Liu et al. (2024) further support this by showing how AI-mediated environments enhance enjoyment and reduce anxiety, particularly in non-judgmental spaces such as self-directed cooking or recipe experimentation. UGT also connects seamlessly with the need for autonomy and control, which is inherent in informal learning, where users dictate pace and content. Therefore, the adoption of AI for cuisine learning among Gen Z aligns directly with UGT's assertion that users actively choose technologies that satisfy their individual cognitive and affective needs.

Informativeness is a core attribute of AI systems, directly impacting trust, satisfaction, and continued engagement. In AI-powered recommendation systems, informativeness pertains to the richness, clarity, and contextual relevance of the information delivered (Bigras et al., 2018). This feature strongly influences user trust, particularly when AI systems provide persuasive and complete explanations (Yang, 2018). Explanations from AI agents have been found to increase system transparency, allowing users to validate outputs and reduce ambiguity (Ghazimatin, 2020; Zanker, 2012). In generative AI contexts, informative output significantly enhances the perceived benefits while mitigating perceived risks, highlighting the importance of balancing information accuracy with usability (Huynh, 2024). Furthermore, informativeness supports perceived value and usability, driving not just first-time use but also continued adoption. Particularly in culinary domains where users seek context-specific guidance—such as dietary substitutions or cooking methods—high informativeness ensures that AI tools are viewed as reliable assistants rather than generic tools. Thus, informativeness serves as a linchpin construct that connects the AI's perceived utility with emotional trust and practical value.

Prior studies involving chatbots, LLMs, and digital assistants in learning and lifestyle contexts consistently underscore the role of informativeness in user satisfaction and learning effectiveness. In educational settings, chatbots have demonstrated moderate-to-significant positive effects on language skills and engagement levels, with informativeness being central to these outcomes (Alemdag, 2025). Jeon et al. (2024) found that participants' perceptions of informative content significantly influenced speaking and listening skills, particularly over long-term interaction with chatbots. Anjulo Lambebo & Chen (2025) similarly noted that perceived usefulness and informativeness contributed to students' continued use of AI communication platforms. In lifestyle interventions involving AI (e.g., sleep, diet, physical activity), informativeness has been critical to behavioral change, with studies confirming statistically significant improvements in health metrics (Singh et al., 2023). In language education, Lyu et al. (2025) also emphasized that features such as AI accessibility and detailed responses increase learning efficacy and user retention. These findings confirm that informativeness, when operationalized correctly, acts as both a cognitive and emotional stimulus for sustained user engagement with AI tools.

Trendiness is a multifaceted concept in digital technology literature, encapsulating the attractiveness of innovation, media visibility, and social appeal. Defined by its alignment with newness and technological evolution, trendiness plays a crucial role in how Generation Z perceives and adopts digital tools. For example, in the digital fashion industry, trendiness has shifted from design functionality to full virtual consumption models, influencing consumer behavior (Chan & Lee, 2023). In infrastructure and digital construction contexts, the adoption of emerging technologies like BIM and automation also reflect how trendiness equates to efficiency and sustainability (Ametepey et al., 2024). Additionally, media representations of trending technologies—such as AI and NFTs—construct public narratives that shape desirability and legitimacy (Echauri, 2024). The TrendNets methodology supports this idea by demonstrating how academic research trends reflect broader societal interests and digital shifts (Katsurai & Ono, 2019). In sum, trendiness is an evolving construct shaped by innovation, peer validation, and contextual utility, and it holds special relevance for technologies targeting younger, innovation-seeking demographics such as Generation Z.

For Gen Z, trendiness influences technology adoption through novelty, peer validation, and social media amplification. As digital natives, Gen Z is highly responsive to the freshness of innovations and the influence of social networks in shaping attitudes. Wells et al. (2010) note that novelty significantly enhances willingness to engage with emerging technologies despite risks. Flecha-Ortiz et al. (2023) found that TikTok users derive both hedonic and utilitarian gratifications from following trends, often driven by parasocial interactions with influencers. Xin & Lim (2025) further emphasized how informational and emotional supports on trending platforms enhance perceived usefulness, often more than traditional content delivery. Verma et al. (2021) also demonstrated that social pressure—especially from peer networks—influences behavioral intentions toward mobile commerce. Interestingly, while Gen Z leans heavily toward digital tools, they still appreciate physical retail spaces, signaling a hybrid appreciation of trend-based experience (Kunc et al., 2024). These findings support the role of perceived trendiness in increasing system desirability and highlight how AI tools that appear trendy can more effectively engage Gen Z in tasks like cuisine learning.

Continuance intention, unlike initial adoption, refers to a user's willingness to keep engaging with a technology over time. Grounded in Expectation-Confirmation Theory, this construct emphasizes the confirmation of prior expectations and satisfaction as primary drivers (Fernández-Guzmán & Bravo, 2018; Zhou, 2011). While initial adoption focuses on features like perceived usefulness and social influence, continuance intention is shaped more by habit, emotional connection, and system reliability (Park & Ok, 2008). Al-Hattami and Almaqtari (2023) showed that continuance intention in digital systems depends heavily on satisfaction derived from information quality and system functionality. In contrast to early models such as TAM, which prioritize cognitive evaluation, continuance intention integrates habitual and affective behaviors that may bypass deliberate decision-making. Thus, for AI in cuisine learning, a high level of satisfaction and emotional ease with the system can foster habitual use, enabling long-term integration of AI tools in users' culinary routines and informal education pathways.

Multiple studies affirm that trust, relevance, and emotional resonance significantly affect continuance intention in AI-powered tools. In education, Mun & Hwang (2024) highlighted that content richness, accuracy, and relevance in tools like ChatGPT directly enhance trust and continued usage. Similarly, Salih et al. (2025) identified hedonic motivation and performance expectancy as predictors of engagement in AI-enabled services. Emotional factors like optimism, loneliness, and perceived intelligence influence continuance, especially in tools used during the COVID-19 period. Facilitating conditions such as ease of access and prior habits also enhance behavioral intention (Cortez et al., 2024). In digital banking, anthropomorphism and perceived intelligence contribute to quality interaction and expectation confirmation, which reinforces ongoing use (Bhatnagr et al., 2024). Among Generation Z, tools like SmartAPPetite demonstrate how food-centric AI apps are well received due to their personalization and utility (Gilliland et al., 2022). Chan & Lee (2023) further observed that Gen Z views generative AI as a productivity enhancer. This convergence of utility, trust, and engagement makes AI-powered cuisine learning tools ideal for sustained adoption by Gen Z.

Hypothesis Formulation

Informativeness is expected to influence the perceived trendiness of AI-based large language models (LLMs) used in culinary learning due to their ability to enhance the learning experience by providing detailed, data-driven insights and personalized recommendations. AI's integration into the food industry, as highlighted by Cui et al. (2025), allows for the exploration and development of new flavor substances, improving product quality and meeting consumer needs through the use of extensive food flavor and ingredient databases. This capability aligns with the broader transformative potential of AI in the food industry, which includes optimizing processes and fostering innovation in personalized nutrition (Zatsu et al., 2024). LLMs, such as ChatGPT, are not merely next-token predictors but possess the ability to engage in human-like communication, which can be leveraged to provide informative and interactive culinary learning experiences (Grzankowski et al., 2025; Roberts et al., 2024). The trendiness of these models is further amplified by their capacity to simulate human conversation and provide tailored feedback, a capability noted in AI-enhanced language learning environments (Handley, 2024). However, perceived trendiness may be tempered by concerns such as misinformation or ethical lapses in AI-generated content (Roberts et al., 2024). Overall, the informativeness offered by AI-based LLMs enhances their educational value and contributes to the perception of trendiness.

H1: Informativeness positively influences the perceived trendiness of AI-based LLMs.

Perceived trendiness significantly influences continuance intention across various digital and AI platforms by enhancing perceived usefulness, enjoyment, and social status, which are central motivators in the Uses and Gratifications Theory (UGT). In the context of digital collaborative learning platforms, Liu et al. (2024) found that perceived usefulness and the alignment between task and technology — often influenced by the platform's trendiness — positively impacted continuance intentions. Similarly, in entertainment and media technologies, such as OTT platforms, trendiness facilitates ongoing usage by enhancing enjoyment, convenience, and even triggering psychological motivators like FOMO. In mobile augmented reality games, the freshness and novelty of content — attributes closely tied to trendiness — drive content and process gratifications, both of which encourage continued engagement (Jang & Liu, 2019). Trendy technologies also enhance user satisfaction in digital health systems, indirectly fostering continuance intention through social influence and enjoyment (Bai & Guo, 2022). Finally, social networking platforms show that usefulness and satisfaction, heightened by perceived trendiness, remain key to retaining user engagement (Gao & Bai, 2014). Thus, trendiness increases the experiential value of AI applications, making them more desirable for continued use.

H2: Trendiness positively influences AI continuance intention.

The link between informativeness and continuance intention has been established across various digital environments, showing that the richness, accuracy, and utility of content drive sustained user engagement. In social commerce, informativeness—marked by the credibility and utility of shared information—enhances perceived usefulness and enjoyment, directly influencing continuance intentions (X. Hu et al., 2022). Similarly, Zhang et al. (2017) observed that in virtual learning communities, high-quality information—measured through completeness, timeliness, and relevance—directly impacts perceived satisfaction and usefulness, which are core to the intent to continue using the platform. Si et al. (2022) found that informativeness in ride-sharing services, expressed as perceived usefulness, significantly boosts continuance intention. This effect varies slightly by demographic variables such as gender. In mobile advertising, informativeness plays a key role in shaping perceived value and satisfaction, both of which lead to continuance intention under the Expectation-Confirmation Model (Hsiao & Chang, 2014). Similarly, in mobile travel platforms, informativeness indirectly influences continued use by fostering enjoyment and user engagement, moderated by the quality of interaction (Chiu et al., 2019). These studies collectively support the notion that informative content in AI-based culinary learning tools increases their perceived utility and drives repeat usage among users.

H3: Informativeness positively influences AI continuance intention.

Trendiness serves as a powerful mediator between informativeness and continuance intention, particularly among Generation Z, who are more inclined toward adopting technologies that align with their preferences for innovation and modernity. Bunea et al. (2024) demonstrate that in AI-assisted online shopping, informativeness influences trendiness perceptions, which then enhance purchase intentions via perceived usefulness and ease of use. This shows that trendiness acts as a bridge, amplifying the impact of informativeness on behavior. Similarly, Mun & Hwang (2024) found that in LLM applications like ChatGPT, the richness and credibility of information directly shape usefulness and trust, which are foundational for continued engagement. These perceptions are particularly important for Gen Z, who evaluate technologies through lenses of novelty and trendiness. Bhatnagr & Rajesh (2024) further confirm that Gen Z's continued use of AI in digital banking is enhanced by anthropomorphic and intelligent characteristics—features closely related to trendiness—which amplify perceived satisfaction and usefulness. Therefore, trendiness can be understood as a cognitive-emotional filter that reinforces the effects of informativeness on behavioral outcomes, particularly among tech-savvy and novelty-driven Gen Z users.

H4: Trendiness mediates the relationship between informativeness and AI continuance intention.

Figure 1 depicts the hypothesized relationships within the study framework. A direct path (solid arrows) represents straightforward influence, while a mediating effect (dashed arrows, H4) indicates the indirect impact of Informativeness on AI Continuance Intention through Trendiness. Arrows represent the hypothesized relationships (H1–H4) between the constructs. Solid lines denote direct effects, and dashed lines show mediating effects, emphasizing the model's structure for testing the interplay between informativeness, perceived trendiness, and intentions to continue using AI systems.

Figure 1. The conceptual framework (Authors)

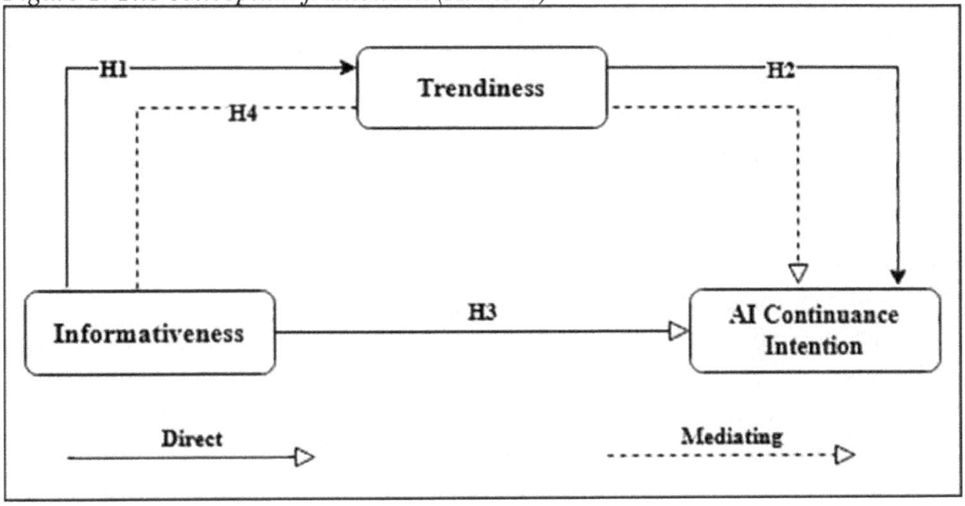

RESEARCH METHODOLOGY

This study was conducted in Northern India and focused on Generation Z individuals, a demographic widely recognized for its digital fluency and receptiveness to emerging technologies. The research specifically targeted users who had prior experience with AI-based large language models (LLMs) in the context of cuisine preparation. To ensure relevance, screening questions were administered to confirm participants' engagement with such tools before they proceeded with the main survey. A structured questionnaire was developed in English and distributed online via email and social media platforms, ensuring accessibility and reach within the Gen Z population. A total of 150 questionnaires were distributed, out of which 119 usable responses were collected, resulting in a satisfactory response rate of 79.3%. The sample size met the item-to-response ratio guideline of at least 10 respondents per item (Hair et al., 2010), considering the 23 items used in the main analysis. This ensured adequate statistical power and model reliability.

The survey instrument comprised 23 items drawn from validated scales used in prior studies. Informativeness and Trendiness were measured using items adapted from Yadav & Rahman (2017), while AI Continuance Intention was measured using the scale by (Giampietri et al., 2018). All items were measured on a five-point Likert scale ranging from 1 (strongly disagree) to 5 (strongly agree), allowing respondents to express varying levels of agreement. Informativeness was measured through items such as "The AI cooking assistant provides accurate recipe instructions," and "The information from the AI is comprehensive." Trendiness included items like "The AI features viral food trends from social media platforms." AI Continuance Intention included items such as "I expect to keep using cooking AI long-term." The constructs were operationalized to capture user perceptions around informational utility, social relevance, and sustained engagement with AI in culinary learning.

Content validity was established by consulting three hospitality and culinary domain professionals, who reviewed the questionnaire for clarity, relevance, and alignment with industry practices. Their suggestions were incorporated into the final version to enhance practical applicability. To ensure the instrument's reliability, a pilot test was conducted using a smaller sample of 30 Gen Z respondents. The pilot testing involved all 23 items, and the Cronbach's alpha values for each construct exceeded the threshold of 0.70 (Hair et al., 2019), confirming internal consistency and scale reliability. Based on pilot feedback, minor linguistic refinements were made to improve item clarity and comprehension. The finalized instrument was then used for full-scale data collection. This rigorous methodological design ensured that the data collected was both reliable and valid for analyzing the relationships among informativeness, trendiness, and AI continuance intention in the domain of AI-assisted cuisine learning. Data analysis was carried with IBM SPSS 23 and AMOS 26.

RESULTS

The study surveyed 119 respondents to understand their interaction with AI-based large language models (LLMs) in cuisine learning. A dominant share (85.7%) of participants reported using ChatGPT, followed by Gemini (9.2%), with minimal usage of alternatives like Deepseek, Grok, Google, and Meta AI. In terms of frequency of use, 34.5% used LLMs daily, while 24.4% used them weekly and 25.2% occasionally. A smaller group used them rarely (15.1%) or never (0.8%). The sample was predominantly young, with 89.1% aged between 15–25 years, and only a marginal share aged above 25. Gender distribution was nearly balanced, with 52.1% females and 47.9% males. A vast majority of the respondents (98.3%) were Indian nationals. Regarding occupational background, 86.6% were students, indicating the study's strong relevance to educational settings. A small portion were self-employed (5%), unemployed (5%), or in other professional roles like teachers and employed individuals. This profile highlights a tech-savvy, youth-dominated, student-centric population engaged with AI tools for culinary learning.

Measurement Model Results

Table 1 presents the measurement model results assessing the reliability and validity of the constructs—Informativeness, Trendiness, and AI Continuance Intention—used in the study. All standardized factor loadings exceed the threshold of 0.70 (Hair et al., 2019), indicating strong item reliability. For instance, loadings range from 0.82 to 0.897, confirming that individual items significantly reflect their respective constructs. Composite reliability (CR) for all constructs also exceeds the recommended minimum value of 0.70 (Fornell & Larcker, 1981), with Informativeness at 0.875, Trendiness at 0.912, and AI Continuance Intention at 0.935, demonstrating high internal consistency. Similarly, the Average Variance Extracted (AVE) values are above the 0.50 threshold (Fornell & Larcker, 1981), confirming convergent validity, with values of 0.701, 0.776, and 0.784, respectively. Cronbach's alpha values, ranging from 0.875 to 0.934, exceed the acceptable benchmark of 0.70 (Hair et al., 2019), ensuring internal reliability of the scales. Furthermore, the overall model fit indices (CMIN/DF = 2.038, GFI = .901, CFI = .966, TLI = .952, IFI = .966, RMSEA = .064, RMR = .044) fall within acceptable ranges, indicating a good fit between the observed data and the hypothesized model (L. Hu & Bentler, 1999). These results collectively confirm that the measurement model is both reliable and valid.

Table 1. Measurement model results

Construct and Items	Standardized Factor Loadings	Composite Reliability	Average Variance Extracted	Cronbach's Alpha
Informativeness (Yadav & Rahman, 2017)		0.875	0.701	0.875
The AI cooking assistant provides accurate recipe instructions.	0.846			
The AI offers useful tips for ingredient substitutions.	0.845			

continued on following page

Table 1. Continued

Construct and Items	Standardized Factor Loadings	Composite Reliability	Average Variance Extracted	Cronbach's Alpha
The information from the AI is comprehensive	0.82			
Trendiness (Yadav & Rahman, 2017)		0.912	0.776	0.91
The AI recommends recipes that are currently trending globally	0.895			
Using an AI cooking assistant is considered trendy among food enthusiasts	0.85			
The AI features viral food trends from social media platforms	0.897			
AI Continuance Intention (Giampietri et al., 2018)		0.935	0.784	0.934
I intend to use AI for weekly meal planning	0.862			
I plan to regularly try AI-generated recipes	0.907			
I'm willing to use AI as my primary cooking assistant	0.867			
I expect to keep using cooking AI long-term	0.904			

Model fit: CMIN/DF: 2.038; GFI: .901; CFI: .966; TLI: .952; IFI: .966; RMSEA: .064; RMR: .044
(Authors' own interpretation)

Table 2 reports discriminant validity using the Fornell-Larcker criterion. Discriminant validity is established when the square root of the AVE of a construct (diagonal values) is greater than its correlations with other constructs (off-diagonal values) (Fornell & Larcker, 1981). The square root of the AVE for Informativeness (0.837), Trendiness (0.881), and AI Continuance Intention (0.885) is higher than their inter-construct correlations. For instance, the correlation between Informativeness and Trendiness is 0.624, while both have lower correlations with AI Continuance Intention (0.500 and 0.735, respectively). These results confirm that the constructs are distinct from each other, fulfilling the requirement of discriminant validity.

Table 2. Discriminant validity using Fornell-Larcker criteria

	Informativeness	Trendiness	AI Continuance Intention
Informativeness	**0.837**		
Trendiness	0.624***	**0.881**	
AI Continuance Intention	0.500***	0.735***	**0.885**

(Author's own interpretation)
Structural Model Results

Table 3 displays the results of hypothesis testing through the structural model. H1, which posits a positive relationship between Informativeness and Trendiness, is supported with a significant path coefficient ($\beta = 0.553$, CR = 6.38, $p < .001$), suggesting that informative AI tools enhance perceived trendiness. H2, asserting that Trendiness positively influences AI Continuance Intention, is also strongly supported (β = 0.858, CR = 6.31, p < .001), indicating that users are more likely to continue using AI tools if they perceive them as trendy. However, H3, proposing a direct effect of Informativeness on AI Continuance Intention, is not supported (β = 0.075, CR = 0.67, p = 0.5), showing that informativeness alone does not significantly drive continued use. The mediation analysis (H4) reveals that Trendiness significantly mediates the relationship between Informativeness and AI Continuance Intention (β = 0.474, p < .05), with the indirect effect being statistically significant. This highlights the pivotal role of trend perception in shaping users' long-term engagement with AI in cuisine learning. These findings support a mediated model where Informativeness enhances Trendiness, which in turn fosters AI Continuance Intention, aligning with contemporary perspectives on technology adoption and user retention.

Table 3. Hypothesis results

Direct Hypothesis	Estimate	S.E.	Bootstrap [95% CI]		C.R.	P	Decision
H1: Informativeness > Trendiness	**0.553**	**0.09**			**6.38**	***	**Supported**
H2: Trendiness > AI Continuance Intention	0.858	0.14			6.31	***	Supported
H3: Informativeness > AI Continuance Intention	0.075	0.11			0.67	0.5	Not Supported
Mediating Hypothesis	**Estimate**	**S.E.**	**Upper**	**Lower**	**C.R.**	**P**	**Label**
H4: Informativeness > Trendiness > AI Continuance Intention	0.474	0.26	0.17	1.001	1.84	*	Supported

***p<.001; **p<.05; SE: Standard error; CR: Critical ratio; P Significance level
(Authors own interpretation)

DISCUSSION AND CONCLUSION

The study findings reveal that the perceived informativeness of AI-powered large language models (LLMs) positively influences users' perceptions of their trendiness. This relationship underscores the growing expectation among Generation Z users for tools that not only provide functional benefits but also align with cultural and digital trends. Consistent with Yadav and Rahman (2017), who highlighted the dual importance of informativeness and trend awareness in technology adoption among youth, our results affirm that informative content enhances users' perceptions of technological modernity. When users feel that AI tools keep them updated with evolving culinary trends and viral food phenomena, they are more likely to view such technologies as innovative and socially relevant. The results suggest that informativeness in LLMs doesn't merely serve a utilitarian function but becomes a springboard for users to connect with broader lifestyle and identity markers. This dimension of trendiness, as a by-product of informative design, opens up new ways to understand digital culinary engagement among young users.

The finding that trendiness has a strong direct influence on AI continuance intention supports the proposition that perceived modernity and alignment with current trends significantly enhance long-term usage behavior. This outcome is corroborated by Giampietri et al. (2018), who emphasized the importance of consumer perceptions and social cues in shaping behavioral outcomes within food-related digital interfaces. As AI applications become increasingly personalized and socially embedded, their trend-related attributes—such as integrating viral recipes or emerging food movements—become instrumental in shaping user retention. Therefore, trendiness acts as both a psychological incentive and a cultural indicator, especially in a generation that values immediacy and novelty in digital content consumption.

Interestingly, informativeness does not have a significant direct effect on continuance intention, indicating that factual and helpful content alone may not suffice to maintain user engagement in AI-mediated cuisine learning. Instead, its effect is channeled through the perception of trendiness, suggesting a mediated relationship. This aligns with Yadav & Rahman (2017) conceptualization, where informativeness in digital services has an indirect effect on behavioral outcomes via mediating factors such as emotional appeal or novelty. The current results thereby underscore the need to blend reliable content with engaging, trend-aligned formats. For LLMs in culinary applications, this implies that success hinges not just on providing accurate recipes, but also on delivering content in ways that feel fresh, social, and aligned with users' digital expectations.

Theoretical Implications

The findings of this study carry significant theoretical implications by extending the scope of the Uses and Gratifications Theory (UGT) within the emerging domain of artificial intelligence-mediated learning experiences, particularly those involving lifestyle knowledge acquisition such as cuisine. Traditionally, UGT emphasizes that users engage with media based on specific gratifications such as entertainment, information, or social interaction (Yadav & Rahman, 2017). This study confirms that informativeness remains a foundational driver for engagement; however, it adds a nuanced layer by positioning perceived trendiness as a vital mediating construct between informativeness and continuance intention. This suggests that traditional gratifications like informational utility must now be viewed in tandem with contemporary social and identity-based motivations. The integration of trendiness broadens the explanatory power of UGT in digital environments where social identity, cultural alignment, and lifestyle signaling are increasingly influencing user engagement. Consequently, this study contributes to the evolving discourse by highlighting the need to recalibrate UGT to reflect generational shifts in media and technology use patterns, particularly among Generation Z.

Beyond validating existing constructs, this study challenges the sufficiency of linear technology adoption frameworks that prioritize functionality over socio-cultural dimensions of technology engagement. Models like the Technology Acceptance Model (TAM) or even extensions of UGT often assume that perceived usefulness or informativeness alone drives usage behavior. However, our findings emphasize the mediating role of trendiness, which reveals that functional gratifications need to be complemented by socially constructed values to sustain user engagement. For Generation Z, perceived innovativeness and social relevance may supersede utilitarian factors in determining long-term usage. This aligns with previous findings by Giampietri et al. (2018) and suggests the need to revisit and extend classical models to include variables like trendiness, aesthetic alignment, and perceived social capital. These constructs better represent the fluid motivations of younger users interacting with AI in informal, lifestyle-centered learning environments. Future theoretical models must account for the interplay between cognitive (informativeness) and affective-social (trendiness) gratifications, thus offering a more holistic and contemporary understanding of technology use.

Practical Implications

The practical implications of this research are far-reaching for developers, marketers, and educational stakeholders aiming to enhance user engagement with AI-powered large language models (LLMs) in the domain of cuisine learning. First, developers must recognize that while functional accuracy and informational depth remain important, the perception of trendiness significantly influences continuance intention. Therefore, AI interfaces designed for cooking guidance should incorporate trend-aware features, such as real-time updates on viral recipes, seasonal food trends, or influencer-curated content. Adding social sharing capabilities, gamification, and aesthetic design that aligns with youth digital cultures can further boost engagement. Marketers can benefit by positioning AI tools not just as informative assistants but as lifestyle enhancers that reflect users' tastes and social identities. Promotional campaigns could showcase how the AI stays updated with contemporary food movements, aligning it with youth aspirations of being innovative and socially in-tune. Ultimately, the blend of trendiness and informativeness offers a compelling user value proposition, particularly for Generation Z consumers seeking both functionality and digital relevance.

For educators and institutions involved in culinary training or digital skill development, this study emphasizes the potential of integrating trendy AI tools into the curriculum. Since Generation Z responds positively to socially relevant and modern learning tools, leveraging AI-powered LLMs as part of experiential or self-paced culinary modules could increase motivation and learning retention. Institutions can develop assignments that involve using AI to explore current food trends, plan balanced meals, or replicate celebrity chef recipes—thereby fostering creativity and engagement. Governmental or non-profit bodies promoting digital literacy, nutrition, or vocational training can also incorporate such tools into outreach programs aimed at young adults. Additionally, by aligning the content with localized trends and cultural cuisines, the effectiveness of AI applications can be increased in diverse demographic settings. For instance, AI models trained on regional ingredients or cooking practices could enhance both relatability and cultural preservation. Overall, these applications underscore how AI, when coupled with trend-conscious design and delivery, can revolutionize informal education in the food and lifestyle domains.

Limitations and Future Scope

This study, while insightful, is limited by its focus on a predominantly student population from Northern India, restricting generalizability to other age groups or regions. Future research could explore cross-cultural comparisons to assess whether the influence of trendiness holds across diverse socio-demographic contexts. Additionally, qualitative studies could delve deeper into how users interpret "trendiness" and what social factors influence their AI tool preferences. Further investigation into other potential mediators like enjoyment or trust would enrich theoretical frameworks. Practically, longitudinal studies could assess whether trendiness continues to influence AI continuance intention over time, particularly as users' novelty-seeking tendencies evolve.

REFERENCES

Ahn, H. Y. (2024). AI-Powered E-Learning for Lifelong Learners: Impact on Performance and Knowledge Application. *Sustainability*, *16*(20), 9066. DOI: 10.3390/su16209066

Alemdag, E. (2025). The effect of chatbots on learning: A meta-analysis of empirical research. *Journal of Research on Technology in Education*, *57*(2), 459–481. DOI: 10.1080/15391523.2023.2255698

Ametepey, S. O., Aigbavboa, C., Addy, H., & Thwala, W. D. (2024). A Bibliometric Review of the Trends of Construction Digitalization Research in the Past Ten Years. *Buildings (Basel, Switzerland)*, *14*(9), 2729. DOI: 10.3390/buildings14092729

Anjulo Lambebo, E., & Chen, H.-L. (2025). Chatbots in higher education: A systematic review. *Interactive Learning Environments*, *33*(4), 2781–2807. DOI: 10.1080/10494820.2024.2436931

Bai, B., & Guo, Z. (2022). Understanding Users' Continuance Usage Behavior Towards Digital Health Information System Driven by the Digital Revolution Under COVID-19 Context: An Extended UTAUT Model. *Psychology Research and Behavior Management*, *15*, 2831–2842. DOI: 10.2147/PRBM.S364275

Bhatnagr, P., & Rajesh, A. (2024). Artificial intelligence features and expectation confirmation theory in digital banking apps: Gen Y and Z perspective. *Management Decision*. Advance online publication. DOI: 10.1108/MD-07-2023-1145

Bhatnagr, P., Rajesh, A., & Misra, R. (2024). Continuous intention usage of artificial intelligence enabled digital banks: A review of expectation confirmation model. *Journal of Enterprise Information Management*, *37*(6), 1763–1787. DOI: 10.1108/JEIM-11-2023-0617

Bigras, E., Jutras, M.-A., Sénécal, S., Léger, P.-M., Fredette, M., Black, C., Robitaille, N., Grande, K., & Hudon, C. (2018). Working with a Recommendation Agent. *Extended Abstracts of the 2018 CHI Conference on Human Factors in Computing Systems*, 1–6. https://doi.org/DOI: 10.1145/3170427.3188639

Boudkouss, H., & Djelassi, S. (2021). Understanding in-store interactive technology use: A uses and gratifications theory (UGT) perspective. *International Journal of Retail & Distribution Management*, *49*(12), 1621–1639. DOI: 10.1108/IJRDM-11-2020-0459

Bunea, O.-I., Corbo , R.-A., Mi u, S. I., Triculescu, M., & Trifu, A. (2024). The Next-Generation Shopper: A Study of Generation-Z Perceptions of AI in Online Shopping. *Journal of Theoretical and Applied Electronic Commerce Research, 19*(4), 2605–2629. DOI: 10.3390/jtaer19040125

Celik, I., Gedrimiene, E., Siklander, S., & Muukkonen, H. (2024). The affordances of artificial intelligence-based tools for supporting 21st-century skills. *Australasian Journal of Educational Technology*. Advance online publication. DOI: 10.14742/ajet.9069

Chan, C. K. Y., & Lee, K. K. W. (2023). The AI generation gap: Are Gen Z students more interested in adopting generative AI such as ChatGPT in teaching and learning than their Gen X and millennial generation teachers? *Smart Learning Environments, 10*(1), 60. DOI: 10.1186/s40561-023-00269-3

Chiu, C.-K., Lin, C.-P., Chen, K.-J., Liu, C.-M., & Ma, H.-C. (2019). Modeling continuance intention towards Mobile Travel Service System (MTSS): A theoretical perspective of motivation and dependency. *Review of Managerial Science, 13*(4), 749–769. DOI: 10.1007/s11846-017-0257-9

Cortez, P. M., Ong, A. K. S., Diaz, J. F. T., German, J. D., & Singh Jagdeep, S. J. S. (2024). Analyzing Preceding factors affecting behavioral intention on communicational artificial intelligence as an educational tool. *Heliyon, 10*(3), e25896. DOI: 10.1016/j.heliyon.2024.e25896

Cui, Z., Qi, C., Zhou, T., Yu, Y., Wang, Y., Zhang, Z., Zhang, Y., Wang, W., & Liu, Y. (2025). Artificial intelligence and food flavor: How AI models are shaping the future and revolutionary technologies for flavor food development. *Comprehensive Reviews in Food Science and Food Safety, 24*(1). Advance online publication. DOI: 10.1111/1541-4337.70068

Echauri, G. (2024). Tech trends: *Wired*'s treatment of emerging technologies (2021-2023). *Journalism*. Advance online publication. DOI: 10.1177/14648849241251966

Fernández-Guzmán, V., & Bravo, E. R. (2018). Understanding Continuance Usage of Natural Gas: A Theoretical Model and Empirical Evaluation. *Energies, 11*(8), 2019. DOI: 10.3390/en11082019

Flecha-Ortiz, J. A., Feliberty-Lugo, V., Santos-Corrada, M., Lopez, E., & Dones, V. (2023). Hedonic and Utilitarian Gratifications to the Use of TikTok by Generation Z and the Parasocial Relationships with Influencers as a Mediating Force to Purchase Intention. *Journal of Interactive Advertising, 23*(2), 114–127. DOI: 10.1080/15252019.2023.2195403

Fornell, C., & Larcker, D. F. (1981). Structural Equation Models with Unobservable Variables and Measurement Error: Algebra and Statistics. *JMR, Journal of Marketing Research, 18*(3), 382. DOI: 10.2307/3150980

Gao, L., & Bai, X. (2014). An empirical study on continuance intention of mobile social networking services. *Asia Pacific Journal of Marketing and Logistics, 26*(2), 168–189. DOI: 10.1108/APJML-07-2013-0086

García-Umaña, A., Beserra, V., & Córdoba, É. F. (2024). Gratifications Associated With the Use of Smartphone and Internet in Students From Ecuador, Spain, and Colombia. *IEEE Access : Practical Innovations, Open Solutions, 12*, 160495–160505. DOI: 10.1109/ACCESS.2024.3488576

Ghazimatin, A. (2020). Explaining Recommendations in Heterogeneous Networks. *Proceedings of the 43rd International ACM SIGIR Conference on Research and Development in Information Retrieval*, 2479–2479. https://doi.org/DOI: 10.1145/3397271.3401451

Giampietri, E., Verneau, F., Del Giudice, T., Carfora, V., & Finco, A. (2018). A Theory of Planned behaviour perspective for investigating the role of trust in consumer purchasing decision related to short food supply chains. *Food Quality and Preference, 64*, 160–166. DOI: 10.1016/j.foodqual.2017.09.012

Gilliland, J. A., McEachern, L. W., Cappuccitti, S., Doherty, S., O'Connor, C., Seabrook, J., Haines, J., & Stranges, S. (2022). SmartAPPetite for youth: development and evaluation of a smartphone app for improving adolescent food literacy and healthy eating. *Proceedings of the Nutrition Society, 81*(OCE5), E167. https://doi.org/DOI: 10.1017/S0029665122002002

Grzankowski, A., Downes, S. M., & Forber, P. (2025). LLMs are not just next token predictors. *Inquiry : a Journal of Medical Care Organization, Provision and Financing*, •••, 1–11. DOI: 10.1080/0020174X.2024.2446240

Hair, J. F., Risher, J. J., Sarstedt, M., & Ringle, C. M. (2019). When to use and how to report the results of PLS-SEM. *European Business Review, 31*(1), 2–24. DOI: 10.1108/EBR-11-2018-0203

Handley, Z. (2024). Has artificial intelligence rendered language teaching obsolete? *Modern Language Journal, 108*(2), 548–555. DOI: 10.1111/modl.12929

Hsiao, W.-H., & Chang, T.-S. (2014). Understanding consumers' continuance intention towards mobile advertising: A theoretical framework and empirical study. *Behaviour & Information Technology, 33*(7), 730–742. DOI: 10.1080/0144929X.2013.789081

Hu, L., & Bentler, P. M. (1999). Cutoff criteria for fit indexes in covariance structure analysis: Conventional criteria versus new alternatives. *Structural Equation Modeling*, *6*(1), 1–55. DOI: 10.1080/10705519909540118

Hu, X., Chen, Z., Davison, R. M., & Liu, Y. (2022). Charting consumers' continued social commerce intention. *Internet Research*, *32*(1), 120–149. DOI: 10.1108/INTR-07-2020-0397

Huynh, M.-T. (2024). Using generative AI as decision-support tools: Unraveling users' trust and AI appreciation. *Journal of Decision Systems*, •••, 1–32. DOI: 10.1080/12460125.2024.2428166

Jang, S., & Liu, Y. (2019). Continuance use intention with mobile augmented reality games. *Information Technology & People*, *33*(1), 37–55. DOI: 10.1108/ITP-05-2018-0221

Jeon, J., Lee, S., & Choi, S. (2024). A systematic review of research on speech-recognition chatbots for language learning: Implications for future directions in the era of large language models. *Interactive Learning Environments*, *32*(8), 4613–4631. DOI: 10.1080/10494820.2023.2204343

Katsurai, M., & Ono, S. (2019). TrendNets: Mapping emerging research trends from dynamic co-word networks via sparse representation. *Scientometrics*, *121*(3), 1583–1598. DOI: 10.1007/s11192-019-03241-6

Kim, D., Kim, S., Kim, S., & Lee, B. H. (2024). Generative AI Characteristics, User Motivations, and Usage Intention. *Journal of Computer Information Systems*, •••, 1–16. DOI: 10.1080/08874417.2024.2442438

Kunc, J., Novotná, M., Reichel, V., Križan, F., & Bilková, K. (2024). Bricks or clicks? Factors influencing shopping behavior of Generation Z. *Equilibrium. Quarterly Journal of Economics and Economic Policy*, *19*(2), 521–548. DOI: 10.24136/eq.2999

Liu, Y.-L. E., Lee, T.-P., & Huang, Y.-M. (2024). Exploring Students' Continuance Intention Toward Digital Visual Collaborative Learning Technology in Design Thinking. *International Journal of Human-Computer Interaction*, *40*(11), 2808–2821. DOI: 10.1080/10447318.2023.2172904

Lyu, B., Lai, C., & Guo, J. (2025). Effectiveness of Chatbots in Improving Language Learning: A Meta-Analysis of Comparative Studies. *International Journal of Applied Linguistics*, *35*(2), 834–851. DOI: 10.1111/ijal.12668

Mun, I. B., & Hwang, K.-H. (2024). Understanding ChatGPT continuous usage intention: The role of information quality, information usefulness, and source trust. *Information Development*. Advance online publication. DOI: 10.1177/02666669241307595

Park, K.-W., & Ok, S.-J. (2008). The Determinants of Continuance Use Intention to Use Web Portal. *Journal of Information Systems*, *17*(2), 49–72. DOI: 10.5859/KAIS.2008.17.2.049

Roberts, J., Baker, M., & Andrew, J. (2024). Artificial intelligence and qualitative research: The promise and perils of large language model (LLM) 'assistance.'. *Critical Perspectives on Accounting*, *99*, 102722. DOI: 10.1016/j.cpa.2024.102722

Ruggiero, T. E. (2000). Uses and Gratifications Theory in the 21st Century. *Mass Communication & Society*, *3*(1), 3–37. DOI: 10.1207/S15327825MCS0301_02

Salih, L., Tarhini, A., & Acikgoz, F. (2025). AI-Enabled Service Continuance: Roles of Trust and Privacy Risk. *Journal of Computer Information Systems*, •••, 1–16. DOI: 10.1080/08874417.2025.2452544

Si, H., Duan, X., Cheng, L., & Zhang, Z. (2022). Determinants of consumers' continuance intention to use dynamic ride-sharing services. *Transportation Research Part D, Transport and Environment*, *104*, 103201. DOI: 10.1016/j.trd.2022.103201

Singh, B., Olds, T., Brinsley, J., Dumuid, D., Virgara, R., Matricciani, L., Watson, A., Szeto, K., Eglitis, E., Miatke, A., Simpson, C. E. M., Vandelanotte, C., & Maher, C. (2023). Systematic review and meta-analysis of the effectiveness of chatbots on lifestyle behaviours. *NPJ Digital Medicine*, *6*(1), 118. DOI: 10.1038/s41746-023-00856-1

Verma, D., Tripathi, V., & Singh, A. P. (2021). From physical to digital: What drives generation Z for mobile commerce adoption? *Journal of Asia Business Studies*, *15*(5), 732–747. DOI: 10.1108/JABS-05-2020-0207

Wells, J. D., Campbell, D. E., Valacich, J. S., & Featherman, M. (2010). The Effect of Perceived Novelty on the Adoption of Information Technology Innovations: A Risk/Reward Perspective. *Decision Sciences*, *41*(4), 813–843. DOI: 10.1111/j.1540-5915.2010.00292.x

Xin, R., & Lim, Y. J. (2025). Information adoption of brand messages among Generation Z in trending topics. *Internet Research*. Advance online publication. DOI: 10.1108/INTR-02-2024-0236

Yadav, M., & Rahman, Z. (2017). Measuring consumer perception of social media marketing activities in e-commerce industry: Scale development & validation. *Telematics and Informatics*, *34*(7), 1294–1307. DOI: 10.1016/j.tele.2017.06.001

Yang, X. (2018). Influence of informational factors on purchase intention in social recommender systems. *Online Information Review*, *44*(2), 417–431. DOI: 10.1108/OIR-12-2016-0360

Zanker, M. (2012). The influence of knowledgeable explanations on users' perception of a recommender system. *Proceedings of the Sixth ACM Conference on Recommender Systems*, 269–272. https://doi.org/DOI: 10.1145/2365952.2366011

Zatsu, V., Shine, A. E., Tharakan, J. M., Peter, D., Ranganathan, T. V., Alotaibi, S. S., Mugabi, R., & Muhsinah, A. Bin, Waseem, M., & Nayik, G. A. (2024). Revolutionizing the food industry: The transformative power of artificial intelligence-a review. *Food Chemistry: X*, *24*, 101867. DOI: 10.1016/j.fochx.2024.101867

Zhang, M., Liu, Y., Yan, W., & Zhang, Y. (2017). Users' continuance intention of virtual learning community services: The moderating role of usage experience. *Interactive Learning Environments*, *25*(6), 685–703. DOI: 10.1080/10494820.2016.1172242

Zhou, T. (2011). An empirical examination of users' post-adoption behaviour of mobile services. *Behaviour & Information Technology*, *30*(2), 241–250. DOI: 10.1080/0144929X.2010.543702

KEY TERMS AND DEFINITIONS

Culinary Innovation: The application of new ideas, technologies, or methods in food preparation and dining, aiming to enhance creativity, quality, and user experience.

Informativeness: The quality of information provided by a system, particularly in terms of richness, clarity, and usefulness; highly informative AI tools are seen as more trustworthy and valuable, supporting user satisfaction and continued use.

Large Language Models (LLM): AI systems trained on massive text datasets to understand, generate, and interact with human language in real time, widely used for tasks like answering questions and creative content generation.

Perceived Trendiness: The degree to which a technology, product, or service is viewed as modern, fashionable, or aligned with current social or cultural trends, influencing adoption and engagement among users—especially youth.

Structural Equation Modelling (SEM): A statistical technique that analyzes relationships between observed and hidden variables, combining regression and factor analysis to test complex models in social science research.

Technological Innovativeness: The extent to which a system or product introduces novel features, functionalities, or user experiences, fostering perceptions of creativity and advancement in the domain (such as AI in culinary learning).

Uses and Gratifications Theory (UGT): A framework that explains how people use media and technology to fulfill specific needs (e.g., information, entertainment) based on their own choices and motivations.

Chapter 14
Sustainable Cooking Solutions Powered by AI and Smart Technologies

Rajashree S. Kini
https://orcid.org/0009-0009-8524-957X
Welcomgroup Graduate School of Hotel Administration, Manipal Academy of Higher Education, Manipal, India

Nagendra Yadav
https://orcid.org/0009-0008-9743-9250
Welcomgroup Graduate School of Hotel Administration, Manipal Academy of Higher Education, Manipal, India

ABSTRACT

This chapter explores the connection of artificial intelligence (AI) and smart technologies with sustainable cooking, in alignment with UN SDGs like Zero Hunger, Clean Energy, Innovation, and Responsible Consumption. The challenges in traditional methods are highlighted with the transformation toward smart cooking. The role of IoT devices, AI meal planning, and energy-efficient tools is mentioned to understand. The real-world experiences are shown in case studies from homes, commercial kitchens, and communities. The chapter also stresses accessibility, ethics, and digital literacy and ends with recommendations for sustainable culinary innovation advancement.

DOI: 10.4018/979-8-3373-4042-5.ch014

INTRODUCTION

Sustainability and the Culinary Industry

In the 21st century, sustainability has emerged as a necessary element in all industries, particularly in food production and consumption. The culinary industry, which includes domestic cooking, commercial kitchens, and food service sectors, is a major stakeholder in environmental sustainability as it has direct influence on energy usage, resource consumption, and waste generation (Milindi et al., 2022). Shukla et al. (2025) provide a forward-looking outlook on rural waste management. The FAO's 2024 report on food loss and wastage indicates the significant global problem of food loss and waste and estimates that 13.2% of food is lost during the supply chain after harvest, and 19% is wasted at retail, food service, and household levels. This leads to significant environmental impact, including 8-10% of greenhouse gas emissions globally, and negatively impacts food security and nutrition (*FAO Policy Series: Food Loss & Food Waste*, 2024). This makes cooking a vital part of interventions.

As cooking activities are resource-intensive with electricity or fuel, water, and perishable goods, they increase carbon footprints with inefficient processes (Goodwin, 2023). The problem will aggravate with the design and function of kitchen appliances, various consumer behavior patterns, and supply chain inadequacies. Adding to that, the social dimensions of cooking relating to nutrition, health, and food security are deeply connected to sustainability principles, which affect vulnerable populations (Dickinson et al., 2022).

Sustainable cooking means not just adapting energy-efficient appliances. It includes a systematic structure in mindful meal planning, waste reduction, energy conservation, and dietary shifts to include plant-based options too (Ibe & Kollur, 2024). This transition needs innovative tools along with strategies that are intelligent, smart, scalable, and also adaptable to different culinary environments, from urban kitchens with high technology usage to rural households with restricted resources.

In sustainable cooking, the circular economy helps in minimizing waste by reusing food resources, composting organic scraps, and designing more durable and repairable kitchen technologies. With this, the approach of the traditional linear 'take-make-dispose' model can be shifted to supporting resource efficiency and environmental resilience (Islam et al., 2024).

The Emergence of AI and Smart Tech in Cooking

Due to globalization and technological advancement, Artificial Intelligence (AI), the Internet of Things (IoT), and other smart technologies have reshaped human interaction with food and cooking environments. When it comes to the concept of 'smart kitchen', the connected technologies like voice-assisted cooking, real-time inventory management, smart ovens with sensor-based automation, and data-driven meal planning apps are considered for convenience, precision, and efficiency (Yu & Sung, 2022).

As one can see, AI is able to simulate human intelligence by machines, which in turn can learn, adapt, and make decisions, based on the given data. In a culinary environment, this can lead to predictive recipe generation, optimized cooking schedules, energy-efficient appliance operation, and dynamic food inventory monitoring (Zohra Dakhia et al., 2025). To give an example, smart refrigerators directly address food waste by tracking the shelf life of ingredients and by suggesting recipes to prevent spoilage. Another example is AI-enabled cooking assistants, such as Samsung's SmartThings or the CookPad app, can personalize meals based on dietary preferences and required calories ("AI-Powered Smart Cooking: Samsung's 'Food' App Helps with Meal Recommendations, Recipes," 2023).

Smart technologies are really needed for sustainability, because of their ability to collect real-time data and behavior pushing. IoT-integrated devices can monitor energy and water use, identify inefficiencies, and provide feedback to users. In domestic, commercial, or institutional kitchens, huge data sets can be analyzed by AI-powered systems to optimize procurement, reduce unnecessary inventory, and forecast demand. This leads to substantial reductions in food waste and enhancement in energy usage (Yeldan et al., 2024).

Chapter Objectives

This chapter seeks to examine how AI and smart kitchen technologies can serve as enablers of sustainable cooking practices across domestic, commercial, and community contexts. The objectives are shown in Figure 1:

Figure 1. Objectives

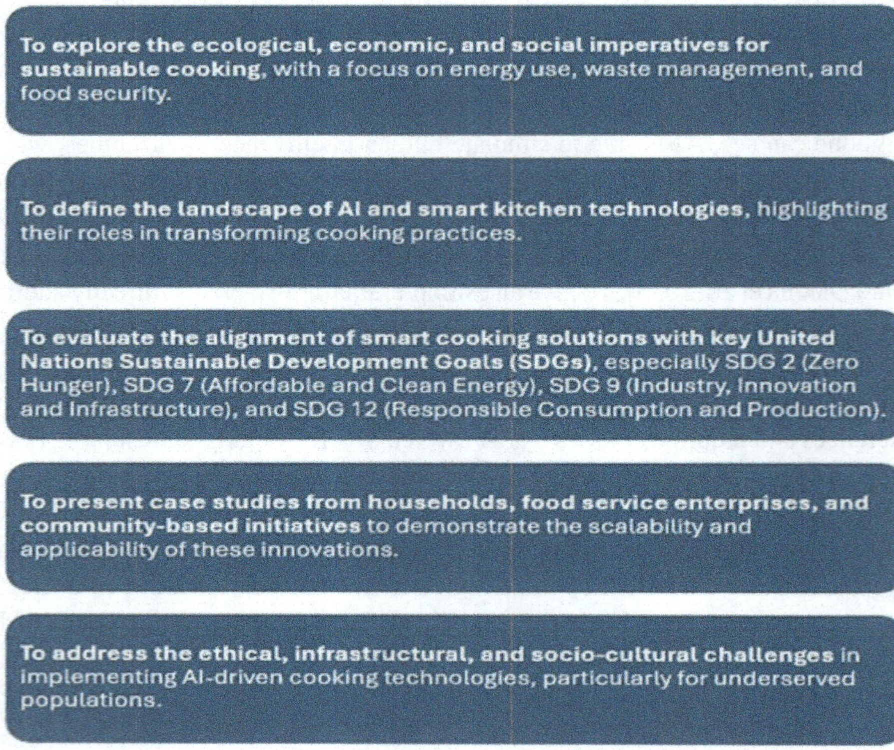

With these objectives, this chapter gives a universal understanding of the connection between digital innovations and their benefits in creating resilient, efficient, and inclusive cooking ecosystems. The chapter contributes to the ongoing concern about the relationship between food, sustainability, and technology. This is an area that has remained underexplored even if it is critically relevant.

Theoretical Underpinning

For the better understand of user adoption and ethical integration of smart cooking technologies, this chapter gives foundation of two key theoretical frameworks: the **Technology Acceptance Model (TAM)** explaining the way users perceive and adopt new technologies, and the **Responsible Innovation Framework** helping in guiding the development of inclusive, ethical, and sustainable technological solutions for sustainable future, as shown in Figure 2:

Figure 2. Two theories

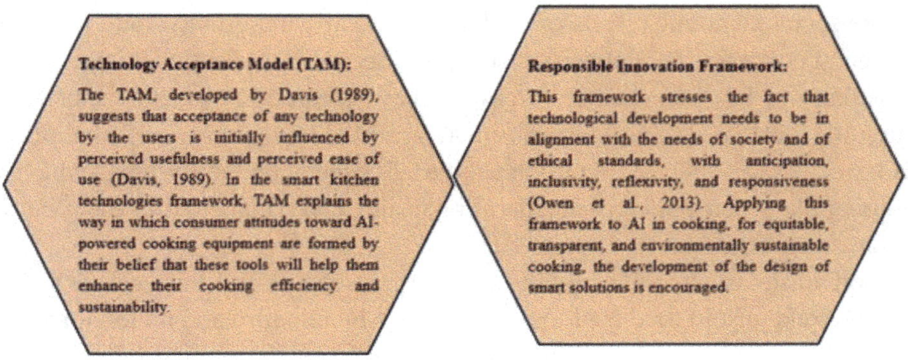

THE NEED FOR SUSTAINABLE COOKING

Environmental Footprint of Traditional Cooking

Traditional cooking methods are a basic activity of society, but they harm the environment in a significant way. As they depend mainly on electricity, gas, or biomass, they are energy-intensive and inefficient. Globally, it can be seen that household cooking accounts for 10% of energy use, as it uses wood and charcoal, and this in developing regions contributes to air pollution and deforestation (IEA, 2022). The kitchen's carbon footprint is increased by poor appliance efficiency and a poor ventilation system.

Another major concern is water usage, as untreated wastewater from the kitchen contains harmful chemicals that can damage ecosystems. It is known that food waste is a critical concern, as 1.05 billion tonnes were wasted globally in 2022, of which 61% from households, due to overbuying, poor storage, and cooking excess food (UNEP, 2023).

As traditional kitchens lack refrigeration, usually in rural areas, it leads to more spoilage. In a linear cooking system, such as buy, cook, consume, discard, the need for and importance of sustainability are overlooked. If there is no awareness and no effort in efficiency improvements, these practices will add to depleting natural resources and affect global food systems (Heydari, 2024).

Socio-Economic Implications

The socio-economic effects of cooking are equally important, along with environmental concerns. Fuel poverty can happen due to inefficient cooking methods, particularly in low-income rural areas, as they lack access to basic technologies (Burguillo et al., 2022). *The integration of AI technologies requires careful adaptation* (Yadav et al., 2024). Over 2.4 billion people still depend on fuel systems, which are polluting like wood and kerosene, harming health, and adding to the labour burden of women and children (WHO, 2024).

Food waste affects the economy of both households and businesses. Spoiled groceries and unsold food lead to huge losses, which in turn add up to global food waste costs around USD 1 trillion annually, still excluding environmental costs (UNFCCC, 2024). The solution is to cut expenses and boost profits by reducing waste and improving efficiency with technology.

Unsustainable cooking mostly results in unhealthy diets. People tend to convenience-based cooking and consume processed foods, do not have portion control, and lack dietary awareness, which leads to obesity and malnutrition (van Kesteren & Evans, 2020). Sustainable cooking promotes nutritious, balanced meals and places less dependence on resource-intensive and animal-sourced products.

The transition from traditional cooking to sustainable cooking must be inclusive. Without equitable access to all, high-tech kitchens may exacerbate the digital divide, which will result in favoring wealthier groups. Governments must initiate low-cost innovations and awareness programs in various areas to ensure fair access and behavioural change (Etim, 2024).

Alignment with UN Sustainable Development Goals (SDGs)

Smart and sustainable cooking is a diverse concept that is connected to several of the UN's Sustainable Development Goals (SDGs) (Figure 3), which support the development of peace, prosperity, and environmental protection:

Figure 3. Related SDGs

- **SDG 2: Zero Hunger**

  One can understand that the AI-enabled cooking systems and smart inventory management tools can reduce food wastage in households and institutions. To enhance health based on balanced diets as per nutritional needs, Personalized meal planning apps can help. Further, AI can be used even in equitable food distribution in community kitchens, and the post-harvest losses can be reduced through predictive demand systems (Fu et al., 2024).

- **SDG 7: Affordable and Clean Energy**

  The traditional stoves can be replaced with induction cookers, solar cookers, and other energy-efficient kitchen appliances, which can help in reducing carbon emissions and air pollution (Mathias et al., 2023). To support clean energy goals in the community, smart devices can be used to monitor and optimize energy usage

- **SDG 9: Industry, Innovation and Infrastructure**

  Smart cooking systems present a very advanced innovation. The sensor-embedded cookware connected kitchen ecosystems, etc., help in showing the good infrastructure and thereby develop sustainable industrialization (Brodny & Tutak, 2023). It is very crucial to invest in digital infrastructure, which is available to both urban and rural communities.

- **SDG 12: Responsible Consumption and Production**

  The optimization of resources and reducing waste is the crux of sustainable and smart cooking. The suggestions based on AI about what is it in the refrigerator, smart portion sizes, and kitchen analytics help users with responsible consumption (Clark et al., 2025). In commercial kitchens, machine learning can be used to optimize purchasing cycles, reduce overproduction, and help institutions meet their sustainability targets.

- **SDG 13: Climate Action**

  There are many climatic benefits in reducing energy use and food waste in cooking. There are AI systems that can analyze the carbon impacts of meal choices and suggest users other greener substitutes, to make environmentally conscious decisions when required. Adding to that, data stored in smart appliances can inform policies and innovations to reduce emissions within the food value chain (Sevda Senturk et al., 2023).

These smart technologies in cooking, if implemented well, can support the SDGs. This can change society's thinking about food preparation, consumption, and waste, beyond considering it as simply using appliances or updating recipes. With responsible application, AI and smart technologies can accelerate the progress of the SDGs and support innovation, inclusion, and resilience in cooking and food consumption.

SMART TECHNOLOGIES ENABLING SUSTAINABLE COOKING

The sustainable goals are necessary for the world to sustain itself in the future. As kitchen activities are basic for all in the world, the application of smart technologies to promote responsible consumption and protection of the environment needs to start with the kitchen itself. It is imperative to include AI, IoT, and machine learning into household and commercial kitchens to reduce wastage of resources, increase nutritional values, and make the decision-making process simple (Isaac Sam Hayford et al., 2023). *Food-related content drives online engagement in hospitality* (Seal et al., 2025). The following are a few major categories of smart technologies used in sustainable cooking that align with the global sustainability goals:

IoT and Sensor-Based Cooking Appliances

The kitchen appliances with Internet of Things (IoT) can collect. Exchange, transform, and analyse data for performance optimization. For example, modern kitchen appliances like smart refrigerators, ovens, dishwashers, and composters have sensors to monitor temperature, humidity, energy usage, and inventory status. These work smart and develop interconnected ecosystems that generate automated responses that can be used in enhancing energy efficiency and minimizing waste (Cooper et al., 2023).

To illustrate, smart refrigerators can detect food items that are getting expired and warn people to use them soon, or can recommend recipes that use those ingredients. This helps in reducing food waste (Sharma et al., 2022). Another example is smart ovens have temperature and humidity sensors that can automatically adjust settings for a few recipes, which in turn will reduce the likelihood of overcooked or underutilized ingredients. It can be used in saving energy as well as food resources (Yu & Sung, 2023). The smart composters can convert food waste into compost for gardens and can support the circular food system. The sensor-based dishwashers can enhance the usage of water and detergent based on the level of dirt and load in commercial kitchens, which can enhance sustainability in the usage of resources (Dolnicar et al., 2024).

These sensor-enabled automations have a dual role: they make it convenient for users to develop sustainable behaviour as they get real-time data feedback. Smart technologies can develop personalized, sustainable solutions by continuously learning user habits, which can be a hallmark of advanced smart technologies.

AI-Driven Meal Planning and Inventory Management

One of the best applications of AI in sustainable cooking is in personalized meal planning and intelligent inventory tracking. AI-based apps and other related domains analyze data based on user preferences, health parameters, and available inventory and recommend meals that minimize waste and maximize nutritional value.

For example, AI tools like Cookpad or Mealime can be used to develop recipes based on leftover food or the ingredients nearing expiry dates (Kansaksiri et al., 2023). This enhances the culture of creativity in the kitchen and helps in reducing food spoilage. The wearable health devices or diet apps used with technology can generate meal options controlling calories, checking for allergies, or supporting health goals, thereby promoting the well-being of the person and the surrounding environment (Zahedani et al., 2023).

Technology-driven inventory management systems can help in analysing patterns of purchase, seasonal availability of items, and can suggest an ideal grocery list of items to reduce overbuying or impulsive purchases (Singh & Adhikari, 2023). AI can also forecast accurate customer demands to support just-in-time procurement of ingredients. This is a demand-responsive planning that can minimize spoilage and reduce unsold items in schools, hospitals, and restaurants (Pal, 2023).

Very interestingly, AI based chatbots or virtual kitchen assistants can help users in real time on preparing meals with limited ingredients or on reusing of leftover food. This reduces the cognitive load on users and also promote responsible consumption (Kim et al., 2025).

Machine Learning for Predictive Maintenance and Energy Efficiency

Another domain where smart technology adds sustainable value is through predictive maintenance and real-time energy optimization. Machine learning algorithms inserted in kitchen appliances can examine operational parameters, identify performance problems, and signal users before failures occur. This extends the life of appliances, reduces repair costs, and limits e-waste—a growing concern in the age of disposable tech (Goyal & Gupta, 2024).

Smart systems can also study usage models to suggest optimum time slots for energy-intensive cooking assignments. For example, AI can coordinate cooking with renewable energy availability from rooftop solar panels or lower-tariff grid times, which reduces both cost and environmental effect (Ukoba et al., 2024). The adaptive planning of this is specifically important in power supply restricted regions or where the cost of energy are unstable.

To prevent overheating or prolonged cooking, a few smart stoves and cooktops can automatically alter heat output to match the exact requirement of a cooking vessel. These modified changes conserve energy and maintain food quality and nutritional integrity (Rasugu Ayub et al., 2021).

In food operations on a large scale, machine learning tools can benchmark energy performance across various branches, which provides comparative analytics to foster constant improvement. To ensure sustainability metrics are part of everyday operations and long-term planning, integration with enterprise resource planning (ERP) systems is needed (Agarwal, 2024).

Integration with Renewable Energy Sources

One of the most promising paths in sustainable cooking is the merging of smart kitchen systems with renewable energy sources. Solar-powered cookers, battery-supported induction systems, and biogas-fed smart appliances are being piloted worldwide, especially in regions with limited grid access (Gordon et al., 2023).

AI plays a vital role in energy source management by prioritizing when to draw power from solar batteries versus the main grid, or by automating shifts in power mode based on weather forecasts. Few systems incorporate AI to forecast solar availability and adjust cooking recommendations accordingly. For example, advising quick-cook meals on cloudy days or scheduling slow-cooked dishes for peak sunshine periods (*World Energy Outlook,* 2024).

Smart microgrids in community kitchens allow several users to share renewable energy resources in a controlled and optimal way. IoT devices help balance load and usage among members, while AI algorithms optimize energy allocation based on user behaviour, appliance needs, and community preferences (Anayo Chukwu Ikegwu et al., 2025).

These hybrid systems, combining AI, IoT, and renewable energy, present a future-ready model of sustainable cooking. They lessen reliance on fossil fuels and reduce greenhouse gas emissions, while also creating opportunities for energy equity in marginalized communities (Marwan Al-Raeei, 2024).

CASE STUDIES AND INDUSTRY APPLICATIONS

When utilised thoughtfully in different areas, the real-world usage of smart technologies in sustainable cooking shows the tangible impact of innovation. Case studies from urban households, commercial kitchens, and community-led programs show good insights about the way AI and IoT can alter cooking behaviour, improve resource efficiency, and encourage environmentally responsible actions (Zhang et al., 2023). This section shows a few experienced cases to show how technological solutions can be efficiently integrated into cooking systems at various scales.

Smart Kitchens in Urban Households

Urban households adopt smart kitchen solutions, which have soon launched in the market, driven by consumer desire for convenience, health-conscious living, and eco-friendly options. In developed countries, tech-savvy consumers prefer AI-enabled kitchen assistants, smart fridges, and connected cooktops.

To give an example, is the incorporation of Samsung's SmartThings Cooking ecosystem in U.S. households (*Samsung SmartThings Connects People to the Things That Matter Most*, 2023). The platform connects appliances, grocery lists, and meal planning apps, and offers personalized recipe suggestions based on existing ingredients and also on dietary goals (*Samsung Announces Global Launch of Samsung Food, an AI-Powered, Personalized Food and Recipe Service*, 2023). This system has been shown to reduce food waste up to 20% per household yearly, according to Samsung's internal reports (Samsung Electronics, 2023). The integration also promotes energy-saving habits by alerting users to peak-hour energy consumption and proposing low-energy cooking options.

The consumers in Scandinavian countries use AI tools like Plant Jammer to get suggestions for plant-based meals using leftover vegetables and pantry items. This helps in reducing food waste and encourages sustainable dietary changes. The gamified user experience of the App, along with its sustainability tips, is an example of changing behaviour through design (Hannu Saarijärvi et al., 2023).

These examples clearly show that smart home ecosystems can assist households to contribute thoughtfully to SDG 12 (Responsible Consumption and Production) and SDG 13 (Climate Action) based on daily actions supported by artificial intelligent systems (*Annual Report 2022 | UNDP*, 2022).

AI in Commercial and Institutional Kitchens

The hotels, restaurants, universities, and hospitals have commercial kitchens, which are intricate environments where huge volumes of food procurement and consumption happen. As a result, there will be a major sustainability challenge and at the same time a strong opportunity for impact. AI and smart technology systems used in these environments can decrease operational ineffectiveness and enhance resource management (Şener & Ulu, 2024).

An AI-powered food waste tracking system, Winnow Vision, is used by commercial kitchens across Europe, Asia, and the Middle East. The system detects and weighs discarded food, using smart cameras and machine learning, which supports chefs and managers with actionable insights on where waste is more and how to decrease it. As per the company, the implementation of Winnow Vision has helped food service operations reduce food waste by up to 50%, which results in average savings of $30,000 per site annually (*How Winnow Is Ending Food Waste with AI*, 2020; Clark et al., 2025).

Similarly, Compass Group, one of the world's largest food service providers, has taken steps to integrate AI-driven forecasting applications and tools to enhance ingredient purchasing and menu planning throughout its worldwide operations. The system analyse consumption patterns and seasonal data, and reduces surplus and ensures that meals are made as per demand, which helps in reducing food waste and improving cost-efficiency (*Compass Group India - Largest Corporate Food & Catering Service Company*, 2023).

The Chef AI domain is helpful in university kitchens as it utilizes past consumption of data and proposes daily menus and portion sizes. The results show a reduction in food waste by 27% over a single semester. This support of artificial intelligence shows that AI can be leveraged for sustainability and also for cost efficiency and improved nutritional results in large-scale institutional settings. (*AI-Driven Recipes and Culinary Delights*, 2023).

These examples strongly align with SDG 9 (Industry, Innovation and Infrastructure) and SDG 2 (Zero Hunger), which show the capacity of smart technologies to enhance economic and ecological performance (Annual Report 2022 | UNDP, 2022).

Sustainable Meal Delivery Platforms

The rise of online meal delivery services has introduced new layers of complexity and opportunity in sustainable cooking. While convenience is the primary driver of such services, concerns have emerged around packaging waste, inefficient routing, and surplus meal production. In response, several tech-driven platforms have emerged that integrate sustainability into their core design using AI and smart logistics (Casciani & D'Itria, 2024).

Olio and Too Good To Go are two mobile apps designed to reduce food waste by connecting consumers with surplus food from grocery stores, restaurants, and even neighbors. AI algorithms match supply with demand in real time and prioritize meals or products nearing expiration. As of 2023, Too Good To Go reported saving over 200 million meals globally from being wasted (Too Good To Go, 2024; Yu et al., 2025).

In India, startups like Rebel Foods are using AI to optimize dark kitchens, centralized food preparation facilities with no dine-in services. AI helps in menu engineering, demand forecasting, and dynamic delivery routing, which cuts fuel emissions and reduces food overproduction. Combined with smart inventory systems, these practices contribute to both environmental and operational sustainability (Lahiri et al., 2024).

Moreover, some platforms are experimenting with AI-based carbon tracking. For instance, Just Eat in the UK has partnered with Carbon Cloud to provide diners with the estimated carbon footprint of each dish (Fernandez et al., 2025). This transparency empowers users to make informed, eco-conscious choices, furthering the goals of SDG 12 and SDG 13 (*Annual Report 2022 | UNDP*, 2022).

Community Kitchens and NGO Innovations Using AI

The application of smart technologies for sustainable cooking is not limited to affluent homes or well-funded corporations. Community kitchens and NGOs around the world are piloting innovative models that bring AI-driven cooking tools to underserved populations (Sammer et al., 2024).

In Nairobi, the non-profit organization Mukuru Clean Stoves equips urban slums with energy-efficient, sensor-enabled cookstoves that monitor cooking duration, temperature, and fuel usage (*Home*, n.d.). Data collected through IoT systems informs community outreach programs, allowing for adjustments in user education and stove design. These interventions have improved indoor air quality and reduced deforestation from charcoal use, supporting both SDG 7 (Affordable and Clean Energy) and SDG 3 (Good Health and Well-being) (*Annual Report 2022 | UNDP*, 2022).

In Brazil, the NGO Gastromotiva operates "social gastronomy" hubs that use AI analytics to match surplus food donations with meal production needs in underserved communities. With the help of machine learning algorithms, the initiative forecasts local demand, optimizes logistics, and ensures food safety standards. As of 2021, the initiative had served over 1 million meals while diverting tons of food from landfills (*Gastromotiva*, 2024; Damasceno et al., 2023).

AI is also aiding disaster response kitchens. The World Food Programme has begun their experiment with AI-powered supply chain tools that help field kitchens in refugee camps to plan meals, optimize rations, and ensure equitable distribution under resource-scarce conditions (Sammer et al., 2024).

The live examples explain well that smart technologies can be made accessible and impactful even in low-resource regions and countries, which aligns with SDG 10 (Reduced Inequalities) and SDG 17 (Partnerships for the Goals) (*Annual Report 2022 | UNDP*, 2022).

CHALLENGES AND ETHICAL CONSIDERATIONS

The integration of Artificial Intelligence (AI) and smart technologies within culinary operations promotes sustainable cooking practices; however, there are many challenges in accessibility and availability. Technological advancements, such as data-driven systems and machine learning, can intensify existing disparities or create new risks if not implemented carefully. This section explores various key barriers - technological, economic, social, and ethical, which need to be addressed to make sure that there is equitable and responsible adoption of smart cooking solutions for sustainable development.

Affordability and Accessibility of Smart Tech

The concern about smart kitchen technologies is their affordability, among other issues. The systems, Advanced appliances, AI-driven cooking apps, and IoT ecosystems are usually priced very high, making them beyond the reach of low-income households or small-scale food businesses. The costs include the initial cost of smart devices such as sensor-enabled ovens, refrigerators, or induction cooktops, and recurring expenses for internet connectivity, updates, and maintenance, which work as a strong entry barrier (Yu & Sung, 2023).

In developing countries, due to cost restrictions, households still rely on traditional biomass stoves, although there are environmental and health risks (Umer Hayyat et al., 2024). Without targeted subsidies, financing models, or inclusive design strategies by the government, smart cooking technologies remain a luxury and privilege of wealthy people rather than equipment used for sustainable development throughout the world.

Additionally, there are infrastructure disparities, such as limited electricity access, unreliable internet, and inadequate digital services, which intensify the challenge of adopting smart kitchen solutions equitably throughout the world (Effoduh, 2024). These structural barriers can be addressed with multistakeholder partnerships, which include governments, NGOs, and the private sector to ensure that sustainability and digital transformation go hand in hand for underserved populations.

Digital Divide and Tech Literacy

Closely related to affordability is the issue of digital literacy. Effective use of AI-powered applications and smart appliances needs a certain degree of technological fluency, which is unequally distributed across different age groups, geographies, and education levels. Previous studies show that to adapt to technology in urban and rural areas, the biggest barrier is digital illiteracy (*UNESCO Global Education Monitoring Report*, 2023).

To give an example, aged adults or individuals without proper formal education find it difficult to navigate voice-assisted meal planners or to configure app settings, or to interpret energy consumption analytics (C. González-González et al., 2025).

Also, language limitations in multilingual societies can restrict user engagement, as interfaces are not localized or culturally adapted in many cases.

This strong digital divide reduces the reach of smart cooking technologies and also enlarges the existing inequalities by enhancing the benefits of AI for the digitally literate people. Thus, educational interventions, user-friendly designs, and multilingual platforms are important and necessary to democratize access and promote people's participation in smart cooking ecosystems universally (Helbing et al., 2023).

Data Privacy and AI Bias in Culinary Recommendations

The smart kitchens are data-driven, and this raises concerns about privacy and algorithmic bias. To function well, many AI-powered cooking applications collect sensitive user data, including dietary habits, household composition, health conditions, and location data. If not handled well, this information can be exploited by third-party agents for targeted marketing, insurance pitches, or any related discriminatory practices (Sapienza & Vedder, 2021).

In addition to this, bias in AI algorithms is a concern in all domains, including culinary applications. For example, recipe recommendation systems may prioritize Eurocentric or affluent dietary patterns due to the biases that are included at the time of training the datasets. This puts the diverse cultural cuisines in the corner and may stress unhealthy eating norms, specifically if 'optimized' for user engagement rather than focusing on sustainability or nutrition (Bidyalakshmi et al., 2024).

Then again, in regions with varied nutritional needs or food access constraints, AI must be trained with localized data and needs to be guided by ethical frameworks to prioritize health, culture, and sustainability. Transparency in algorithm design, consent-based data collection, and the right to opt out must be strictly included in all smart kitchen applications to protect user autonomy (Khatiwada et al., 2024).

Lifecycle Sustainability of Smart Devices

While smart kitchen technologies promote efficiency during use, their entire lifecycle, including production, transportation, and disposal, must be considered to assess true sustainability. Many smart appliances rely on rare earth metals, complex circuits, and non-recyclable components that contribute to environmental degradation when disposed of improperly (Oladunni Bola Abogunrin-Olafisoye et al., 2025).

The rapid pace of innovation also results in short product life cycles, pushing consumers to upgrade frequently and leading to mounting electronic waste (e-waste). The United Nations reports that over 53.6 million metric tons of e-waste were generated globally in 2019, with only 17.4% being recycled (UNITAR, 2024). If smart kitchen technologies are to advance sustainable development, manufacturers must be held accountable for product longevity, repairability, and recyclability.

Circular economy principles, such as modular design, take-back programs, and second-life applications, should be integrated into the development of smart kitchen ecosystems. Governments can play a role by mandating extended producer responsibility (EPR) and offering incentives for environmentally responsible innovation (Lifset et al., 2023).

Cultural Resistance and Behavioral Inertia

Sustainable cooking involves both a technological and behavioural challenge. The traditional cooking methods are culturally significant and socially included, and individuals are attached to them. For example, in a few cultures, the process of cooking on an open flame is considered identity, celebration, and community. With this, the introduction of AI-based cooking methods can be understood as impersonal, sterile, and threatening to cultural continuity (Seaman et al., 2024).

Similar to this, in highly digital societies also, users resist change because of habit, mistrust in AI, or thinking about losing control over food decisions. For example, users may feel uncomfortable with automated grocery purchases or AI-generated meal plans that lack human emotion, intuition, and improvisation (Selter et al., 2023).

To overcome these behavioural hesitations, smart cooking technologies must be designed with empathy, allowing users to gradually adapt to them and giving an option for manual override of the automatic actions. Co-designing technologies with community insights and incorporating feedback features can lead to greater acceptance and trust. Highlighting user empowerment, more than automation, will help align AI tools with cultural and emotional values around food with people (Schaper et al., 2023).

FUTURE OUTLOOK AND POLICY RECOMMENDATIONS

The future of sustainable cooking is strongly connected to AI, smart technologies, and global sustainability goals (Marwan Al-Raeei, 2024). As these tools are developed and used, they get integrated into how people cook, eat, and engage with food systems. Yet, for integration to be successful, it needs thoughtful strategies to balance innovation, inclusivity, and regulation. This section discusses key opportunities and policy recommendations for stakeholders - governments, industry, educators, and communities- to support the adoption of smart cooking solutions.

Scaling Sustainable AI-Driven Cooking Solutions

This era moves with rapid advancements in AI system performance, affordability, and kitchen integration, with luxury trends like AI-powered autonomous meal-preparation robots, blockchain-enabled food supply chain traceability etc. (Parvez et al., 2024). The scalable application happens with interoperability between different tools, devices and systems. Open-source AI models and recognised standards help in cross-market innovation and reduce vendor lock-in. To support greater adoption, the cloud-based platforms can centralize the recipe management, energy analytics, and food inventory tracking (Freitas et al., 2023). It can be complemented by necessary investments in smart infrastructure such as energy-efficient housing, renewable microgrids, and digital food marketplaces to enable smooth AI-integrated kitchen operations (Qayyum et al., 2024). The innovations, when integrated with sustainability principles, can significantly reduce waste, enhance nutrition, and improve resource efficiency in cooking, with public-private partnerships having a major role in financing and implementing the systems at scale, which is helpful for developing and underdeveloped countries.

Policy Support for Smart Kitchen Technologies

To accelerate the adoption of sustainable cooking technologies, the government needs to develop policies. Regulatory frameworks need to support research and development in green kitchen technologies, creating incentives and benefits to consumers and businesses that base their actions on sustainability.

Figure 4. The suggested key policies can be (Google Images)

- **Subsidies and tax incentives** for purchasing energy-efficient appliances and renewable-powered cooking systems.
- **Grants for local innovation hubs** that develop culturally relevant, low-cost smart cooking solutions tailored to regional needs.
- **Standards and labeling schemes** that certify appliances based on lifecycle sustainability, data privacy compliance, and AI transparency.
- **Integration of smart cooking into national digitalization and energy access plans**, particularly in alignment with SDGs 7, 9, and 12.

Policymakers should also update building codes to include smart kitchen readiness, ensuring that new constructions are equipped for intelligent energy and appliance management (Hernández et al., 2024). It is very important that the policies need to be inclusive and equity-focused, and need to ensure that the rural and marginalized populations are not left behind in the smart cooking revolution, which is developed for society itself.

Education and Awareness Initiatives

Technology innovation itself is not enough; it should be integrated with raising awareness through education and also with user competence (Legi et al., 2023). Sustainable cooking is a combination of the right tools and knowing how and why to use them in different settings. Education initiatives regarding awareness with customised content and delivery need to target multiple audiences, such as students, household cooking, professional chefs, and community leaders.

To empower individuals and to support them in adopting responsible cooking habits, there is a need to integrate sustainability and digital literacy into school curricula, culinary training programs, and community workshops.(Gema Ika Sari et al., 2024).

Syllabus revision in culinary schools and hospitality programs should include modules on AI, sustainability, and smart kitchen design. For example, students should made to learn how to use data analytics for menu optimization or practice zero-waste cooking using IoT-based monitoring systems, which is embedded in syllabus (Buchan et al., 2024).

Public campaigns using social media, community centers, and digital influencers can help to normalize sustainable cooking behaviors. Storytelling, gamification, and participatory design approaches may enhance engagement and foster a sense of ownership among users.

Incentivizing Sustainable Behavior through AI Nudging

One of AI's most powerful capabilities lies in its potential to influence behavior through personalized, real-time nudges (Figure 5). Just as recommendation engines influence what we watch, wear, or buy, similar systems can encourage individuals to make environmentally responsible food choices (Moshtaghian et al., 2023).

Figure 5. Influence of AI-driven cooking apps

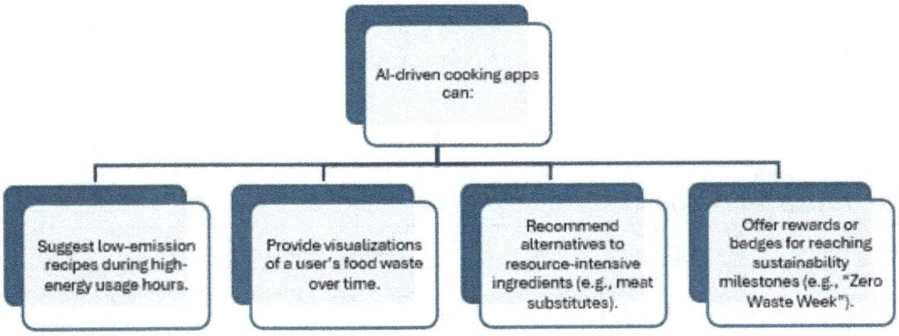

These micro-interventions, or 'digital nudges,' can significantly alter consumption patterns when embedded into daily routines. Research shows that such interventions are most effective when they are contextual, timely, and aligned with user values (Ali et al., 2024).

However, ethical considerations must be addressed to ensure nudges are transparent and non-coercive (Craigon et al., 2022). Users should retain full control over their preferences, with clear options to customize or disable AI suggestions, to avoid paternalism or privacy violations.

Governments and app developers can collaborate to establish 'Sustainable Design Guidelines' for AI in cooking, similar to frameworks already being explored in other sectors like finance and healthcare (Gurung et al., 2024), to standardize best practices and encourage ethical innovation.

The future of sustainable cooking lies at the intersection of intelligent technology, inclusive policy, and empowered communities. With supporting ecosystems for responsible adoption of AI and smart kitchen tools, societies can support environmental goals along with enhancing human well-being. This section concludes that the path forward is not only technological, but it includes cultural, educational, and political aspects.

PRACTICAL IMPLICATIONS OF THE STUDY

The connection between Artificial Intelligence and smart technologies in cooking has theoretical as well as major real-world applications. As established throughout this study, these innovations can transform individual behavior, institutional operations, and government policy related to food systems and sustainability. The practical implications shown in Table 1 below deliver actionable information for households, industry professionals, educators, policymakers, and technology developers to develop responsible consumption, reduce negative impact on the environment, and support the larger goals of the circular economy and the Sustainable Development Goals (SDGs).

Table 1. Practical implications

Implication Area	Description
1.Household Empowerment	At the household level, AI-based meal planners, inventory apps, and smart appliances help reduce food waste, conserve energy, and encourage eco-friendly behaviour.
2.Commercial & Institutional Efficiency	To optimize procurement, reduce overproduction, and achieve sustainability targets, hotels, hospitals, NGOs, and large kitchens can use predictive analytics and IoT systems.
3. Policy & Access Equity	Policymakers can develop policies to support comprehensive technology implementation through subsidies, digital infrastructure, and community-level smart cooking initiatives.
4. Culinary Education & Skill Development	Hospitality programs can include AI, sustainability, and digital literacy concepts in the curriculum to prepare students as chefs and professionals for tech-smart and green kitchen operations.
5. Data-Informed Public Policy	Data from smart kitchens (e.g., energy use, food waste, carbon tracking) can be used by the government and other related authorities to develop national policies on food security, climate change, and responsible consumption.
6. Green Tech Innovation Catalyst	For sustainable and inclusive cooking, entrepreneurs can design sustainable kitchen technologies in connection with circular economy principles, by focusing on accessibility, durability, user inclusivity, and awareness.

CONCLUSION

Sustainable cooking is not a luxury or a trend; it is mandatory. With pressure of climatic changes, resource scarcity, and growing urbanization on global food system, the cooking strategies must change. This chapter has developed connection between Artificial Intelligence (AI) and smart technologies, and it proves that they are not just futuristic conveniences, but are essential tools and systems for building resilient, inclusive, and ecologically responsible cooking ecosystems for sustainable development. Starting from smart kitchens in urban homes to AI-assisted community meal programs, the advanced technologies support scalable solutions which align with the United Nations Sustainable Development Goals (SDGs), SDG 2 (Zero Hunger), SDG 7 (Affordable and Clean Energy), SDG 9 (Industry, Innovation and Infrastructure), and SDG 12 (Responsible Consumption and Production).

This chapter has shown the multifaceted impact of traditional cooking methods, which will affect energy waste, food loss, and greenhouse gas emissions. This chapter examines the way in which digital tools, ranging from IoT-enabled appliances and AI-based meal planning apps to sensor-integrated composters and predictive maintenance systems, can strongly enhance efficiency and lessen ecological footprints. Real-world experiences and case studies have shown these impacts in regular practice, giving evidence of tangible reductions in waste and an increase in user engagement, nutritional outcomes, and operational efficiency.

Along with the advantages, there are many challenges as well. The significant barriers to adoption are affordability, accessibility, and digital literacy. The chapter has highlighted the importance of designing inclusive technologies and policies to include underserved communities in the digital transformation of cooking. Along with that, ethical issues like data privacy, algorithmic bias, and electronic waste are addressed to make sure that smart kitchen ecosystems support sustainability in the whole lifecycle.

The future of sustainable cooking is based on the integration of technology with culture, of policy with innovation, and of environmental goals with regular daily habits. While making regulatory frameworks, policymakers should consider ethical aspects too. Developers should give importance to user-centric and eco-conscious design in technologies. Educators should redesign culinary learning to include sustainable cooking and AI literacy. The consumers must enhance their knowledge, learning, tools, and confidence to make smarter, greener, and sustainable choices in the kitchen.

In the long run, this chapter advocates for a shift in the mindset of people, from considering AI and smart technologies as a luxury addition to recognizing them as a crucial ecosystem for sustainable living. The kitchen is usually seen as a personal or private space, and hence a powerful link in global sustainability networks. By changing how we cook, we can change how we live, consume, and care for the planet.

FINAL THOUGHT

Sustainability starts at home, and with intelligent integration of technology, our kitchens can become microcosms of planetary well-being. As we move towards a more digitally connected and environmentally conscious societies, the question isn't whether AI belongs in our kitchens, but how we can make sure that it serves everyone, everywhere, in creating a healthier and more sustainable future.

REFERENCES

Abogunrin-Olafisoye, O.B., Adeyi, O., Adeyi, A. J., Oke, E. O., & Oluranti Ojediran, G. (2025). Sustainable Waste Electrical and Electronics Equipment Recycling in Nigeria, Policies and Emerging Technologies. *Environmental Quality Management, 34*(3). Advance online publication. DOI: 10.1002/tqem.70020

Agarwal, P. (2024). Harnessing the Power of Enterprise Resource Planning (ERP) and Customer Relationship Management (CRM) Systems for Sustainable Business Practices. *International Journal of Computer Trends and Technology, 72*(4), 102–110. DOI: 10.14445/22312803/ijctt-v72i4p113

AI-Driven Recipes and Culinary Delights. (2023). Chefai.net. https://www.chefai.net/

AI-powered smart cooking: Samsung's "Food" app helps with meal recommendations, recipes. (2023, September 1). *The Economic Times.* https://economictimes.indiatimes.com/magazines/panache/ai-powered-smart-cooking-samsungs-food-app-helps-with-meal-recommendations-recipes/articleshow/103273473.cms?from=mdr

Al-Raeei, M. (2024). Artificial intelligence for climate resilience: Advancing sustainable goals in SDGs 11 and 13 and its relationship to pandemics. *Discover Sustainability, 5*(1). Advance online publication. DOI: 10.1007/s43621-024-00775-5

Al-Raeei, M. (2024). The smart future for sustainable development: Artificial intelligence solutions for sustainable urbanization. *Sustainable Development, 33*(1). Advance online publication. DOI: 10.1002/sd.3131

Ali, S. H., & Rahman, F., Aakanksha Kuwar, Khanna, T., Nayak, A., Sharma, P., Sarika Dasraj, Auer, S., Rouf, R., Patel, T., & Dhar, B. (2024). Rapid, Tailored Dietary and Health Education Through A Social Media Chatbot Microintervention: Development and Usability Study With Practical Recommendations. *JMIR Formative Research, 8*(1), e52032. DOI: 10.2196/52032 PMID: 39652870

Annual Report 2022 | UNDP. (2022). UNDP. https://annualreport.undp.org/2022/index.html

Bidyalakshmi, T., Jyoti, B., Mansuri, S. M., Srivastava, A., Mohapatra, D., Kalnar, Y. B., & Indore, N. (2025). Application of artificial intelligence in food processing: Current status and future prospects. *Food Engineering Reviews, 17*(1), 27–54.

Buchan, M. C., Bhawra, J., & Katapally, T. R. (2024). Navigating the digital world: Development of an evidence-based digital literacy program and assessment tool for youth. *Smart Learning Environments, 11*(1). Advance online publication. DOI: 10.1186/s40561-024-00293-x

Burguillo, M., Barisone, M., & Juez-Martel, P. (2022). Which cooking and heating fuels are more likely to be used in energy-poor households? Exploring energy and fuel poverty in Argentina. *Energy Research & Social Science*, *87*, 102481. DOI: 10.1016/j.erss.2021.102481

Casciani, D., & D'Itria, E. (2024). Fostering Directions for Digital Technology Adoption in Sustainable and Circular Fashion: Toward the Circular Fashion-Tech Lab. *Systems*, *12*(6), 190. DOI: 10.3390/systems12060190

Clark, Q. M., Kanavikar, D. B., Clark, J., & Donnelly, P. J. (2025). Exploring the potential of AI-driven food waste management strategies used in the hospitality industry for application in household settings. *Frontiers in Artificial Intelligence*, *7*. Advance online publication. DOI: 10.3389/frai.2024.1429477 PMID: 39917546

Compass Group India - largest corporate food & catering service company. (2023, February 15). Compass Group. https://compass-group.co.in/

Cooper, A., Lion, R., Rodriguez-Sierra, O.E., Jeffrey, P., Thomson, D., Peters, K., & Christopher, L., Mona, Wistrand, L., van, & E. van Herpen. (2023). Use-up day and flexible recipes: Reducing household food waste by helping families prepare food they already have. *Resources, Conservation and Recycling*, *194*, 106986–106986. DOI: 10.1016/j.resconrec.2023.106986

Craigon, P. J., Sacks, J., Brewer, S., Frey, J., Mendoza, A. G., Jacobs, N., Kanza, S., Manning, L., Munday, S., Wintour, A., & Pearson, S. (2022). Ethics by Design: Responsible Research & Innovation for AI in the Food Sector. *Journal of Responsible Technology*, *13*, 100051. DOI: 10.1016/j.jrt.2022.100051

Dakhia, , Z., Russo, M., & Merenda, M. (2025). AI-Enabled IoT for Food Computing: Challenges, Opportunities, and Future Directions. *Sensors (Basel)*, *25*(7), 2147–2147. DOI: 10.3390/s25072147 PMID: 40218659

Damasceno, R. R., Bonvini, O., Martins, N. de S., Bastos, R. M. S., & Garcez, T. C. da N. A., Carneiro, A. P. de G., & Costa, E. de A. (2023). How can gastronomy impact the social change of a vulnerable population assisted by social programs? The case of Brazilian social gastronomy. *International Journal of Gastronomy and Food Science*, *32*, 100728. DOI: 10.1016/j.ijgfs.2023.100728

Dickinson, A., Wills, W., Kapetanaki, A. B., Ikioda, F., Godfrey-Smythe, A., & Halliday, S. V. (2022). Food security and food practices in later life: A new model of vulnerability. *Ageing and Society*, *42*(9), 1–26. DOI: 10.1017/S0144686X20002020

Dolnicar, S., & Greene, D., Siamak Layeghy, & Portmann, M. (2024). Does every hotel room need a minifridge? Empirical evidence from consumer self-reports and an automatic sensor-based system measuring electricity consumption and guest use. *Annals of Tourism Research Empirical Insights*, *5*(2), 100134–100134. DOI: 10.1016/j.annale.2024.100134

Effoduh, J. O. (2024). Africa's Energy Poverty in An Artificial Intelligence (AI) World: Struggle for Sustainable Development Goal 7. *Journal of Sustainable Development Law and Policy (The)*, *15*(3), 32–63. DOI: 10.4314/jsdlp.v15i3.2

Electronics, S. (2023). *Samsung electronics sustainability report 2023*. Samsung Electronics. https://www.samsung.com/global/sustainability/media/pdf/Samsung_Electronics_Sustainability_Report_2023_ENG.pdf

Etim, E. (2024). Leveraging public awareness and behavioural change for entrepreneurial waste management. *Heliyon*, *10*(21), e40063–e40063. DOI: 10.1016/j.heliyon.2024.e40063 PMID: 39553649

FAO Policy Series: Food Loss & Food Waste. (2024). PolicySupport. https://www.fao.org/policy-support/policy-themes/food-loss-and-food-waste/fao-policy-series--food-loss---food-waste/en

Fernandez, R. R., Longhurst, J., & Barnes, J. (2025). *Local policies and interventions to reduce a city's carbon footprint using plant-based diets: Bristol as a case study*. Deleted Journal., DOI: 10.1088/2976-601x/adad76

Freitas, N., Araújo, S. O., Alemão, D., Ramos, J., Guedes, M., Gonçalves, J., Peres, R. S., Rocha, A. D., & Barata, J. (2023). Cloud-Based Machine Learning Application for Predicting Energy Consumption in Automotive Spot Welding. *Processes (Basel, Switzerland)*, *11*(1), 284. DOI: 10.3390/pr11010284

Gastromotiva. (2024). Gastromotiva.org/En/. https://gastromotiva.org/en/

González-González, C., Violant-Holz, V., Colombo-Ruano, L., & Rodríguez-Silva, C. (2025). Designing ethical and inclusive voice assistants through co-creation with older adults. *Behaviour & Information Technology*, •••, 1–23. DOI: 10.1080/0144929x.2025.2514245

Goodwin, L. (2023). The Global Benefits of Reducing Food Loss and Waste. *World Resources Institute*, *1*(1). https://www.wri.org/insights/reducing-food-loss-and-food-waste

Gordon, J. A., Balta-Ozkan, N., & Nabavi, S. A. (2023). Divergent consumer preferences and visions for cooking and heating technologies in the United Kingdom: Make our homes clean, safe, warm and smart! *Energy Research & Social Science, 104*, 103204. DOI: 10.1016/j.erss.2023.103204

Goyal, S., & Gupta, S. (2024). A Comprehensive Review of Current Techniques, Issues, and Technological Advancements in Sustainable E-Waste Management. *E-Prime - Advances in Electrical Engineering Electronics and Energy, 9*, 100702–100702. https://doi.org/DOI: 10.1016/j.prime.2024.100702

Gurung, S., Ritacco, J., Chen, L. A., & Lai, J. (2024). A Case Study on the Home Cooking Movement–Legalization, Market, and Competition. *Applied Economics Teaching Resources (AETR), 6*(4). umn.edu. https://scholar.google.com/scholar?hl=en&as_sdt=0%2C5&as_ylo=2023&as_yhi=2025&q=Governments+and+AI+in+cooking&btnG=

Hayyat, U., Khan, M. U., Farooq, M., Sultan, M., Amjed, M. A., Liu, G., & Alkhedher, M. (2024). Recent developments and challenges in biomass cookstove. *Energy Reports, 12*, 2193–2208.

Helbing, D., Mahajan, S., Fricker, R. H., Musso, A., Hausladen, C. I., Carissimo, C., Carpentras, D., Stockinger, E., Argota Sanchez-Vaquerizo, J., Yang, J. C., Ballandies, M. C., Korecki, M., Dubey, R. K., & Pournaras, E. (2023). Democracy by Design: Perspectives for Digitally Assisted, Participatory Upgrades of Society. *Journal of Computational Science, 71*, 102061. DOI: 10.1016/j.jocs.2023.102061

Hernández, J. L., de Miguel, I., Vélez, F., & Vasallo, A. (2024). Challenges and opportunities in European smart buildings energy management: A critical review. *Renewable & Sustainable Energy Reviews, 199*, 114472. DOI: 10.1016/j.rser.2024.114472

Heydari, M. (2024). Cultivating sustainable global food supply chains: A multifaceted approach to mitigating food loss and waste for climate resilience. *Journal of Cleaner Production, 442*, 141037–141037. DOI: 10.1016/j.jclepro.2024.141037

home. (n.d.). Mukurustoves.org. Retrieved June 7, 2025, from https://mukurustoves.org/

How Winnow is Ending Food Waste with AI. (2020, April 6). Food Tank. https://foodtank.com/news/2020/04/how-winnow-is-using-artificial-intelligence-to-help-kitchens-end-food-waste/

Ibe, K. K., & Kollur, S. P. (2024). Challenges towards the adoption and use of sustainable cooking methods: A comprehensive review. *Sustainable Environment, 10*(1). Advance online publication. DOI: 10.1080/27658511.2024.2362509

IEA. (2022). *Access to clean cooking – SDG7: Data and Projections – Analysis.* IEA. https://www.iea.org/reports/sdg7-data-and-projections/access-to-clean-cooking

Ikegwu, A. C., Obianuju, O. J., Nwokoro, I. S., Kama, M. O., & Ebem, D. U. (2025). Investigating the Impact of AI/ML for Monitoring and Optimizing Energy Usage in Smart Home. Artificial Intelligence Evolution, 30-43.

Hayford, I. S., Ofori, E. K., Gyamfi, B. A., & Gyimah, J. (2023). Clean cooking technologies, information, and communication technology and the environment. *Environmental Science and Pollution Research International, 30*(48), 105646–105664.

Islam, N. F., Gogoi, B., Saikia, R., Yousaf, B., Narayan, M., & Sarma, H. (2024). Encouraging circular economy and sustainable environmental practices by addressing waste management and biomass energy production. *Regional Sustainability, 5*(4), 100174. DOI: 10.1016/j.regsus.2024.100174

Kansaksiri, P., Panomkhet, P., & Tantisuwichwong, N. (2023). Smart Cuisine: Generative recipe & ChatGPT powered nutrition assistance for sustainable cooking. *Procedia Computer Science, 225*, 2028–2036. DOI: 10.1016/j.procs.2023.10.193

Khatiwada, P., Yang, B., Lin, J.-C., Mugurusi, G., & Underbekken, S. (2024). A Reference Design Model to Manage Consent in Data Subjects-Centered Internet of Things Devices. *IoT, 5*(1), 100–122. DOI: 10.3390/iot5010006

Kim, H., Choi, S., & Shin, H. H. (2025). Artificial intelligence in the kitchen: Can humans be replaced in recipe creation and food production? *International Journal of Contemporary Hospitality Management, 37*(5). Advance online publication. DOI: 10.1108/ijchm-04-2024-0549

Lahiri, S., Bose, I., & Majumdar, A. (2024). Rebel Foods' Cloud Kitchen Technologies: Food for Thought? *Communications of the Association for Information Systems, 54*(1), 155–179. DOI: 10.17705/1cais.05407

Legi, H., Damanik, D., & Giban, Y. (2023). Transforming Education Through Technological Innovation In The Face Of The Era Of Society 5.0. *Educenter: Jurnal Ilmiah Pendidikan, 2*(2), 102–108. DOI: 10.55904/educenter.v2i2.822

Lifset, R., Kalimo, H., Jukka, A., Kautto, P., & Miettinen, M. (2023). Restoring the incentives for eco-design in extended producer responsibility: The challenges for eco-modulation. *Waste Management (New York, N.Y.), 168*, 189–201. DOI: 10.1016/j.wasman.2023.05.033 PMID: 37311386

Milindi, P. S., De Lieto, F., & Chopra, S. S. (2022). Food waste-Energy-Water-Emissions (FEWE) nexus in the food service sector: Comparing a restaurant meal from imported ingredients versus an imported ready-to-eat meal. *Journal of Cleaner Production, 380*, 134871. DOI: 10.1016/j.jclepro.2022.134871

Moshtaghian, H., Bolton, K., & Rousta, K. (2023). Upcycled food choice motives and their association with hesitancy towards consumption of this type of food: A Swedish study. *British Food Journal, 126*(1). Advance online publication. DOI: 10.1108/bfj-09-2022-0757

Pal, S. (2023). Advancements in AI-Enhanced Just-In-Time Inventory: Elevating Demand Forecasting Accuracy. *International Journal for Research in Applied Science and Engineering Technology, 11*(11), 282–289. DOI: 10.22214/ijraset.2023.56503

Parvez, M. O., Hossain, M. S., & Islam, M. S., & Kayode Kolawole Eluwole. (2024). Customers' acceptance of AI-powered robot chefs in casual dining restaurants. *Journal of Hospitality and Tourism Insights, 8*(4). Advance online publication. DOI: 10.1108/jhti-04-2024-0347

Qayyum, F., Jamil, H., & Ali, F. (2024). A Review of Smart Energy Management in Residential Buildings for Smart Cities. *Energies, 17*(1), 83. DOI: 10.3390/en17010083

Rasugu Ayub, H., Jakanyango Ambusso, W., Muriuki Manene, F., & Mongeri Nyaanga, D. (2021). A Review of Cooking Systems and Energy Efficiencies. *American Journal of Energy Engineering, 9*(1), 1. DOI: 10.11648/j.ajee.20210901.11

Saarijärvi, , H., Sparks, L., Närvänen, E., Erkkola, M. Fogelholm, M. & Nevalainen, J. (2023). From transactions to transformations: Exploring transformative food retailing. *International Review of Retail, Distribution and Consumer Research, 34*(1), 1–18. DOI: 10.1080/09593969.2023.2213423

Sammer, M., Seong, K., Olvera, N., Gronseth, S. L., Anderson-Fletcher, E., Jiao, J., Reese, A., & Kakadiaris, I. A. (2024). AI-FEED: Prototyping an AI-Powered Platform for the Food Charity Ecosystem. *International Journal of Computational Intelligence Systems, 17*(1). Advance online publication. DOI: 10.1007/s44196-024-00656-9

Samsung Announces Global Launch of Samsung Food, an AI-Powered, Personalized Food and Recipe Service. (2023). News.samsung.com. https://news.samsung.com/global/samsung-announces-global-launch-of-samsung-food-an-ai-powered-personalized-food-and-recipe-service

Samsung SmartThings Connects People to the Things That Matter Most. (2023, August 31). News.samsung.com. https://news.samsung.com/global/samsung-smartthings-connects-people-to-the-things-that-matter-most

Sapienza, S., & Vedder, A. (2021). Principle-based recommendations for big data and machine learning in food safety: The P-SAFETY model. *AI & Society*. Advance online publication. DOI: 10.1007/s00146-021-01282-1

Sari, G. I., Winasis, S., Pratiwi, I., Nuryanto, U. W., & Basrowi, N. (2024). Strengthening digital literacy in Indonesia: Collaboration, innovation, and sustainability education. *Social Sciences & Humanities Open*, *10*, 101100–101100. DOI: 10.1016/j.ssaho.2024.101100

Schaper, M.-M., Charlotte Smith, R., van Mechelen, M., Aki Tamashiro, M., & Ole Sejer Iversen. (2023). Co-designing sustainable practices for emerging technologies education. *International Journal of Technology and Design Education*, *34*(3). Advance online publication. DOI: 10.1007/s10798-023-09857-3

Seal, P. P., Shukla, A., Bhutia, L. D., & Yadav, N. (2025). The role of food in food blogs: A netnographic analysis of Indian food bloggers' posts on social media. *Asia-Pacific Journal of Innovation in Hospitality and Tourism*, *14*(1), 115–136.

Seaman, A. N., Franzidis, A., & Nelson, M. (2024). Considering invasive alien species as a food source: Current motivations and future implications for controlling through consumption. *Geographical Review*, *115*(1-2), 1–18. DOI: 10.1080/00167428.2023.2299785

Selter, J.-L., Fota, A., Wagner, K., & Schramm-Klein, H. (2023). Aspects driving customers' intention to use automated purchasing processes. *International Journal of Retail & Distribution Management*, *51*(9/10). Advance online publication. DOI: 10.1108/ijrdm-10-2022-0397

Şener, E., & Ulu, E. K. (2024). Culinary Innovation. *Tourism (Zagreb)*, *72*(3). Advance online publication. DOI: 10.37741/t.72.3.4

Sharma, A., Kuthiala, T., Thakur, K., Thatai, K. S., Singh, G., Kumar, P., & Arya, S. K. (2022). Kitchen waste: Sustainable bioconversion to value-added product and economic challenges. *Biomass Conversion and Biorefinery*, *15*(2). Advance online publication. DOI: 10.1007/s13399-022-02473-6

Shukla, A., Yadav, N., Khunasathitchai, K., Bakshi, I., & Sharma, N. (2025). Waste Management Outlook and Future Directions in Rural Touristic Areas. In Albattat, A., Singh, A., Tyagi, P., & Haghi, A. (Eds.), *Solid Waste Management and Disposal Practices in Rural Tourism* (pp. 495–522). IGI Global Scientific Publishing., DOI: 10.4018/979-8-3693-9621-6.ch020

Singh, N., & Adhikari, D. (2023). AI in Inventory Management: Applications, Challenges, and Opportunities. *International Journal for Research in Applied Science and Engineering Technology*, *11*(11), 2049–2053. DOI: 10.22214/ijraset.2023.57010

Too Good To Go. (2024). *Save Food - Help The Planet*. Www.toogoodtogo.com. https://www.toogoodtogo.com/en-us

Ukoba, K., Olatunji, K. O., Adeoye, E., Jen, T.-C., & Madyira, D. M. (2024). Optimizing renewable energy systems through artificial intelligence: Review and future prospects. *Energy & Environment*, *35*(7). Advance online publication. DOI: 10.1177/0958305x241256293

UNESCO issues urgent call for appropriate use of technology in education | Global Education Monitoring Report. (2023). Www.unesco.org. https://www.unesco.org/gem-report/en/articles/unesco-issues-urgent-call-appropriate-use-technology-education

UNFCCC. (2024, September 30). *Food loss and waste account for 8-10% of annual global greenhouse gas emissions; cost USD 1 trillion annually | UNFCCC*. Unfccc.int. https://unfccc.int/news/food-loss-and-waste-account-for-8-10-of-annual-global-greenhouse-gas-emissions-cost-usd-1-trillion

UNITAR. (2024, March 20). *Global e-Waste Monitor 2024: Electronic Waste Rising Five Times Faster than Documented E-waste Recycling*. United Nations Institute for Training and Research. https://unitar.org/about/news-stories/press/global-e-waste-monitor-2024-electronic-waste-rising-five-times-faster-documented-e-waste-recycling

van Kesteren, R., & Evans, A. (2020). Cooking without thinking: How understanding cooking as a practice can shed new light on inequalities in healthy eating. *Appetite*, *147*, 104503. DOI: 10.1016/j.appet.2019.104503 PMID: 31707071

WHO. (2024, October 16). *Household Air Pollution and Health*. Who.int; World Health Organization: WHO. https://www.who.int/news-room/fact-sheets/detail/household-air-pollution-and-health

World Energy Outlook 2024. (2024). https://iea.blob.core.windows.net/assets/c036b390-ba9c-4132-870b-ffb455148b63/WorldEnergyOutlook2024.pdf

Yadav, N., Shukla, A., Khunasathitchai, K., & Kaur, R. (2024). AI Adaptions: Challenges and Strategies for the Hotel and Tourism Industry. In Talukder, M., Kumar, S., & Tyagi, P. (Eds.), *Hotel and Travel Management in the AI Era* (pp. 19–38). IGI Global Scientific Publishing., DOI: 10.4018/979-8-3693-7898-4.ch002

Yeldan, G., Yılmaz, G., & Kayatürk, G. (2024). AI-Driven Optimization of Order Procurement and Inventory Management in Supply Chains. *The European Journal of Research and Development*, *4*(3), 46–56. DOI: 10.56038/ejrnd.v4i3.605

Yu, M., Secondi, L., Laureti, T., & Palumbo, L. (2025). Saving food surplus and developing new business models: exploring the potential of "Too Good To Go" at territorial level using web-scraped data. *Big Data Research*, 100536–100536. https://doi.org/DOI: 10.1016/j.bdr.2025.100536

Yu, Y., & Sung, T.-J. (2022). Perceived Values to Evaluate Smart Product-Service Systems of Smart Kitchen Appliances. *Engineering Management Journal*, *35*(3), 1–15. DOI: 10.1080/10429247.2022.2075210

Yu, Y., & Sung, T.-J. (2023). A value-based view of the smart PSS adoption: A study of smart kitchen appliances. *Service Business*, *17*(2). Advance online publication. DOI: 10.1007/s11628-023-00529-9

Zahedani, A. D., Veluvali, A., McLaughlin, T., Aghaeepour, N., Hosseinian, A., Agarwal, S., Ruan, J., Tripathi, S., Woodward, M., Hashemi, N., & Snyder, M. (2023). Digital health application integrating wearable data and behavioral patterns improves metabolic health. *NPJ Digital Medicine*, *6*(1), 1–15. DOI: 10.1038/s41746-023-00956-y PMID: 38001287

Zhang, Z., Yu, J., & Tian, J. (2023). Community Participation, Social Capital Cultivation and Sustainable Community Renewal: A Case Study from Xi'an's Southern Suburbs, China. *Journal of the Knowledge Economy*, *15*(3). Advance online publication. DOI: 10.1007/s13132-023-01536-x

Chapter 15
AI-Driven Transformation in the Hospitality Industry Enhancing Guest Services and Operational Performance: A Review of Literature

Pankaj
https://orcid.org/0009-0001-1518-6296
Maharishi Markandeshwar University, Mullana, India

Abhishek Ghai
https://orcid.org/0000-0002-0994-9760
Maharishi Markandeshwar University, Mullana, India

ABSTRACT

The study analyses the role played by artificial intelligence (AI) in improving experience of consumers and operational efficiency at hospitality industry. In concern, it is identified that the hospitality industry operating through traditional approaches face certain issues like that of security in data recording, cash flow management and so on that requires an advanced technological implementation for generating successive outcomes. In relation, the varied benefit of AI in eliminating existing issues at the hospitality firms and generating operational efficiency is evaluated where market trends and other aspects are achieved by AI application to better manage the business through technological concerns. Simultaneously, it is observed that AI application in the specific industry generates problem of accountability,

DOI: 10.4018/979-8-3373-4042-5.ch015

high cost, liability and so on that requires proper improvement. In responses, recommendations like adaptation of lean management approach, and so on are cited for improving application of AI in hospitality industry for generating the desired outcomes respectively.

INTRODUCTION

The latest computer technology is artificial intelligence, which utilises human intelligence in performing activities to ensure an automatic decision-making process (Korteling et al., 2021). In this regard, AI is the combination of various technologies that allows running complicated procedures by independently evaluating the orders of humans. The hospitality industry is a broad term defined to include dining, accommodation, tourism, and event management, among other various forms of services aimed at satisfying people most efficiently. This paper will discuss the role of AI in the hospitality industry as a method of adding ability in efficiency and overall customer experiences. To comprehend the circumstances and problems of the sphere today, the notion of the hospitality sphere and the importance of the involvement of the customer in this sphere are addressed. The role of AI and its importance are presented in the attempt to evaluate how it contributes towards enhancing operations and consumer experiences within the hospitality industry.

In the past two decades, the hospitality industry in the world has metamorphosed tremendously due to the emergence of new convenient and comfortable technologies, evolving customer preferences and competitive environments among other activities. The application of technology in the hospitality industry is not viewed by the World Travel and Tourism Council (WTTC, 2023) as optional any more, which is seen as a survival measure and a growth tool. It has been driven by artificial intelligence, which has enabled institutions to offer services that are faster, more personalised and culminate in effective operations. More than performing monotonous tasks, AI enables companies with decision-making power by providing predictive analytics, which lets hotels surmise the required expectations and align their resources in the process (Gretzel et al., 2015). This makes sense with the larger industry transition towards service approaches that are experience-based, where loyalty and enjoyment of guests are driven by the ability to provide flawless, personalised interactions throughout the customer experience. In the past, hospitality businesses largely depended on manual management of the processes, including check-ins, handwritten concierge services and paper reservations. On the one hand, these practices promoted personal interaction, but on the other hand, it inhibited scalability, consistency, and operational efficiency (Ivanov & Webster, 2019). These limitations have been resolved with the introduction of AI-powered systems that incorporate such features as natural language recognition, computer, and machine learning models. As an example, AI-based chatbots used by Marriott International have decreased the response rate on guest requests by more than 60 percent, which translates into measurable guest satisfaction indexes.

Due to the post-pandemic environment, the hospitality sector has adopted AI more quickly, stimulated by rising customer demand for efficient, sanitary, and frictionless services (Kaushal & Srivastava, 2020). AI technologies during COVID-19, such as robotic room service delivery, voice-activated room systems, and automated cleaning schedules, did not just make the property a simpler thing to operate, but seeing such technologies on the property demonstrated to the guest that the property is serious about ensuring the health regulations are enforced. The technology change is supposed to have long-term implications. According to industry experts, the use of AI in a significant number of hospitality enterprises, over 70 percent, will grow to handle vital tasks in the industry, such as revenue management, guest services, and monitoring sustainability by 2030 (Statista, 2024). Due to these shifts, the timeliness and necessity of the topic addressed in the present study, namely, how AI can enhance the experience of the guests and improve the performance of the operations, are deliberate. The study aims to inform strategic decision-making among hospitality managers, technology developers, and policymakers since it follows an industry challenge and customer expectations lens and the technological capabilities lens to analyse AI applications to deliver an actionable insight into AI applications (Ivanov & Webster, 2019).

RESEARCH BACKGROUND AND PROBLEM STATEMENT

The hospitality sector consists of various businesses that work together to make consumers' experiences enjoyable and worthwhile in the marketplace. It includes activities such as providing shelter, food, cultural experiences and so on, ensuring that visitors' purposes are well organized when they go to a new place. Consequently, it is clear that customer satisfaction and proper management of business operations are crucial for achieving consistent success in these businesses. However, the hospitality industry struggles with accuracy in operational aspects like safeguarding customer data, managing cash flow and keeping administrative records. Likewise, it fails to meet modern consumer demands and enhance their business experiences through market trend analysis, high-quality customer service management, technological innovations and more. This shows that traditional hospitality practices are inadequate for achieving quality results and need digital advancements to meet contemporary consumer expectations and stay competitive in the marketplace (Fatima & Elbanna, 2020).

The traditional hospitality industries are facing operational issues in managing consumer experiences that require the proper application of advanced technologies like artificial intelligence to eliminate the existing issues. However, there are challenges in effectively using artificial intelligence in hospitality businesses. Achieving good results requires knowing the right methods for applying this advanced technology Du et al, (2025). The hospitality industry around the world has been under more and more pressure to use new technologies because of changing customer expectations, a competitive market, and problems in the job market. According to the "United Nations World Tourism Organisation" (UNWTO, 2023), digital transformation has become one of the core components of longer-term competitiveness and sustainable recovery of the tourism-related industries. In this regard, AI is highly significant as it allows individuals to analyse the data in real-time, predict the future demand and automate the tasks which previously required the efforts of a large number of people. These competencies are of great significance to large hotel chains, which receive thousands of visitors daily. Minor issues on the processes may also cause large cost overruns and dissatisfied guests. There are issues about operations of the hospitality industry, but it also exists as a structural gap to the speed at which there is implementation of new technology is implemented in developed and developing economies. One of the market participants to use the AI-based solutions was luxury brands and large hotel chains, such as the Hilton AI-powered concierge known as Connie and the Accor voice-enabled guest services. However, SMEs or the small or medium-sized enterprises in this hospitality sector lack the funds and technical knowledge to operate such systems (Tussyadiah, 2020). In the smaller businesses, the digital divide issues face the businesses in terms of personalised services, operational efficiency, and brand differentiation. To close this gap, it is important to invest in AI infrastructure and training programs that will help hospitality workers learn how to use AI effectively in their roles.

Moreover, the implementation of AI in the hospitality sector raises some impressive data security, cybersecurity, and ethical decision-making concerns. The hospitality industry is responsible for managing heavy amounts of sensitive data relating to their guests' names and other personal identifications, payment methods and behavioural preferences, among others. This data may negatively affect the fame and reputation of a business and lead to legal conflicts upon its mismanagement (Gursoy et al., 2019). The regulatory rules, such as the EU regulation "General Data Protection Regulation" (GDPR), and the Indian regulation, the "Digital Personal Data Protection Act" (DPDPA, 2023), enact strict rules on data management. The hospitality industry uses systems that must have strong security and notifications of compliance at the initial stage. These concerns were paramount in terms of developing trust among guests since it determines their behaviour concerning the need to interact with services powered by AI.

LITERATURE REVIEW

Hospitality Industry and the Importance of Guest Experience

Achmad & Yulianah (2022) anticipated that the hospitality industry refers to the vast businesses concerned with providing economic activities that satisfy people's needs during their travel to specific places. In contrast, it was argued that the hospitality industry combines consumer service businesses such as food and beverage service, lodging, and event management. The integration of AI technologies requires careful adaptation (Yadav et al., 2024). In addition, it was interpreted that the hospitality industry has a primary purpose to provide a memorable experience to consumers during their visit to the organisation. On the other hand, contemplated that the hospitality industry expresses those businesses that have a corporate social responsibility to offer sustainable tourism services to people visiting different places, such that their satisfaction leads to an increase in business sales. It was interpreted that the concept of guest experience indicates the satisfaction level of consumers with a business process.

In this regard, Kim & Kim (2022) evaluated that guest experience was vital in the hospitality sector since the effectiveness of a guest has the potential to influence other customers to visit the business space and make the organisation successful in their operations to gain regular and increased revenue. From a different perspective, it was considered that the guest experience is important for the hospitality sector since it guarantees high-quality business process marketing for a more dependable growth of brand image. Apart, Achmad & Yulianah (2022) commented that guest experience is important because it can highlight the changes required for the business to attain sustainability and can appropriately guide revenue growth of the firm in the competitive marketplace. Social media has redefined the role of food in cultural expression (Seal et al., 2025).

Buhalis and Sinarta (2019) examined the transition within the hotel sector from service delivery to real-time co-creation, highlighting how immediate interactions and personalisation are increasingly influencing the visitor experience. Their study demonstrated how technology, especially artificial intelligence (AI), makes the idea of "nowness service" possible by providing immediate, situation-specific solutions that raise satisfaction. This supports the idea that businesses in the hospitality sector must employ AI technologies to adapt to the demands of their clients in real time, increasing client retention and service excellence.

According to Walls et al. (2011), the guest involvement was a multifaceted concept with social, emotional, and physical components. According to their findings, positive word-of-mouth and repeat business are significantly influenced by emotional engagement. Today's consumers look for staged, memorable experiences in addition to goods and services. Every element of a trip, including behavioural, cognitive, and emotional contributions, contributes to the total experience in tourism. A positive or negative experience is inherent in even the most basic offerings. In hospitality research, the meaning of "experience" needs to be clearly explained and defined because many different definitions exist. AI supports these dimensions by enabling hotels to monitor and respond in real time to guests' emotions through tools like sentiment analysis and automated feedback systems. This shows that AI can turn experiential quality measures into practical actions, helping hotels improve their competitive position.

Sakdiyakorn et al. (2021) reviewed the life history of 125 Thai university students. The research examined the extent to which the group mentality of Generation Z (Gen Z) influences individual values and approaches to work. Important values such as universalism, achievement, benevolence, self-direction, and security were formed by some common and formative experiences and thus have significance on the aspect of attracting, retaining, and engaging Gen Z in the hospitality sector. Understanding Gen Z's motivations is essential given the industry's reliance on young workers and the ongoing difficulties of precarious employment and early disillusionment. The study filled gaps in primarily Western-focused generational studies by applying collective consciousness theory, using a qualitative narrative approach, and providing a geographically specific perspective.

Ariza-Montes et al. (2017) examined, according to Schwartz's (1992) theory of fundamental human values, this study explored how the authority and responsibility levels in various hospitality roles, such as restaurant managers, chefs and cooks, waiters and bartenders, kitchen assistants, and other hospitality staff, correlate with their personal value profiles. Logistic regression models were employed to test hypotheses regarding the connection between job roles and the conservation value dimensions of self-transcendence, self-enhancement, and openness to change. The findings, which aligned with the proposed hypotheses, indicated that higher-status positions like managers and chefs/cooks exhibited stronger self-enhancement and openness to change values, while lower-status positions were more aligned with self-transcendence and conservation values. By applying Schwartz's value framework to the hospitality industry, the research filled a gap in the literature, illustrating how value orientations differ significantly across hierarchical levels, thereby affecting leadership styles, job compatibility, and organisational outcomes.

CURRENT PERFORMANCE OF THE HOSPITALITY INDUSTRY

Fatima & Elbanna (2020) focused on highlighting how the hospitality industry continues to be primarily fixed on traditional models of delivering services, often lacking integration with the latest technologies and the trend of sustainable operations. They found out that the adoption of outdated systems impeded the ability of the industry to adapt to the modern needs of consumers and the dynamism of the market. The authors emphasised that to enhance operational resilience and customer satisfaction, it is imperative to digitalise operations, in particular, through AI. The hospitality industry at present operates mostly based on traditional approaches and provides basic consumer services of food, lodging and entertainment without focusing much on sustainability and market trends for a better understanding the consumer behaviour and patterns. According to Shukla and Sharma (2020), service quality drives customer retention.

EHL (2024) analysed the hospitality industry's financial performance over some years by the EHL Hospitality Business School (2024). The COVID-19 pandemic significantly reduced global revenues in 2019 and 2020. According to their findings, the recovery has been uneven, with hotels that use more sophisticated technology recovering more quickly than those that still use manual procedures. This strengthens the argument for incorporating operational systems powered by AI to lessen future disruptions. It is observed that the global revenue of the hospitality industry has declined significantly between the period of 2019 to 2020 as the COVID-19 pandemic has hit the business aspects, and due to a lack of innovative strategies, most of the firms faced a negative influence of the same.

Figure 1. Global revenue of hospitality firms over the years (Ehl, 2024)

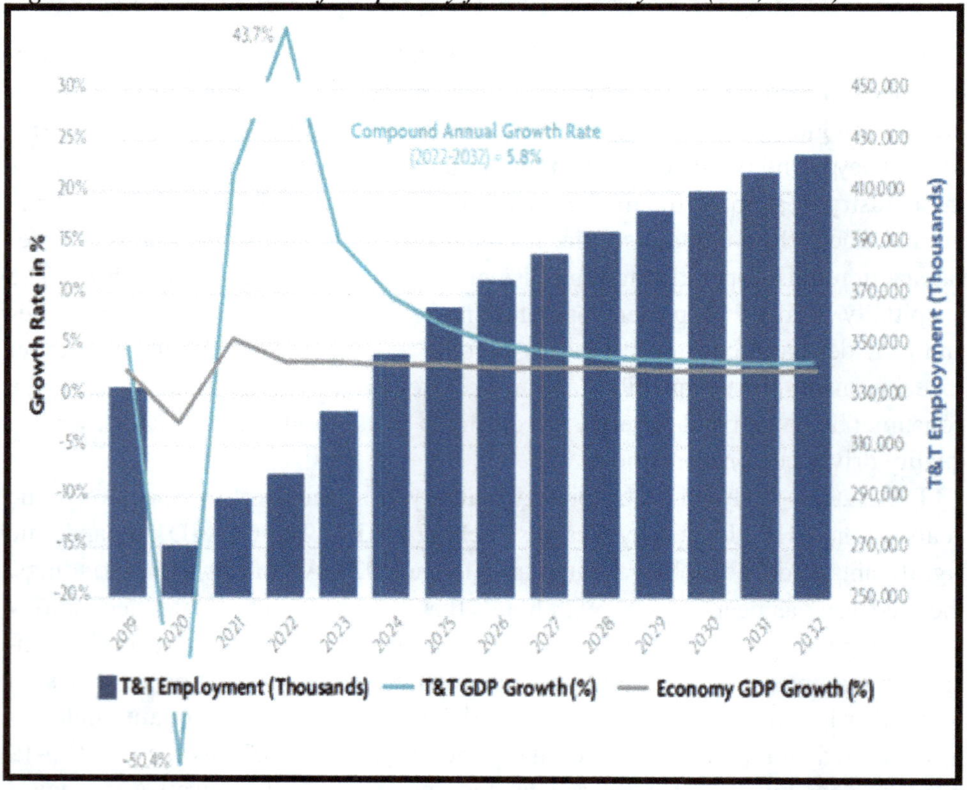

BCG (2021) revealed that even though the hospitality industry is adopting digital technology more and more, only around 25% of businesses make investments in long-term digital value creation. According to the report, hotels run the risk of operational inefficiencies if hotels only concentrate on trend-based technologies without integrating them with long-term sustainability objectives. When paired with sustainability objectives, AI has the potential to boost the economy and the environment. It was contemplated that the hospitality industry at present has invested heavily in digital transformation of businesses to a certain extent; however, only 25% of hospitality firms across the globe have invested in sustainable value generation aspects of their organisation, hampering the potential of meeting current business trends and consumer needs.

Figure 2. Investment data of hospitality firms on digitalisation (BCG, 2021)

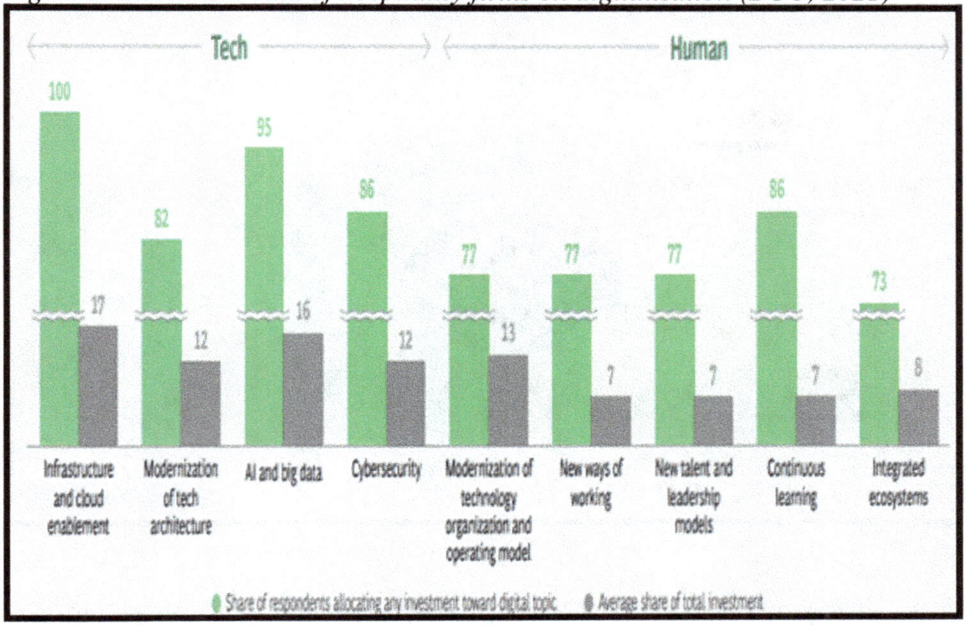

Ghani et al. (2022) determined that a major issue facing the international hospitality sector was high employee turnover. The operational consistency and service quality are directly impacted by unstable workforce structures. By improving employee engagement and optimising staffing levels, AI-driven workforce management solutions like performance analytics and automated scheduling can assist in overcoming this difficulty. Researchers have assimilated that currently the global hospitality industry is facing issues in retaining its skilled employees due to high turnover affecting adequate management of the workforce, respectively.

Prakash et al. (2023) investigated how hospitality operations affect the environment, pointing out that the industry was responsible for about 8% of greenhouse gas emissions worldwide. To monitor and lower carbon footprints while preserving service quality, their study underlined the necessity of green technologies, such as AI-powered energy management systems. The global hospitality industry at present fails to meet the modern demand of operational sustainability.

Figure 3. Carbon footprint in the global hospitality industry (Revnomix, 2022)

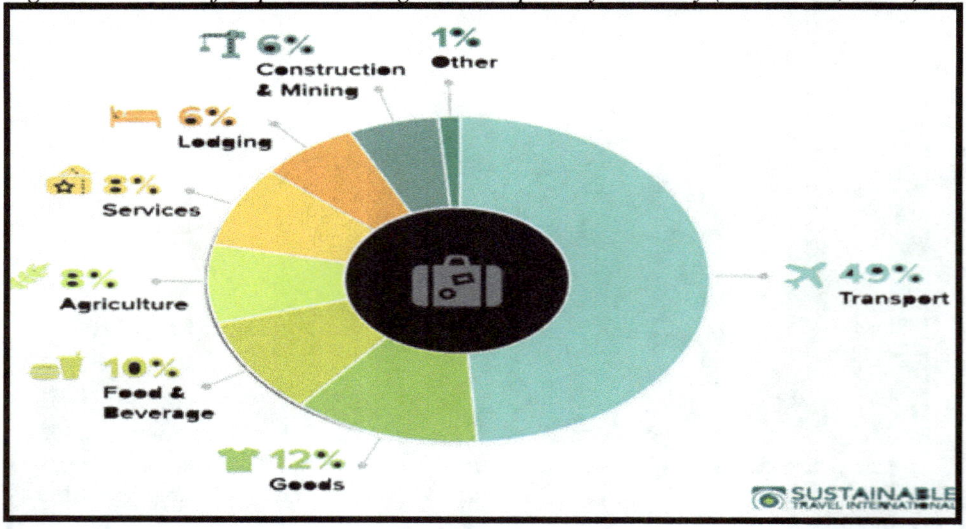

Ghazi (2023) examined how the use of clean technologies, energy-efficient designs, and renewable energy, the hotel industry improved its carbon efficiency by 27% between 2014 and 2019, lowering daily emissions per room from 17.5 kg to 12.9 kg. Reductions in the US, Canada, the Euro Area, and the UK varied from 16% to 47%; the UK led the way because it transitioned from coal to renewable energy. In comparison to 2019, major hotel chains, including Hilton, IHG, Hyatt, and Marriott, reported an additional 17% gain in efficiency by 2021. This increase was primarily due to continuous sustainability initiatives, but it was also partially caused by decreased demand.

ARTIFICIAL INTELLIGENCE AND ITS SIGNIFICANCE

Korteling *et al.* (2021) interpreted that AI is a suitable advanced technology that adheres to human intelligence in carrying out activities automatically, so that the burden on humans is relieved to a certain extent. When comparing artificial intelligence (AI) with human intelligence in complex tasks, AI stands out for its ability to quickly process large amounts of data, identify patterns, and make consistent decisions without getting tired or showing bias. These features were essential in the hospitality industry for tasks like predictive demand forecasting, personalised guest recommendations, and real-time room pricing, which allow for quicker and more accurate decision-making.

Xu *et al.* (2021) assimilated that AI is an innovative technology of the modern era that enables varied advanced functions by analysing common human interpretations, ensuring a scientific paradigm to recommend technical ideas relatively. Apart, it is conveyed that artificial intelligence is a modern technology used for various purposes in the modern world, such as guidance on delicate wings, translating languages, providing digital security to data, tracking trends and so on. AI is a new way of thinking about science that could change whole industries by automating difficult tasks like translating languages, recognising images, and predicting behaviour. In the hospitality industry, their research shows how AI can make it easier for guests to communicate in multiple languages, speed up check-in by recognising faces, and predict what customers want to give them personalised experiences.

Verma et al. (2021) emphasised that AI has distinct importance, due to which it was widely accepted in business operations to provide a technological direction to operations for enhancing accuracy and efficiency reliably. Researchers investigated the ways to overcome the lack of understanding of how artificial intelligence can assist in marketing, in particular, how it might analyse the behaviour of people and foretell trends. Such types of insights enable the dynamic pricing techniques, tailored advertisement campaigns, and customised service opportunities in the hospitality industry. In their research, they reveal that customer relationship management systems may significantly enhance brand positioning and guest loyalty through marketing tools made possible by Artificial intelligence. In this regard, researchers also added that AI is also high in quality when it comes to predicting the needs of the future because AI-based application automatically examines current market trends and consumer patterns, significantly aiding in the satisfaction of marketing aspects of business processes. Separately, it was demonstrated that AI has numerous other advantages, including those related to the enhanced decision-making process, the personalisation of consumer needs, the attention paid to sustainability issues in processes, etc., that apply to the contemporary needs and tendencies, respectively.

Galaz et al. (2021) demonstrated the fact that artificial intelligence played an important role in the systematic reduction of risks and errors and the recording of digital and repetitive data. Different AI Contributions in the Hospitality Operations. Researchers investigated the way AI could assist in reducing systemic risks and help things become more sustainable. Artificial intelligence can identify vulnerabilities within operations, utilise resources more effectively and assist a business to transition to low-carbon models. Due to the presence of the hospitality sector, AI-based energy management software and waste reduction analytics can help hotels save money and keep up with environmental standards.

Polignano et al. (2024) said this breakneck evolution of Large Language Models (LLMs) led to their ability to be used with many more tasks, including text generation, decision generation, and customer support. Nevertheless, it is difficult to open, comprehend and trust them because of their complexity. Explainable AI (XAI) has become necessary to deal with the dangers of bias, false information, and harm, especially in important areas like finance, healthcare, and law enforcement. XAI for LLMs faces unique difficulties due to the nuanced, context-dependent nature of human language and the massive scale of these models. Existing methods, such as LIME and SHAP, offer local and feature-based interpretability but often struggle with the computational demands and intricacy of deep neural networks. The XAI 2024 workshop highlighted advances and future directions, emphasising scalable, task-oriented, and human-in-the-loop techniques to detect and mitigate biases, hallucinations, and other risks, ensuring trustworthy, accountable AI systems.

VARIOUS CONTRIBUTIONS OF AI IN OPERATIONS OF THE HOSPITALITY INDUSTRY

Merkle (2023) stated that artificial intelligence (AI) has the potential to benefit the hotel sector since the cutting-edge technology may increase operational efficiency for companies and give customers a sense of security when using their databases. To improve workplace productivity, scientific approaches should be used, according to the scientific philosophy of management. Similarly, the use of AI in the hospitality sector adopts a particular theoretical viewpoint since it enhances the security system by offering features like facial recognition, automated data surveillance, and so forth, ensuring that only restricted access to individualised business data is guaranteed.

Ruel & Njoku (2021) looked into how AI applications were changing the security systems in the hospitality industry, especially through technologies like biometric authentication, facial recognition, and anomaly detection systems. AI-powered surveillance systems can keep an eye on and limit access to private areas, which protects the data of both guests and the business. Through big hotel chains, the tools can be used to prevent people from reaching places that they should not and to monitor and record their actions as they take place. These features are critical in an environment where the security of the personal details of the guests and their physical security were what determine their trust. Artificial intelligence contributes significantly to security as it helps develop a digital security barrier that may regulate and protect access to data.

Limna (2023) has studied the possible impact of AI on the efficiency of hospitality work by automating the organisational processes related to it to a significant extent, including watching over the inventory and creating housekeeping plans, as well as organising the front office. The study has shown that AI-based systems may keep all the operation records at the same location, therefore, making the decisions of the managers more beneficial because they are presented with the real-time information. One of them is that predictive maintenance algorithms can plan the services to be performed on the equipment even before the machine malfunctions to lessen the downtimes and repairs. This integration not only makes things work more smoothly but also makes guests feel happy because there are fewer interruptions in services. The authors argued that AI can benefit the hospitality sector by improving company operations, such as managing administrative records, inventory data, and housekeeping schedules efficiently.

Sony and Naik (2020) used socio-technical systems theory to explain how AI coordinates hospitality operations. Their study showed that AI platforms make it easier for departments to share information, like housekeeping, food and drink, and reservations, which ensures that service delivery goes smoothly. In practice, this could mean that AI combines booking data with housekeeping schedules to make sure that rooms are cleaned right after guests check out to get the most guests in and out. The Systems theory of management states that an organisation is an integrated part of distinct systems that require working together to achieve a common goal. Consequently, it is evaluated that the specific theoretical perspective is adopted by AI as its implementation in the hospitality industry centralises administrative records, making it much easier to accurately record and access business-related data. AI contributes to the hospitality industry by reducing operational costs of the firm as it can accurately track utilisation of resources and eliminate any sort of error in the process, so that turnover and wastage rates can be reduced.

Saleh et al. (2021) looked into how AI can make hotel financial reporting better, especially when it comes to managing income statements, keeping track of expenses, and predicting cash flow. Their research demonstrated that AI systems can accurately detect spending irregularities, enhance budget distribution, and forecast future revenue patterns. Hotels can cut down on mistakes made by people, make sure people follow the rules, and give managers more time to make strategic decisions by automating financial analysis. For instance, AI algorithms can find seasonal changes and changes in when goods are bought to avoid having too much or too little of something. AI contributes to a decline in operational costs of the hospitality industry by appropriately determining the financial statements, like income statements, cash flow statements, financial trends of expenses and so on, such that the business could be properly guided on the economic concerns automatically. Subsequently, AI can easily understand the complex financial cash flow of hospitality businesses and regulate the operations in the best possible manner with the identification of performance indicators.

Mariani & Borghi (2024) contradicted that AI has a highly beneficial contribution to operational aspects, as the specific advanced technology can guide in improving operational efficiency by assuring high productivity in managing the distinct processes. For instance, AI can develop the productivity rate of repeated tasks in hospitality firms, such that employees can be engaged in more efficient tasks to achieve successive outcomes reliably. In addition, the customer service by different employees could be tracked by an AI system and identify the human resources hampering operational progress, so that quality services could be generated over the marketplace. It measured the impact of AI-enabled automation in enhancing labor productivity of workers because it frees up human resources in the most redundant and repetitive tasks and deploys them towards other more valuable tasks. Illustratively, AI has the capability of automation of check in tasks, releasing front desk agents to spent more time in interacting closely with individual guests. Also, in their paper, they have shown how AI-based performance monitoring could identify weak-performing areas of performance and bottlenecks in the services so that training interventions could be provided on those areas. These measures not only can set the efficiency of the operations of the employees but also boost the morale of employees upon engaging in more appealing and guest-encountering activities.

IMPACTS OF AI ON GUEST EXPERIENCES IN THE HOSPITALITY INDUSTRY

Pulignano et al. (2024) pointed out, AI can be highly efficient in changing the service delivery in case it is used in a manner that makes it complementary to human interaction instead of substituting it. The study based on human relations conceptualised that guests are usually interested in the human warmth and understanding, as well as the convenience brought by the technology. The authors suggested incorporating a two-level strategy, implying that hotels should use a hybrid solution, relying on AI to resolve routine and data-driven tasks, leaving human staff to perform on high-empathy needs like resolving complaints, delivering personalised greetings and so on. To give an example, the check-in process could be handled by AI and a staff member could personally guide the guests to their rooms, combining the effective efficiency and the personal touch which is vital to the hospitality business. The influence of AI in the hospitality sector has a significant role in guest experience since the marketplace appreciates the eco-friendly AI technology, which ensures sustainability of the business context relatively. According to the Human Relations theory of management, an organisation can be improved through the satisfaction of its employees and stakeholders and through the incorporation of their interpersonal and social relations in the friendliest way.

Limna and Kraiwanit (2023) developed a preliminary analysis of the role played by AI-powered tools such as ChatGPT in the international hospitality customer support. As per their study, with AI-powered chat systems, the hotels will be able to offer 24/7 services in various languages and will be able to resolve customer queries promptly without the time lag that accompanies human-operated service desk systems. The authors stated that even AI systems can get trained on answering frequently asked questions, dealing with modifications in reservations, and coming up with personalised recommendations on which facilities the hotels have to offer. This kind of automation, in addition to shortening wait time, also enhances perceptions of responsiveness among the visitors, which is an influential element of overall ratings in satisfaction levels. Besides, the hotels can store the recordings of interaction in previous interactions, which could be offered by the hotel by merging the AI with the CRM platforms, such that subsequent interactions could be more suited to the preferences of specific guests. The AI-chatting tools, if employed, can ensure that consumers no longer require waiting in a queue and instead, can provide them the services they require at any hour of any day or night. The most important way in which an AI can improve the consumer experience is by guaranteeing omnichannel facilitation to individuals to enable communication with the business according to their preferred mode.

Saxena et al. (2024) discovered that the customised contact enhances customer loyalty considerably. To personalise the services offered, the authors described how the use of AI algorithms could analyse the trend of previous bookings, food preferences, and participation in leisure activities. As an illustration, an already registered customer who may have booked a spa package will automatically receive a personalised offer of promotion, i.e., spa discount on the next visit. As well, the research constructed that personalisation enhances the ties that customers develop with the brand-emotionally, highly encouraging repeat business and positive electronic word-of-mouth (eWOM). Crucially, these personalisation initiatives are scalable, so even big hotel chains can continue to provide a customised experience without overworking their employees. AI has a greater impact on the hospitality industry because it personalises consumer demands by analysing their data and predicting the pattern of consumer experience, so that it can accordingly be guided to promote the business brand. In other words, AI impacts consumer experience in the business space by proactively suggesting products for their necessities during their time of visit to the organisation, so that the customer is digitally engaged in the marketplace. This can assure the fulfilment of modern consumer demands and enhance their business experiences.

Bisoi et al. (2020) asserted that AI assures quality improvement in consumer services that influence the experiences of consumers in the business process. In other words, AI with its technological interface can trace the performance of workers and detect improvement areas so that service activities are accordingly modified to ensure that the words of customers are valued for improving their trust. Hotels and restaurants could also employ the use of AI to monitor and enhance the quality of service. Their study implied that AI is capable of monitoring the performance of the employees in real-time and identifying areas where service is lacking, which may end up leaving guests dissatisfied. As an example, through point-of-sale solutions based on AI, it will be possible to track the average time frame it takes to fill room service orders as well as notify the managers when a time is entering abnormal limits so that they intervene immediately. The software can analyse guest reviews posted on different online sources, which helps hotels focus on and solve problems. Based on their findings, the authors have concluded that computer-aided monitoring helps hotels in quality management of service delivery, which is greatly essential in a competitive business environment to ensure a strong brand reputation.

AI is efficient because of its low-tech interaction with customers and high standards of customer service. This will be demonstrated, such as voiceovers are AI tools that can easily interpret and understand voice notes of various individuals, thus answering their demands to achieve satisfaction. Moreover, Hyari et al. (2023) assumed that AI has analytical tools that monitor patterns in the market and help hospitality companies revolutionise services to satisfy the needs of customers and meet their sustainable technological needs. In this regard, AI technology would be useful in facilitating improved services to the customers because, in its entire use, the technology does not result in any pollution, hence, in line with the emerging tendency of sustainable business conduct and customer loyalty to the business. Researchers sought to understand how AI is contributing towards the provision of environmentally conscious visitor experiences in luxury hotels. Their analysis determined that energy management systems powered by AI would allow them to regulate their water consumption, air conditioning, and lighting in rooms in real-time using occupancy data to maximise the comfort of every visitor and minimise environmental damage. As an example, AI sensors will be able to automatically adjust the temperature of the room, before the arrival of the visitor, based on their preferences and identified on their prior visits. Such a combination increases the operational performance, besides appealing to environmentally friendly tourists who are typically focusing on sustainable lodgings. Green AI applications enhance the corporate social responsibility and corporate profile of a hotel and can help it appeal to an increasingly valuable segment of consu mers: the so-called green consumers.

WAYS TO IMPROVE THE APPLICATION OF AI IN THE HOSPITALITY INDUSTRY

It is noted that despite all the advantages that may be offered by the role of AI in hospitality sector to enhance operational efficiency and consumer experience; it has been noticed that there are some concerns related to AI and concerns of AI, i.e. liability issue, accountability issue, cost etc. At the same time, Sharma et al. (2021) focused on the issue of the liability aspect of the AI usage, particularly in those cases when using bad algorithms might lead to a low quality of service or dissatisfied customers. Through their research, the way to reduce liability risk is to train AI systems using representative, diverse and high-quality data. This should also focus on incorporating previous booking history, customer reviews and service records into the AI learning algorithms in the hospitality sector to enable it to evade biased decision making and give better recommendations. Trained on the multilingual data, e.g., a chatbot can equally react to the visitors of diverse linguistic backgrounds and mitigate the likelihood of complaints due to being misunderstood. The liability pledge in the use of AI in hospitality can be strengthened by learning to perfect high-quality and diverse information that may tend to integrate well into operational processes and systems to detect and eliminate outside access and behaviour.

Lv et al. (2022) realised that the hospitality industry could implement the use of AI technology to enhance the performance of business operations by establishing the issue of accountability by conducting audits periodically. The audits were to be used to detect whether there are any discriminatory patterns in AI activities and ensure that biased decisions are never produced. The authors analysed the potential of deploying systematic reviews and transparency in facilitating the discussion of responsibility issues of AI, e.g., decision-making processes not decipherable. To make sure that unreasonable biases or tendencies in the products of AI-based decisions are not observed, it was suggested to conduct audits of the AI performances as a matter of routine, and such audits should involve third parties. Such industries as the hospitality industry may be left with an audit, which may lead to maintaining fair and non-discriminatory rates on rooms through the use of AI-powered pricing mechanisms. Moreover, the study found that the strategy of lean management could be implemented in the hospitality business to facilitate the introduction of AI technology. This was one of the ways through which the businesses were able to get rid of the redundant costs, thus raising their financial ability to make good investments in AI solutions. Furthermore, the study demonstrated that further enhancement of AI implementation was possible with the creation of customer-service robots that could carry out various human tasks with precision, attain consumer expectations, and eliminate human error in the hospitality work environment.

Ko et al. (2022) proposed that advancements in AI are necessary, particularly in managing data privacy, as the rise of deepfakes raises significant concerns. The risks of AI require its control to aid in the prevention of data breaches. The researchers underlined the significance of privacy-by-design frameworks concerning the AI implementation at a time when data misuse and deepfake technologies have become of considerable concern. The given practice would make the AI systems, which deal with visitor-related information in the hospitality sector, compliant with GDPR and any other data protection laws through automated data destruction, the adoption of a robust access control system, and an in-built encryption feature. As an example, to reduce breach risks, facial recognition check-in systems powered by AI should save biometric data temporarily on the device used instead of sending it to cloud-based servers. Practices in this category, which ensure they do not infringe on visitor privacy, also improve the brand since they show a policy of responsible data management.

Du et al. (2024) carried out two successive studies to understand the possibility of consumer resistance to this innovation, emphasising the young generation who were likely to adopt new technologies first. In the first study, a theoretical framework was proposed that combined the TPB (Theory of Planned Behaviour) with the TAM (Technology Acceptance Model), and its empirical validation was examined. The results showed that the attitudes, subjective norms, and perceived behavioural control positively affected the intentions that people possessed to use unmanned smart hotels. Based on these results, the second study determined the mediation of the role of perceived security by usage intentions. As it was discussed in the paper, it is essential to take into consideration the security implications when thinking about technology adoption and provide a thorough insight into the perception of the average consumer against the introduction of unmanned smart hotels.

Huang et al. (2024) found that the role of increasing business efficiency within the hospitality industry with the help of AI technology may be reinforced with the strengths that include the problem of accountability, resolvable via frequent auditing aimed at detecting the biasing tendencies and drive to make objective decisions. It examined the aspects of systematic reviews and transparency, resolving the concern regarding the accountability of AI, including the lack of transparency in the decision-making process, by suggesting a regular AI performance audit, including a third-party assessment. The paper also revealed how the lean management strategy could potentially assist the hospitality businesses with the AI technology integration via minimising the irrelevant costs, and raising the funds to invest into the AI technology. It also demonstrated that AI applications can be improved by creating customer-service robots that will be able to precisely do different tasks that humans can do, including meeting customer expectations and the eradication of human error in the hospitality business.

DISCUSSION

The current state of the hospitality industry reflects a sector transforming due to the dual pressures of post-pandemic recovery and swift technological progress. Variations in operational flexibility, technological preparedness, and market adaptability led to a sharp decline in revenue from 2019 to 2020, with recovery being uneven (EHL, 2024). Several companies that followed the traditional pattern in services have found it difficult to keep up with a modernised demand as data management, cash flow, and service standardisation remain inefficient (Fatima & Elbanna, 2020). Artificial intelligence (AI) has proven to be the key in fighting these obstacles through automation of administrative processes, coordination and integration of the business, as well as improving the responsiveness of customer service (Limna, 2023). Artificial intelligence-powered chatbots, facial recognition check-in tools, predictive analytics, or dynamic pricing analytics can enable hospitality companies to provide non-contact, highly customised services fulfilling the changing visitor demands in terms of previous convenience, security, and customised experiences (Saxena et al., 2024). In addition to the efficiency improvement of its operations, AI implementation can contribute to sustainability by helping with predictive maintenance, energy efficiency, waste management strategies and efficient energy consumption, sparking an increased interest among travellers who are environmentally aware (Hyari et al., 2023). It has possible effects that are magnified by the combination with well-known management theories. As an example, the scientific management style achieves efficiency and standardisation of the processes, the system theory maximises the information circulation among the departments, and human relations theory makes sure that the technology is just a supplement to the human factor (Sony & Naik, 2020). Among major adoption-related challenges provided in the literature, there is the issue of initial high investment, lack of competent workers, responsibility, and the possibility of ethical issues handled by algorithms and data privacy (Lv et al., 2022). These are complex issues and tackling them needs a multipronged approach, such as regular AI performance audit to be transparent and fair in automated decision-making, privacy-by-design to protect sensitive guest information and lean management ethos to prioritise resources. On the whole, the discussion renders AI as a strategic imperative as opposed to being another technological tool. It can help fuel operational resiliency, business competitiveness, and sustainable hospitality growth industries in a world where hospitality businesses must increasingly find equilibrium between efficiency, personalisation, and ethical responsibility systems of operation in an extremely competitive and highly digitalised global marketplace.

CONCLUSION

Artificial intelligence has emerged as the key ingredient that can help ensure long-term competitiveness, as opposed to being a secondary add-on, as indicated by the current performance analysis of the hospitality sector and recent academic discoveries. Overall, strong fundamentals of the industry notwithstanding, existing intra-industry operational inefficiencies, labour issues and changing consumer expectations require an adaptive and futuristic technology strategy. Artificial intelligence presents solutions to these issues by helping automate routine workloads, supporting data-driven decisions to increase efficiency and quality of visitor services, and scaling personalised visitor experiences. Moreover, its use in sustainability initiatives, including waste minimisation and energy efficiency, is aligned with market demands to take up initiatives in sustainability. But the way to successful integration of AI is replete with pitfalls: to bring out successful integration of AI, then deliberate governance frameworks, an upskilled workforce, and ethical protections need to be in place to take care of budget limitations, competency gaps, accountability, and data privacy concerns. Incorporating the deployment of AI into the existing management theories avoids neutralising the interpersonal nature of hospitality as a service, keeping the balance between service efficiency and high humanism.

In the future, the most competitive hospitality enterprises will be those that leverage AI as a comprehensive driver of innovation, resilience, and sustainable growth, rather than merely a cost-cutting tool. These businesses will adapt to shifting market conditions and consistently deliver superior guest experiences in an increasingly digital global economy. Additionally, the early adopters stand a good chance of increasing their brand equity because they can then be considered innovative, ecologically friendly, and safe. The end-to-end AI enabled a possibility of brand new types of service, including immersive virtual concierge services, predictive tour planning and real-time operations adjustments based on sentiment analysis of the guests. Over the next decade, the competitive edge in the hospitality industry will be redefined by the fusion of technology and service culture. Human hospitality professionals will focus on empathy, creativity, and relationship-building, while AI handles complexity and data-driven decision-making. This intelligent balance will ultimately shape the industry's future. By addressing both current operational and experiential needs, this collaboration will also protect hospitality companies from the uncertainties of a rapidly changing global landscape.

REFERENCES

Achmad, W., & Yulianah, Y. (2022). Corporate social responsibility of the hospitality industry in realising sustainable tourism development. *Enrichment: Journal of Management, 12*(2), 1610–1616.

Al-Hyari, H. S. A., Al-Smadi, H. M., & Weshah, S. R. (2023). The impact of artificial intelligence (AI) on guest satisfaction in hotel management: An empirical study of luxury hotels. *Geo Journal of Tourism and Geosites, 48*, 810–819.

Ariza-Montes, A., Arjona-Fuentes, J. M., Han, H., & Law, R. (2017). Employee responsibility and basic human values in the hospitality sector. *International Journal of Hospitality Management, 62*, 78–87. DOI: 10.1016/j.ijhm.2016.12.001

Bcg.com. (2021). *TRAVEL AND TOURISM: What the Data Tells Us About Digital Transformation, by Industry*. https://media-publications.bcg.com/BCG-Travel-and-Tourism-What-the-Data-Tells-Us-About-Digi

Bisoi, S., Roy, M., & Samal, A. (2020). Impact of artificial intelligence in the hospitality industry. *International Journal of Advanced Science and Technology, 29*(5), 4265–4276.

Buhalis, D., & Sinarta, Y. (2019). Real-time co-creation and nowness service: Lessons from tourism and hospitality. *Journal of Travel & Tourism Marketing, 36*(5), 563–582. DOI: 10.1080/10548408.2019.1592059

Du, H., Li, J., Kam, K., & King, C. (2024). Artificial intelligence in hospitality services: Examining consumers' receptivity to unmanned smart hotels. *Journal of Hospitality and Tourism Insights, 8*(11). Advance online publication. DOI: 10.1108/jhti-06-2024-0548

Ehl.edu. (2024, March 28). *Hospitality Industry: All Your Questions Answered*. Hospitalityinsights.ehl.edu. https://hospitalityinsights.ehl.edu/hospitality-industry

Fatima, T., & Elbanna, S. (2020). Balanced scorecard in the hospitality and tourism industry: Past, present and future. *International Journal of Hospitality Management, 91*, 102656.

Galaz, V., Centeno, M. A., Callahan, P. W., Causevic, A., Patterson, T., Brass, I., & Levy, K. (2021). Artificial intelligence, systemic risks, and sustainability. *Technology in Society, 67*, 101741.

Ghani, B., Zada, M., Memon, K. R., Ullah, R., Khattak, A., Han, H., & Araya-Castillo, L. (2022). Challenges and strategies for employee retention in the hospitality industry: A review. *Sustainability*, *14*(5), 2885.

Ghazi, S. (2023, March 14). *Hotel carbon efficiency improved 27% from 2014-2019*. Oxford Economics. https://www.oxfordeconomics.com/resource/hotel-carbon-efficiency-improved-27-from-2014-2019/

Gretzel, U., Sigala, M., Xiang, Z., & Koo, C. (2015). Smart tourism: Foundations and developments. *Electronic Markets*, *25*(3), 179–188. DOI: 10.1007/s12525-015-0196-8

Gursoy, D., Chi, O. H., Lu, L., & Nunkoo, R. (2019). Consumers acceptance of artificial intelligence (AI) device use in service delivery. *International Journal of Information Management*, *49*(1), 157–169. DOI: 10.1016/j.ijinfomgt.2019.03.008

Huang, A., Ozturk, A. B., & Zhang, T., Efren, & Haney, A. (2024). Unpacking AI for hospitality and tourism services: Exploring the role of perceived enjoyment on future use intentions. *International Journal of Hospitality Management*, *119*, 103693–103693. DOI: 10.1016/j.ijhm.2024.103693

Ivanov, S. H., & Webster, C. (2019). *Robots, Artificial Intelligence, and Service Automation in Travel, Tourism and Hospitality*. Emerald Publishing Limited., DOI: 10.1108/9781787566873

Kaushal, V., & Srivastava, S. (2020). Hospitality and Tourism Industry amid COVID-19 Pandemic: Perspectives on Challenges and Learnings from India. *International Journal of Hospitality Management, 92*(1), 102707. NCBI. https://doi.org/DOI: 10.1016/j.ijhm.2020.102707

Kim, Y. J., & Kim, H. S. (2022). The impact of hotel customer experience on customer satisfaction through online reviews. *Sustainability*, *14*(2), 848.

Ko, H., Lee, S., Park, Y., & Choi, A. (2022). A survey of recommendation systems: Recommendation models, techniques, and application fields. *Electronics (Basel)*, *11*(1), 141.

Korteling, J. H., van de Boer-Visschedijk, G. C., Blankendaal, R. A., Boonekamp, R. C., & Eikelboom, A. R. (2021). Human versus artificial intelligence. *Frontiers in Artificial Intelligence*, *4*, 622364.

Limna, P. (2023). Artificial Intelligence (AI) in the hospitality industry: A review article. *International Journal of Computing Sciences Research*, *7*, 1306–1317.

Limna, P., & Kraiwanit, T. (2023). The role of ChatGPT on customer service in the hospitality industry: An exploratory study of hospitality workers' experiences and perceptions. *Tourism and Hospitality Management*, *29*(4), 583–592.

Lv, H., Shi, S., & Gursoy, D. (2022). A look back and a leap forward: A review and synthesis of big data and artificial intelligence literature in hospitality and tourism. *Journal of Hospitality Marketing & Management*, *31*(2), 145–175.

Mariani, M. M., & Borghi, M. (2024). Artificial intelligence in service industries: Customers' assessment of service production and resilient service operations. *International Journal of Production Research*, *62*(15), 5400–5416.

Merkle, J. A. (2023). *Management and ideology: The legacy of the international scientific management movement*. Univ of California Press.

Polignano, M., Musto, C., Pellungrini, R., & Setzu, M. (2024, November 26). *XAI.it 2024: An Overview on the Future of AI in the Era of Large Language Models*. ResearchGate; unknown. https://www.researchgate.net/publication/386135144_XAIit_2024_An_Overview_on_the_Future_of_AI_in_the_era_of_Large_Language_Models

Prakash, S., Sharma, V. P., Singh, R., & Vijayvargy, L. (2023). Adopting green and sustainable practices in the hotel industry operations analysis of critical performance indicators for improved environmental quality. *Management of Environmental Quality*, *34*(4), 1057–1076.

Pulignano, V., Grimshaw, D., Domecka, M., & Vermeerbergen, L. (2024). Why does unpaid labour vary among digital labour platforms? Exploring socio-technical platform regimes of worker autonomy. *Human Relations*, *77*(9), 1243–1271.

Revnomix.com. (2022, November 21). *Ultimate Guide to Sustainability in the Hospitality Industry*. Revnomix. https://revnomix.com/ultimate-guide-to-sustainability-in-the-hospitality-industry/

Ruel, H., & Njoku, E. (2021). AI is redefining the hospitality industry. *Journal of Tourism Futures*, *7*(1), 53–66.

Sakdiyakorn, M., Golubovskaya, M., & Solnet, D. (2021). Understanding Generation Z through Collective consciousness: Impacts for Hospitality Work and Employment. *International Journal of Hospitality Management*, *94*(102822). Advance online publication. DOI: 10.1016/j.ijhm.2020.102822

Saleh, M. M. A., Jawabreh, O. A., Al Om, R., & Shniekat, N. (2021). Artificial intelligence (AI) and the impact of enhancing the consistency and interpretation of financial statements in the classified hotels in Aqaba, Jordan. *Academy of Strategic Management Journal*, *20*, 1–18.

Saxena, S. K., Gupta, V., & Kumar, S. (2024). Enhancing Guest Loyalty in the Hotel Industry Through Artificial Intelligence-Driven Personalisation. In *New Technologies in Virtual and Hybrid Events* (pp. 335–350). IGI Global.

Seal, P. P., Shukla, A., Bhutia, L. D., & Yadav, N. (2025). The role of food in food blogs: A netnographic analysis of Indian food bloggers' posts on social media. *Asia-Pacific Journal of Innovation in Hospitality and Tourism*, *14*(1), 115–136.

Sharma, R., Kumar, A., & Chuah, C. (2021). Turning the black box into a glass box: An explainable machine learning approach for understanding hospitality customers. *International Journal of Information Management Data Insights*, *1*(2), 100050.

Shukla, A., & Sharma, N. (2020). An analysis of casual dining outlet: A Himachal-based study. *International Journal of Advanced Science and Technology*, *29*(3 Special Issue), 1546–1557.

Sony, M., & Naik, S. (2020). Industry 4.0 integration with socio-technical systems theory: A systematic review and proposed theoretical model. *Technology in Society*, *61*, 101248.

Tussyadiah, I. (2020). A review of research into automation in tourism: Launching the Annals of Tourism Research Curated Collection on Artificial Intelligence and Robotics in Tourism. *Annals of Tourism Research*, *81*(1), 102883. DOI: 10.1016/j.annals.2020.102883

Verma, S., Sharma, R., Deb, S., & Maitra, D. (2021). Artificial intelligence in marketing: Systematic review and future research direction. *International Journal of Information Management Data Insights*, *1*(1), 100002.

Walls, A. R., Okumus, F., & Wang, Y. (Raymond), & Kwun, D. J.-W. (2011). An epistemological view of consumer experiences. *International Journal of Hospitality Management*, *30*(1), 10–21. DOI: 10.1016/j.ijhm.2010.03.008

Xu, Y., Liu, X., Cao, X., Huang, C., Liu, E., Qian, S., & Zhang, J. (2021). Artificial intelligence: A powerful paradigm for scientific research. *The Innovation*, *2*(4).

Yadav, N., Shukla, A., Khunasathitchai, K., & Kaur, R. (2024). AI Adaptions: Challenges and Strategies for the Hotel and Tourism Industry. In Talukder, M., Kumar, S., & Tyagi, P. (Eds.), *Hotel and Travel Management in the AI Era* (pp. 19–38). IGI Global Scientific Publishing., DOI: 10.4018/979-8-3693-7898-4.ch002

Chapter 16
Assessing the Role of Augmented Reality in Shaping the Future of the Hotel Industry:
A Literature-based Perspective

Sahil Saini
https://orcid.org/0009-0003-4326-5076
Maharishi Markandeshwar Univertsity, India

Abhishek Ghai
https://orcid.org/0000-0002-0994-9760
Maharishi Markandeshwar University, India

Megha Gupta
https://orcid.org/0009-0003-6385-0563
Kurukshetra University, Haryana, India

ABSTRACT

Augmented reality gamified experiences may entertain and engage visitors. A hotel might implement an augmented reality treasure hunt that directs guests to various locations inside the facility. Product visualization Augmented Reality (AR) is a technology that overlays computer-generated visuals over a user's perception of the actual environment, so creating a composite vision. Room customizations - Augmented reality applications for room customization enable customers to digitally alter the décor of their hotel accommodations, including furniture, artwork, and lighting. Augmented reality (AR) is transforming travel and our engagement with the world. This research seeks to examine the transformational effects of technological

DOI: 10.4018/979-8-3373-4042-5.ch016

breakthroughs in the hospitality sector which reflects the technology is transforming the visitor experience and the operational framework of hotels, resorts, and other hospitality venues. Augmented reality enhances comprehension of how technology innovations are transforming guest experiences and operational methodologies in the hotel industry.

INTRODUCTION

Augmented Reality (AR) and its Growing Role in Service Industry

Hospitality industry has experienced substantial growth in recent years driven by rising global tourism and business travel and changing business preferences and technology plays an important role in streamlining expansion growing with every technological advancement and that comes along with competitive dynamics (Shabani et al., 2019). Augmented Reality (AR) Is shaping the hospitality industry AR technology or due to travel constraints (Shivekar, 2023). Utilizing AR technology enables individuals to visually experience a location without physically visiting there (Shabani et al., 2019). Augmented Reality technology can provide customized experiences according to individual tastes and requirements. This customization can enhance tourist accessibility, especially for individuals with physical impairments, budgetary limits, or time constraints.

Usage of Website-Based Augmented Reality in Indian Hotels

Augmented reality may provide immersive and interactive experiences for prospective tourists, including virtual tours, 360-degree movies, and three-dimensional models of destinations. Augmented Reality for Entertainment and Education - Augmented Reality can augment the enjoyment and educational experience of travellers by including engaging components, such as games, quizzes, or narratives, into tourism events (Shivekar, 2023).

Product visualization Augmented reality (AR) is a technology that superimposes computer-generated images on a user's view of the real world, thus providing a composite view. In the hospitality industry, AR is being used to create immersive and engaging experiences for guests. Therefore, (Tussyadiah et al., 2017) study provided empirical evidence to confirm the effectiveness of VR in shaping consumers attitudes and behaviour.

Table 1. The VR experience centre features

Feature	Description	Implementation Details
VR Experience Center	Physical VR center located in the hotel lobby	Provides immersive 8K three-dimensional environment that guests can explore interactively
3D Property Mapping	Complete virtual representation of the hotel	Includes 3D mapping of every facet of the hotel property
360° Virtual Tours	Comprehensive virtual exploration	Enables clients to experience a 360° tour of various hotel spaces from a single point in the lobby
Banqueting Space Visualization	Virtual event space planning	Showcases the hotel's 51,000 sq.ft. banqueting space with multiple décor setup visualizations to assist with decision-making
WAVE Technology	WhatsApp Virtual Experience	Exclusive technology allowing global distribution of virtual experiences through WhatsApp and other messaging platforms
Virtual Booking Process	VR-guided booking solutions	Allows prospective clients to survey different services and get information through live chat before booking

Usage of Website-Based Augmented Reality Globally

Benefits of using AR in the hospitality industry - Enhanced guest experience, AR can create more engaging and memorable experiences for guests. Increased guest satisfaction, AR can help guests find their way around, learn about the local area, and make informed decisions about their stay. Improved marketing and sales, AR can be used to create engaging marketing materials that can help to attract new guests. Differentiation from competitors, AR can help hotels and resorts to stand out from the competition. Augmented reality (AR) empowers travel operators to improve their services by providing streamlined navigation, immediate destination information, and overcoming linguistic obstacles. The user may examine the frigid terrain through 360° photos and videos, as well as access relics from the structures tourists can explore famous landmarks, interactive museums, and natural wonders from the comfort of their own homes. They can view 3D models of iconic buildings like the Eiffel Tower or the Taj Mahal and even take virtual walking tours through bustling city streets or far-flung wilderness. An example of this is the travel app Antarctic Heritage Trust AR, which takes users on an expedition to Antarctica. Here a user can explore the icy landscape with 360° images and videos and access artefacts from the buildings (Impact of Technology in Hospitality Sector 2019). (Velina Kazandzhieva, 2019)

Table 2. Current applications of AR/VR in hospitality

Application			Description
Virtual Hotel Tours			**Allowing guests to take virtual tours of hotels and rooms before booking, providing more immersive and realistic experiences compared to traditional photos**
Enhanced Booking Processes			Streamlining and improving the room booking experience through immersive visualization
Virtual Amenity Previews			Offering virtual tours of hotel amenities to help guests better understand available facilities
Competitive Differentiation			Creating unique experiences that distinguish hotels from competitors, increasing customer engagement and potentially driving revenue

Comparative Analysis with Other Technologies

This comprehensive analysis investigated augmented reality (AR) and virtual reality (VR) within the realms of tourism and hospitality. Staff Training: Hotels utilize augmented reality to instruct staff members on new protocols or enhance their comprehension of the hotel's amenities, resulting in superior client service.

Pre-Travel Research and Exploration

Tourists can digitally investigate places prior to finalizing their travel plans. Augmented reality applications provide previews of forthcoming experiences, including accommodations and regional points of interest. Shukla et al. (2025) highlight policy-level solutions to reduce waste in rural tourism zones.

Novelty and Enjoyment

Tourists like AR since it introduces an aspect of exhilaration and exploration to their travels. Augmented reality is utilized throughout the vacation experience, encompassing pre-trip study and on-site investigation. *The integration of AI technologies requires careful adaptation* (Yadav et al., 2024).

Content Quality

High-resolution imagery and precise information are essential.

Seamless Integration

Augmented Reality must effortlessly merge with the tangible surroundings.

Intuitive augmented reality applications improve user experience. As per the article by (Jalilvand & Ghasemi, 2024)

Collaborations: Alliances with local artists, historians, and narrators enhance AR content.

Ongoing Updates: Consistently updating AR material maintains its relevance and engagement.

Objective

This study aims to explore and critically evaluate the role of Augmented Reality (AR) in transforming the hotel industry by conducting a thorough analysis of existing literature. It focuses on understanding the diverse applications of AR, its benefits and challenges and the broader implications for enhancing both customer experience and operational efficiency within the hospitality sector. Additionally, the study seeks to examine how AR technologies contribute to increased customer engagement and satisfaction in hotel services, while also identifying the operational and marketing advantages associated with their implementation. Ultimately, the research aims to assess how AR can drive innovation and strengthen the future competitiveness of hotel businesses.

LITERATURE REVIEW & CONCEPTUAL BACKGROUND

Definition and Types of AR

The intriguing realm of augmented reality (AR) and its applications in travel and hospitality. Augmented reality (AR) is a technology that superimposes digital information—such as photographs, movies, or 3D models—onto the physical environment. In contrast to virtual reality (VR), which fully immerses viewers in artificial situations, augmented reality (AR) enriches the real-world context by including interactive components. Envision utilizing AR glasses or a smartphone to get further information on a historical landmark, navigate to local attractions, or see virtual characters overlaid in your environment.

Augmenting Tourist Experiences

Tourism management acknowledge that augmented reality may enhance the tourist experience. By delivering AR-enhanced material, companies may furnish visitors with dynamic and unforgettable experiences at destinations, landmarks, and cultural places. The essay emphasizes that augmented reality substantially influences consumer experiences in the tourist industry by augmenting engagement, personalizing interactions, functioning as a potent marketing instrument, safeguarding cultural assets, and converging with virtual reality to provide immersive experiences. (Jalilvand & Ghasemi, 2024)

Advantages of Implementing AR in the Hospitality Sector

1. Improved guest experience; AR may facilitate more captivating and unforgettable interactions for patrons. Augmented Reality enhances guest happiness by assisting them in navigation, providing information about the local region, and facilitating educated decision-making over their stay.
2. Enhanced marketing and sales; augmented reality may be utilized to develop captivating promotional materials that attract new patrons.
3. Augmented reality can assist hotels and resorts in distinguishing themselves from rivals. Particular Challenges Hotels May Encounter with Augmented Reality.

AR, MR, VR and How AR Differs

This article highlights augmented reality's significance as an effective marketing communication instrument. It can entice prospective visitors by offering captivating previews of activities, therefore affecting their decision-making process. The immersive quality of augmented reality fosters a profound emotional bond with the place, hence increasing the probability of revisiting.

Augmented Reality (AR) may significantly contribute to the preservation of cultural heritage by recreating the original aesthetics of historical locations and provide educational information regarding them. This enhances the visitor experience while fostering understanding and respect of cultural heritage.

Types: Marker-Based, Location-Based, Projection-Based, Superimposition-Based AR

Augmented reality is the overlay of computer-generated imagery onto a real-world environment. Certain experts consider augmented reality (AR) a subclass of virtual reality (VR), especially a form of mixed reality where real-world environments are augmented with digital material.

Applications in T&H: Augmented Reality (AR) has been acknowledged for its prospective uses across several domains, such as management, marketing, and heritage conservation. It serves as a marketing communication instrument and enhances the visitor experience through interactive and engaging experience as briefly explained in (Calisto & Sarkar, 2024).

Evolution of AR in Hospitality

VR and AR technologies are two major applications that are examples of these innovations. AR is defined as a technology that provides a "unified view" of a computer-generated image over the real world and in real time (McMillan et al., 2017), while VR technology refers to a computer-generated environment that simulates an actual event (Fernandez, 2017). A comprehensive literature analysis was conducted to elucidate the notion of Augmented Reality within the service industry, encompassing sectors such as tourism, hospitality, transportation, and dining establishments. The evaluated literature was categorized into the following components. The paper by (Jingen Liang & Elliot, 2021) examines the role of augmented reality (AR) in the tourist sector, highlighting its enhancement of real-world experiences through digital means. The paper systematically traces the path from first research in AR design and development to contemporary emphasis on user adoption, employing bibliometric analysis to uncover emergent subjects and research clusters. The selected clusters encompass AR design and development, user acceptability, user experience and behavioural intention, AR implementation and administration, and gamification in AR. The article highlights the increasing interest in augmented reality applications within tourism, providing a thorough review of the present landscape, research methodology, and prospective future avenues for utilizing AR to enhance visitor experiences in the tourist industry. The paper authored by (Ivasciuc, 2020a) document is an article that discusses how augmented reality (AR) can be used in the hospitality industry, especially in the context of the COVID-19 pandemic. The article reviews the literature on the impact of the pandemic on the tourism sector, the benefits and challenges of AR and the features and functionalities of the AR Media application, a social network based on these technologies developed by the Future Web project. *The growth of casual dining reflects changing urban lifestyles* (Shukla & Sharma, 2020).The article proposes several ways that the AR Media application can contribute to increasing the competitiveness of tourism companies, such as creating an in-site social network, offering a "try before you buy" opportunity, improving customer service and security, and facilitating payment and information.

Timeline or Milestones of AR use in the Hospitality Sector

The article concludes that AR are promising tools for innovation and adaptation in the hospitality industry, and that the AR Media application can offer valuable experiences and solutions for both tourists and businesses. (Ivasciuc, 2020b) examines the application of augmented reality (AR) within the hotel sector, particularly in relation to the COVID-19 pandemic. The paper examines the literature about the pandemic's effects on the tourist industry, the advantages and obstacles of augmented reality (AR), and the characteristics and capabilities of the AR Media application, a social network utilizing these technologies developed by the Future Web project. The article suggests multiple methods by which the AR Media application can enhance the competitiveness of tourism enterprises, including the establishment of an onsite social network, provision of a "try before you buy" option, enhancement of customer service and security, and facilitation of payment and information. The article concludes that augmented reality represents a promising tool for innovation and adaptation within the hospitality sector, and that the AR Media application can deliver valuable experiences and solutions for both tourists and businesses. The study undertaken by (Paulo et al., 2018) seeks to comprehend the adoption of Mobile Augmented Reality in Tourism (MART) by synthesizing the Task Technology Fit (TTF) model with the Unified Theory of Acceptance and Usage of Technology (UTAUT2). The research investigates the effects of task and technology attributes on Task-Technology Fit (TTF), analysing their influence on user performance expectations, behavioural intentions, and use behaviour. Research concerning several facets of Augmented Reality conducted by (Saxena, 2023).

Applications of AR in the Hotel Industry

Younger demographics frequently demonstrate more excitement and adaptability in adopting AR, owing to their familiarity with digital interfaces and gaming experiences (Ivarson et al., 2024) Conversely, elder demographics may exhibit more prudent attitudes or encounter obstacles associated with technology proficiency (Zhang, 2023)

Nonetheless, elderly users have demonstrated heightened interest in augmented reality, especially within hospital settings for rehabilitation or assisted living applications (Chung et al., 2015).

BENEFITS OF AR FOR THE HOTEL INDUSTRY

The essay posits that intelligent technologies, such as augmented reality (AR), have the potential to improve visitor experiences (Chang et al., 2022). (Shabani et al., 2019) examine e-marketing and its vital significance in the success of enterprises, goods, or services, with an emphasis on the tourist and hospitality sector. It underscores the distinctive attributes of this business, such as perishability, seasonality, and consumer-centricity, hence accentuating the necessity for successful marketing techniques. The article presents Augmented Reality (AR) as a revolutionary technology for e-marketing, illustrating effective applications in travel, including AR museums, Inamo restaurants in London, and Kuoni Travel. Augmented Reality offers cost-effectiveness, enhanced client engagement, and the capacity to provide memorable experiences, rendering it an invaluable asset for marketers in the tourist and hospitality industries. The conclusion highlights the prospective influence of augmented reality (AR) on this sector and the necessity for more study and innovation. The article by (Mariani & Borghi, 2021a) examines the effects of artificial intelligence (AI) manifested through service robots within the hotel and tourist business. This study examines the impact of AI-enabled service interactions, particularly those involving service robots, on customer assessments and satisfaction levels. The research suggests utilizing online reviews as a data source to evaluate the impact of consumers' perceptions of service robots on their evaluations of AI-driven hospitality services. This study seeks to examine the prevalence of robot-related content in online reviews and its relationship with consumer happiness, as shown by review ratings. This study examines the convergence of AI in services, data analytics, and electronic Word-of-Mouth (e-WOM), enhancing the comprehension of client responses to service robots inside enhanced hospitality experiences. (Drexler & Beckman Lapré, 2019) examine the incorporation of artificial intelligence (AI) and robots into the hotel sector, emphasizing the effects of digital transformation. It examines instances of robotic implementations in hotels and restaurants, highlighting the necessity for a measured approach in the use of technology. The management ramifications, encompassing the re-evaluation of conventional methods and the issues associated with staff turnover, are emphasized. The paper further examines overarching social issues, like privacy and data protection, within the framework of AI and robots. The augmented reality approach focuses on integrating AI and robotic components into the hospitality experience, perhaps improving customer interactions via augmented reality interfaces or applications. (Pinto et al., 2022a) examines the effects of the COVID-19 pandemic on tourism in Porto, Portugal, which has emerged as a prominent destination in Europe. The study seeks to empirically evaluate the adoption of Mobile Augmented Reality in Tourism (MART) in Porto, focusing on factors influencing adoption, their significance, and the moderating effects of age and gender. The paper by (Kumar Bisoi et al., 2020a) on Augmented Reality (AR) and Artificial Intelligence (AI) presents substantial insights into the applications of AR, including immersive experiences such as virtual tours and supplementary information for guests.

CHALLENGES FACED BY HOTELS IN THE IMPLEMENTATION OF WEBSITE-BASED AUGMENTED REALITY

Hotels are facing difficulties in integrating augmented reality and beacon technologies into their operational systems. This technology facilitates services for guests, including digitally guided tours, previews of in-room environments (e.g., decor, facilities, and amenities), instant translation services for signs and written materials, interactive restaurant menus featuring dish previews, critic reviews, food allergy information, and interactive trivia games related to on-property points of interest. While utilizing these services, HSP can promote new offerings or gather data to enhance visitor preferences (Kansakar et al., 2017) Interoperability Data Management - The hotel sector has a deficiency in standardization. Numerous HSPs are creating proprietary solutions grounded on their measurements and methodology to meet the technical service requirements of contemporary visitors. *Social media has redefined the role of food in cultural expression* (Seal et al., 2025. *The growth of casual dining reflects changing urban lifestyles* (Shukla & Sharma, 2020. This has resulted in a varied array of implementations aimed at delivering a comparable range of services. The Back of House (BoH) management systems in hotels must effectively handle the flood of diverse visitor data from several sources. To deliver tailored services to visitors, BoH management systems must evaluate guest preference profiles in conjunction with data on the surrounding environment obtained from IoT devices and sensors. (Kansakar et al., 2017) Security and Privacy - To deliver highly tailored services, Hospitality Service Providers (HSPs) must monitor visitor preferences, behaviours, and whereabouts. Hospitality service providers must guarantee the right utilization and storage of guest data to safeguard individuals from physical, financial, and social dangers. A secure hotel network inhibits hackers from accessing visitor data. (Kansakar et al., 2017). (Wu & Kim, 2022)

Challenges connected to age-specific disparities in AR utilization encompass cognitive capabilities, visual sharpness, and motor proficiency. Senior users may encounter challenges in navigating AR interfaces or understanding intricate interactions. Creating straightforward interfaces and adjustable features help mitigate these issues, enhancing the accessibility of AR across all age demographics (Peleg-Adler et al., 2018) investigate the effects of the COVID-19 pandemic on Indonesia's hotel sector, highlighting the significance of crisis management and technology resilience in relation to augmented reality, purchase decisions, and guest satisfaction levels. It examines the function of artificial intelligence and robotics, emphasizing discussions on human vs robotic work. The study, derived from interviews conducted in Greater Jakarta, delineates obstacles and possibilities.

CONSUMER BEHAVIOUR AND PERCEPTION

Alshammare et al. (2022) examine the effect of augmented reality (AR) and website quality on purchase intention within the Indonesian textile industry, specifically focusing on the housing sector. The research used the SOR (Stimulus-Organism-reaction) framework, wherein website quality and augmented reality serve as stimuli, consumer satisfaction functions as the organism, and purchase intention represents the reaction.

The direct impact of augmented reality and website quality on purchase intention is deemed minimal; however, a considerable indirect effect exists via customer satisfaction mediation. The research underscores the capability of augmented reality to deliver a more immersive and pleasurable experience, hence enhancing consumer happiness and, in turn, purchase intention. (Djakasaputra et al., 2023) examine the impact of website quality and augmented reality (AR) on purchase intention within the textile industry's housing sector in Indonesia, employing the Stimulus-Organism-Response (SOR) theory.

THEORETICAL FRAMEWORKS IN EXISTING LITERATURE

The principal obstacle encountered by consumers utilizing augmented reality in retail pertains to technological impediments. Users may have compatibility issues with their devices or suffer challenges in using AR applications, hindering their engagement. Moreover, apprehensions around data privacy and security persist. Consumers may exhibit reluctance to utilize augmented reality owing to concerns around the disclosure of personal information or the risk of privacy violations while interacting with these immersive technologies (Alam et al., 2021) .Awkward interfaces or interruptions in the purchasing process might reduce the attractiveness of AR apps, impacting user happiness and their propensity for extended use (Wu & Kim, 2022) .

The study of (Saxena, 2023) examined the increasing utilization of Augmented Reality (AR) in tourism, particularly via mobile applications. It seeks to comprehend how the amalgamation of augmented reality with mobile technology fulfils the requirements of tourists. The study centres on two principal frameworks: the Unified

Theory of Acceptance and Use of Technology (UTAUT) and the Task Technology Fit (TTF) model. The study underscores the importance of augmented reality in improving visitor experiences and decision-making via smartphone applications. (Özkul & Kumlu, 2019) examine the development and influence of augmented reality (AR) apps within the tourist sector. The text commences with a historical review, emphasizing the function of augmented reality in captivating and engaging visitors via improved management, marketing, and advertising strategies. The notion of augmented reality, which is the incorporation of computer images into the physical environment, is examined, highlighting its diverse uses across several sectors. This study conducted by (Anand et al., 2023) examines the quality characteristics of augmented reality-based mobile applications for smart tourism and their influence on consumer satisfaction and intention to reuse.

RESEARCH GAPS AND FUTURE DIRECTIONS

Approach

Given the uniqueness of the AR media application, we deemed it appropriate to conduct a thorough examination of the literature concerning emerging technologies, including augmented reality and face recognition technology. The primary conclusions will be utilized to assess the influence of these developing technologies on AR Media.

Outcomes and Deliberations

The application of Augmented Reality (AR) in tourism and hospitality has seen substantial development from 2010 to 2024.

1. Initial Adoption (2010-2015): The primary applications aimed at improving visitor experiences at heritage sites and museums using rudimentary AR overlays that offered historical information.
2. Expansion and Innovation (2016-2020): Augmented reality technology advanced significantly, incorporating interactive components and gamification in amusement parks and urban excursions.
3. The COVID-19 pandemic (2020-2022) expedited the integration of augmented reality, providing virtual excursions and experiences to mitigate the limitations imposed by travel restrictions.

4. Mainstream Integration (2023-2024): Augmented Reality is extensively incorporated into tourist strategy, augmenting real-world experiences with digital material, expanding accessibility, and customizing visitor interactions.

The research underscores AR's impact on education, information tailoring, interactivity and entertainment in tourism. Notable examples, such as Kuoni Travel and Thomson Cruises, demonstrate AR's potential for personalized and memorable experiences. The literature review emphasizes smartphones as prevalent AR devices, and the study aims to qualitatively explore tourism-specific AR value dimensions, focusing on tourism suppliers. The overarching goal is to understand AR's perceived value comprehensively before widespread implementation, considering the significant investments involved, and ensure its successful adoption for long-term profitability and business success. The finding from a book titled "Hospitality & Tourism Sectors Global Perspectives and Challenges", edited by Kumar et al. (2023). This article written by (Flavián et al. 2021) delves into the impact of Augmented Reality technologies, including VR headsets, smartphones, and desktop computers, on the early stages of the hotel customer journey. The Spherical Fuzzy COPRAS MCDM Model is a systematic approach for objectively selecting AR solutions based on organizational needs. Augmented Reality Satisfaction Level or purchase decision of guests and users According to Nayyar et al. (2018) the applications of Augmented Reality (AR) in the 21st century, highlight its role in tourism and hospitality. AR integrates virtual information with the real world, offering benefits such as effective planning, entertainment, education, and marketing in the tourism industry. Examples like Pokémon GO demonstrate AR's popularity. The summary concludes by underscoring the potential of AR in hotel management, marketing, and staff training, indicating its significant impact on the hospitality sector. Putri, Situmorang, and Mehrotra et al. (2023) explore how the COVID-19 pandemic has impacted Indonesia's hotel industry, emphasizing the importance of crisis management and technological resilience. It discusses the role of artificial intelligence and robotics, highlighting debates about human vs. robotic labour. The research, based on interviews in Greater Jakarta, identifies challenges and opportunities. In terms of augmented reality (AR), the article suggests that while intelligent technologies, including AR, could enhance guest experiences. The study contributes insights to the discourse on technology adoption in hospitality, calling for more research on AR applications in hotels. The conclusion emphasizes the potential impact of AR on this industry and the need for further research and innovation. According to an article written by (Mariani & Borghi, 2021b) explores the impact of Artificial Intelligence (AI) in the form of service robots on the hospitality and tourism industry. The article written by (Drexler & Beckman Lapré, 2019) explores the integration of artificial intelligence (AI) and robotics in the hospitality industry, focusing on the impact of digital transformation. It discusses examples of robotic applications in hotels and restaurants, emphasizing the need for a balanced approach in adopting technology. The management implications, including the reconsideration of traditional models and the challenges related to employee turnover, are highlighted. The article also addresses broader societal concerns such as privacy and data protection in the context of AI and robotics. In the augmented reality perspective, the emphasis would be on incorporating AI and robotic elements into the hospitality experience, potentially enhancing customer interactions through augmented reality interfaces or applications. The article published by (Pinto et al., 2022b) explores the impact of the COVID-19 pandemic on tourism in Porto, Portugal, which has become one of Europe's attractive destinations. The study aims to empirically test the adoption of Mobile Augmented Reality in Tourism (MART) in Porto and addresses research questions related to factors influencing adoption, the importance of these factors, and the moderating role of age and gender. The UTAUT-3 framework is employed, considering constructs such as performance expectancy, effort expectancy, social influence, facilitating conditions, hedonic motivation, price value, habit, and personal innovativeness. The study uses a quantitative approach with a partial least square regression (SEM-PLS) analysis to estimate the model. The conceptual framework highlights the role of UTAUT-3 in predicting MART adoption, considering age and gender as moderators. The study positions MART as a tool for enhancing the tourism experience and aiding in the industry's recovery post-pandemic. The paper on Augmented Reality (AR) and Artificial Intelligence (AI) by (Kumar Bisoi et al., 2020b) jointly enhance the hospitality industry.

CONCLUSION

This study examines the emerging role of Augmented Reality (AR) and related immersive technologies within the Indian hospitality sector, with a specific focus on hotels and resorts. It is based largely on secondary data and case evidence, recognizing that adoption in India remains at an early stage. The scope is limited to applications that directly influence guest experiences, marketing strategies, and operational practices in hospitality, rather than the wider tourism industry. Given the pace of technological change, the findings should be viewed as a snapshot of current developments, requiring continuous updates to remain relevant. The success of initiatives such as Crowne Plaza Greater Noida's VR Experience Centre demonstrates the growing potential of immersive technologies in hospitality. As costs decline and consumer awareness increases, AR/VR applications are expected to expand in areas such as remote property exploration, immersive marketing campaigns, event and conference planning, staff training, and personalized guest services. These tools can strengthen customer engagement and operational efficiency, positioning hotels more competitively in a digital-first market. However, future adoption will depend on addressing key challenges, including high initial investments, technical expertise, guest readiness, and integration with existing hotel systems. Overall, immersive technologies hold strong promise to shape the future of tourism and hospitality management.

Augmented Reality (AR), together with VR and MR, is gradually reshaping the hospitality sector by creating more engaging guest experiences and smarter ways of running operations. In India, adoption is still at an early stage, but applications such as virtual tours, interactive navigation, product visualization, and room personalization already show how these technologies can add real value for both customers and businesses. As AR tools become easier to access and implement, hotels and resorts that embrace them will be better positioned to stand out, build stronger customer connections, and remain competitive in a rapidly evolving market. Immersive technologies are not just a trend—they represent a significant step toward the future of tourism and hospitality management.

REFERENCES

Alam, S. S., Susmit, S., Lin, C.-Y., Masukujjaman, M., & Ho, Y.-H. (2021). Factors affecting augmented reality adoption in the retail industry. *Journal of Open Innovation*, *7*(2), 142.

Alshammare, G. I., Halim, M. S. B. A., & Alsheikh, G. A. A. (2022). Online Booking Services Assisted By Technology To Improve Customer Loyalty In Jordanian Five-Star Hotels. *International Journal of Professional Business Review*, *7*(3). Advance online publication. DOI: 10.26668/businessreview/2022.v7i3.0551

Anand, K., Arya, V., Suresh, S., & Sharma, A. (2023). Quality Dimensions of Augmented Reality-based Mobile Apps for Smart-Tourism and its Impact on Customer Satisfaction & Reuse Intention. *Tourism Planning & Development*, *20*(2), 236–259. DOI: 10.1080/21568316.2022.2137577

Calisto, M. de L., & Sarkar, S. (2024). A systematic review of virtual reality in tourism and hospitality: The known and the paths to follow. *International Journal of Hospitality Management*, *116*. Advance online publication. DOI: 10.1016/j.ijhm.2023.103623

Chang, H. Y., Binali, T., Liang, J. C., Chiou, G. L., Cheng, K. H., Lee, S. W. Y., & Tsai, C. C. (2022). Ten years of augmented reality in education: A meta-analysis of (quasi-) experimental studies to investigate the impact. *Computers & Education*, *191*. Advance online publication. DOI: 10.1016/j.compedu.2022.104641

Chung, N., Han, H., & Joun, Y. (2015). Tourists' intention to visit a destination: The role of augmented reality (AR) application for a heritage site. *Computers in Human Behavior*, *50*, 588–599. DOI: 10.1016/j.chb.2015.02.068

Djakasaputra, A., Pramezwary, A., & Lemy, D. M. (2023). SEIKO: Journal of Management & Business Influence of Website Quality and Augmented Reality to Purchase Intention with Satisfaction as Mediation. *SEIKO: Journal of Management & Business*, *6*(1), 730–743. DOI: 10.37531/sejaman.vxix.688

Drexler, N., & Beckman Lapré, V. (2019). For better or for worse: Shaping the hospitality industry through robotics and artificial intelligence. *Research in Hospitality Management*, *9*(2), 117–120. DOI: 10.1080/22243534.2019.1689701

Fernandez, M. (2017). Augmented virtual reality: How to improve education systems. *Higher Learning Research Communications*, *7*(1), 1. DOI: 10.18870/hlrc.v7i1.373

Flavián, C., Ibáñez-Sánchez, S., & Orús, C. (2021). Impacts of technological embodiment through virtual reality on potential guests' emotions and engagement. *Journal of Hospitality Marketing & Management*, *30*(1), 1–20. DOI: 10.1080/19368623.2020.1770146

Impact of Technology in Hospitality sector. (2019). www.jetir.org

Ivarson, E., Erlandsson, V., Faraon, M., & Khatib, S. (2024). Augmented reality and gamification in higher education: Designing mobile interaction to enhance students' motivation and learning. *E-Learning and Digital Media*. Advance online publication. DOI: 10.1177/20427530241239981

Ivasciuc, I. S. (2020a). Augmented reality and facial recognition technologies. Building bridges between the hospitality industry and tourists during pandemic. *Bulletin of the Transilvania University of Brasov. Series V, Economic Sciences*, •••, 75–92. https://webbut.unitbv.ro/index.php/Series_V/article/view/219

Ivasciuc, I. S. (2020b). Augmented reality and facial recognition technologies. Building bridges between the hospitality industry and tourists during pandemic". Bulletin of the Transilvania University of Brasov Series V Economic Sciences, 13(62)(2), 75–92. https://doi.org/DOI: 10.31926/but.es.2020.13.62.2.8

Jalilvand, M. R., & Ghasemi, H. (2024). Augmented reality technology in tourism and hospitality research: a review from 2010 to 2024. Journal of Science and Technology Policy Management. https://doi.org/DOI: 10.1108/JSTPM-04-2024-0136

Jingen Liang, L., & Elliot, S. (2021). A systematic review of augmented reality tourism research: What is now and what is next? *Tourism and Hospitality Research*, *21*(1), 15–30. DOI: 10.1177/1467358420941913

Kansakar, P., Munir, A., & Shabani, N. (2017). Technology in Hospitality Industry: Prospects and Challenges. https://arxiv.org/abs/1709.00105

Kumar Bisoi, S., Roy, M., Samal, A., Bisoi, S., & Professor, A. (2020a). Impact of Artificial Intelligence in the Hospitality Industry. *International Journal of Advanced Science and Technology*, *29*(5), 4265–4276. https://www.researchgate.net/publication/343180745

Kumar Bisoi, S., Roy, M., Samal, A., Bisoi, S., & Professor, A. (2020b). Impact of Artificial Intelligence in the Hospitality Industry. *International Journal of Advanced Science and Technology*, *29*(5), 4265–4276. https://www.researchgate.net/publication/343180745

Mariani, M., & Borghi, M. (2021a). Customers' evaluation of mechanical artificial intelligence in hospitality services: A study using online reviews analytics. *International Journal of Contemporary Hospitality Management*, *33*(11), 3956–3976. DOI: 10.1108/IJCHM-06-2020-0622

Mariani, M., & Borghi, M. (2021b). Customers' evaluation of mechanical artificial intelligence in hospitality services: A study using online reviews analytics. *International Journal of Contemporary Hospitality Management*, *33*(11), 3956–3976. DOI: 10.1108/IJCHM-06-2020-0622

McMillan, K., Flood, K., & Glaeser, R. (2017). Virtual reality, augmented reality, mixed reality, and the marine conservation movement. *Aquatic Conservation*, *27*, 162–168. DOI: 10.1002/aqc.2820

Nayyar, A., Mahapatra, B., Le, D., & Suseendran, G. (2018). Virtual Reality (VR) & Augmented Reality (AR) technologies for tourism and hospitality industry. [). www.sciencepubco.com/index.php/IJET]. *IACSIT International Journal of Engineering and Technology*, *7*(2), •••.

Özkul, E., & Kumlu, S. T. (2019). Augmented reality applications in tourism. *International Journal of Contemporary Tourism Research*, *3*(2), 107–122. https://dergipark.org.tr/en/pub/ijctr/issue/51052/625192

Paulo, M. M., Rita, P., Oliveira, T., & Moro, S. (2018). Understanding mobile augmented reality adoption in a consumer context. *Journal of Hospitality and Tourism Technology*, *9*(2), 142–157. DOI: 10.1108/JHTT-01-2017-0006

Peleg-Adler, R., Lanir, J., & Korman, M. (2018). The effects of aging on the use of handheld augmented reality in a route planning task. *Computers in Human Behavior*, *81*, 52–62. DOI: 10.1016/j.chb.2017.12.003

Pinto, A. S., Abreu, A., Costa, E., & Paiva, J. (2022a). Augmented Reality for a New Reality: Using UTAUT-3 to Assess the Adoption of Mobile Augmented Reality in Tourism (MART). *Journal of Information Systems Engineering & Management*, *7*(2). Advance online publication. DOI: 10.55267/iadt.07.12012

Pinto, A. S., Abreu, A., Costa, E., & Paiva, J. (2022b). Augmented Reality for a New Reality: Using UTAUT-3 to Assess the Adoption of Mobile Augmented Reality in Tourism (MART). *Journal of Information Systems Engineering & Management*, *7*(2). Advance online publication. DOI: 10.55267/iadt.07.12012

Putri, R. N. P., Situmorang, R., & Mehrotra, A. (2023). Human vs. Robot: The application of smart technology in the hotel industry. *Jurnal Manajemen Dan Kewirausahaan*, *25*(2), 142–151. DOI: 10.9744/jmk.25.2.142-151

Saxena, S. (2023). "Unveiling The Digital Marketplace: A Comprehensive Analysis of The Purchase Decision-Making Process Among Online Consumers in Jammu and Kashmir" (Vol. 5, Issue 2).

Seal, P. P., Shukla, A., Bhutia, L. D., & Yadav, N. (2025). The role of food in food blogs: A netnographic analysis of Indian food bloggers' posts on social media. *Asia-Pacific Journal of Innovation in Hospitality and Tourism*, *14*(1), 115–136.

Shabani, N., Munir, A., & Hassan, A. (2019). E-Marketing via Augmented Reality: A Case Study in the Tourism and Hospitality Industry. *IEEE Potentials*, *38*(1), 43–47. DOI: 10.1109/MPOT.2018.2850598

Shivekar, R. (2023). A Study Of Digital Marketing Strategies In Tourism Management. https://www.researchgate.net/publication/375926420

Shukla, A., & Sharma, N. (2020). An analysis of casual dining outlet: A Himachal-based study. *International Journal of Advanced Science and Technology, 29*(3 Special Issue), 1546–1557.

Shukla, A., Yadav, N., Khunasathitchai, K., Bakshi, I., & Sharma, N. (2025). Waste Management Outlook and Future Directions in Rural Touristic Areas. In Albattat, A., Singh, A., Tyagi, P., & Haghi, A. (Eds.), *Solid Waste Management and Disposal Practices in Rural Tourism* (pp. 495–522). IGI Global Scientific Publishing., DOI: 10.4018/979-8-3693-9621-6.ch020

Tussyadiah, I. P., Wang, D., Jung, T. H., Metropolitan, M., Claudia, M., & Dieck, T. (2017). Virtual Reality, Presence, and Attitude Change: Empirical Evidence from Tourism School of Hotel and Tourism Management Faculty of Business and Law 20.

Velina Kazandzhieva, G. I. H. F. (2019). Impact of Technology in Hospitality sector. www.jetir.org

Wu, Y. F., & Kim, E. Y. (2022). Users' Perceptions of Technological Features in Augmented Reality (AR) and Virtual Reality (VR) in Fashion Retailing: A Qualitative Content Analysis. *Mobile Information Systems, 2022*. Advance online publication. DOI: 10.1155/2022/3080280

Yadav, N., Shukla, A., Khunasathitchai, K., & Kaur, R. (2024). AI Adaptions: Challenges and Strategies for the Hotel and Tourism Industry. In Talukder, M., Kumar, S., & Tyagi, P. (Eds.), *Hotel and Travel Management in the AI Era* (pp. 19–38). IGI Global Scientific Publishing., DOI: 10.4018/979-8-3693-7898-4.ch002

Zhang, M. (2023). Older people's attitudes towards emerging technologies: A systematic literature review. In Public Understanding of Science (Vol. 32, Issue 8, pp. 948–968). SAGE Publications Ltd. https://doi.org/DOI: 10.1177/09636625231171677

Chapter 17
Exploring Consumer Experiences in Wine Tourism Through Sentiment Analysis:
An Assessment of Tripadvisor Reviews About Wineries in France, Portugal, and Italy

Laura Siegl
Faculty of Economics, Universidade do Algarve, Portugal

Ema M. Q. Baptista
Faculty of Economics, Universidade do Algarve, Portugal

Daniele Ruvo
University of Naples Parthenope, Italy

Célia M. Q. Ramos
https://orcid.org/0000-0002-3413-4897
CinTurs, ESGHT, Universidade do Algarve, Portugal

ABSTRACT

This study examines TripAdvisor reviews from three wineries in Europe to analyze customer perceptions and preferences in wine tourism. The investigation aims to understand the impact of online reviews on winery reputations and customer satisfaction in the context of wine tourism experiences. The study utilizes an advanced data analytics tool, namely Power BI, to analyze the development and demographics of the

DOI: 10.4018/979-8-3373-4042-5.ch017

reviews and Orange Data Mining to conduct sentiment analysis and text mining. By examining reviews from wineries in France, Portugal, and Italy, this work identifies key themes, sentiments, and preferences expressed by customers, providing actionable insights for wineries to enhance visitor experiences and strategic decision-making in the wine tourism industry using an artificial intelligence approach.

INTRODUCTION

Wine tourism has evolved from a niche activity for connoisseurs into a dynamic and multifaceted experience that attracts a wide range of visitors, including professionals, enthusiasts, and curious travelers. Today, it encompasses not only wine tasting but also cultural, natural, and social experiences, often integrated with events, gastronomy, and outdoor activities. Key drivers of this form of tourism include authenticity, tradition, cultural heritage, nature, learning, health, safety, and sustainability (Garibaldi, 2024).

In parallel with this evolution, the digital transformation of tourism has introduced new tools for understanding visitor behavior. In particular, the rise of electronic Word of Mouth (eWOM) through platforms like TripAdvisor has created vast amounts of user-generated content that can be analyzed to extract insights into tourist perceptions and satisfaction. This is where Artificial Intelligence (AI) plays a crucial role.

This chapter explores how AI-driven techniques such as text mining, sentiment analysis, and machine learning can be applied to wine tourism research. Using tools like Power BI and Orange, which integrate Business Intelligence and data mining capabilities, we analyze 432 TripAdvisor reviews from three wineries located in Portugal, Italy, and France. The aim is to uncover patterns in visitor sentiment and perception, and to assess whether there are significant differences in the average sentiment score across countries. Additionally, the study investigates whether the size of the winery influences the evaluations received.

By combining traditional tourism research with AI-based analytical methods, this work contributes to a deeper understanding of wine tourist experiences and offers practical insights for winery managers and tourism professionals.

The chapter is structured as follows: the next section presents the state of the art; this is followed by a description of the methodology; the results are then outlined and discussed; and finally, the conclusion summarizes the findings and suggests future research directions.

BACKGROUND

The research problem addressed in this work focuses on how online reviews and user-generated content on social media platforms like TripAdvisor influence the reputation and perceived quality of wineries in Portugal, France, and Italy.

Online Reviews

Online reviews, also known as electronic word of mouth (eWOM), are voluntary evaluations or feedback consumers share on digital platforms about their experiences with products or services. These reviews are often considered more trustworthy and objective than the information provided by the companies themselves. As a result, they are a crucial part of the customer experience and decision-making process. Customers often seek online reviews to reduce perceived risks and gain confidence in their choices, relying on the experiences of others who have consumed the service or product (Fu et al., 2022).

In the context of the wine industry, online reviews are perceived to have a relatively high reliability and relevant information for management. This is because they influence those who plan to visit a winery and shape their overall expectations for the experience, particularly regarding the menu and service (Fu et al., 2022).

Online Reputation Management

The reputation of wine regions and wineries is crucial for attracting tourists, significantly influencing wine tourism and overall perception. Managers should, therefore, maintain winery landscapes, enhance reputation, and ensure fulfilling visitor experiences (Wu et al., 2024).

Understanding online destination image is vital, as it impacts tourists' decision-making and helps develop effective online marketing strategies. Social media significantly shapes wine consumer behavior, presenting essential opportunities for wineries (Wu et al., 2024). Analyzing online customer reviews provides a basis for developing relevant marketing strategies and gaining insights into the hospitality and travel industry (Fu et al., 2022). Managers should monitor online reviews to identify areas for improvement and use them as marketing tools to generate revenue. Text mining and semantic network analysis of reviews offer reliable customer satisfaction assessments, helping compare performance with competitors and devise long-term strategies (Fu et al., 2022).

Winery operators should focus on creating and maintaining a unique competitive advantage, considering tourists' satisfaction during their visit, which influences their overall experience and future intentions (Gao et al., 2022). Knowledgeable and hospitable winery staff are crucial for shaping a wine destination's image and enhancing tourism experiences. Excellent service and effective marketing strategies can leave lasting impressions on visitors (Wu et al., 2024).

Wine regions should market the winescape by emphasizing natural landscapes as brand elements. Promotional messages should balance core wine tourism products, such as wine tasting and buying, with hedonic experience elements. For repeat visitors, marketing efforts should focus on wine-related aspects, while for first-time visitors, the natural environment should be emphasized (Bruwer & Lesschaeve, 2012).

Wine Tourism

Wine tourism, a growing segment of gastronomy tourism, significantly impacts regional economies by combining elements of wine, food, tourism, and arts (Hospitality ON, 2024). This form of tourism involves visiting wine-producing regions and wineries, often including activities like winery visits, wine tastings, vineyard tours, and other wine-related experiences, catering to both casual enthusiasts and connoisseurs (Brown & Getz, 2005). Its development enhances visitor experiences and contributes to these regions' economic and cultural growth (Hospitality ON, 2024).

Today's wine tourism attracts wine enthusiasts and visitors interested in cultural events, heritage, gastronomy festivals, agricultural activities, and educational seminars, benefiting both wineries and wine regions (Alebaki et al., 2022). Experiences are central to the tourism industry, often stimulating sensations such as relaxation, pleasure, fun, escapism, and discovery, which generate memorable events (Amaral et al., 2024; Gao et al., 2022).

In wine tourism, experiences are considered hedonic, focusing on the search for pleasure related to wine and other aspects like quality food, entertainment, interaction with people, and enjoying the landscape (Amaral et al., 2024). Studies of wine tourism experiences take a holistic approach, often focusing on establishments or events associated explicitly with wine tourism, such as wineries, wine festivals, wine routes, or destinations (Gao et al., 2022).

Wineries in Different Countries

The "World's Best Vineyards" recently released their 2023 ranking of the top 100 wine tourism destinations globally, underscoring the increasing prominence of wine tourism. This sector, often grouped with food tourism, is gaining substantial media attention, reaching a diverse audience (Pereira, 2023).

The growing interest in wine tourism brings new responsibilities for stakeholders, including wine producers, tourism operators, and regional and national authorities. Ensuring high-quality experiences for tourists is crucial for generating satisfaction.

Wine tourism can be a distinctive and value-creating product, especially in inland regions where it often becomes a key tourist asset alongside scenic and natural values. It is also compatible with preserving existing natural and cultural values, contributing to sustainability across environmental, economic, and social pillars (Pereira, 2023).

Regional Highlights

Europe

Europe dominates the rankings with 52 out of the top 100 destinations, followed by the Americas (30) and Oceania (8). Most European wineries are in southern countries, particularly France (17), Portugal (9), Italy (7), and Spain (6). Outside Europe, Argentina (11) and Chile (10) have a significant presence (Pereira, 2023).

Portugal, Italy, and France will be the countries addressed in this paper due to their significant roles in the global wine tourism market. Each of these countries has a rich winemaking heritage, diverse wine offerings, and vibrant tourism industries, making them key players in the sector (Lima & Pereira, 2024).

Portugal

Wine tourism in Portugal is integrated into broader tourism activities. Studies show that 15-20% of visitors engage in wine tourism as a primary activity. In regions like the North and Alentejo, wine tourism is a primary action, while in the Algarve, Madeira, and Lisbon, it is often a secondary attraction. The Azores have seen significant growth, particularly with the UNESCO recognition of Pico Island's vineyard landscape (Lima & Pereira, 2024).

The analyzed winery Quinta dos Vales, located in the Western Algarve, combines wine and art. Founded in 2007 by Karl Heinz Stock, today's estate offers tours, tastings, and accommodations. In 2017, the estate expanded to include “The Winemaker Experience” and real estate under the brand “The Vines” (Quinta dos Vales, 2024).

Italy

Italy's wine tourism has grown rapidly, and it is recovering well post-pandemic. Tuscany is the leading region, especially with areas like Montepulciano and Chianti. Wine tourism in Italy is often integrated into broader tourism experiences. Despite high visitor numbers, Italy's market value and revenues are still below France and the

USA. Revenue growth in 2022 was led by winery visits (78.8%), accommodations (32.5%), and catering (27.5%) (Wine News, 2023).

The analyzed Icario Winery in Montepulciano, Tuscany, is known for its "Vino Nobile di Montepulciano". Founded in 1999, it emphasizes the family's devotion, quality, art, and respect for the territory (Icario, 2018).

France

France remains the leading wine producer worldwide, with over 45 million litres in 2023 (World Population Review, 2024). The country has over 11,000 wineries open to visitors, attracting 10 million wine tourists annually and generating 5.2 billion euros in revenue (Hospitality ON, 2024). Popular destinations include Bordeaux, Champagne, and regions hosting wine events like Fête le Vin (Hospitality ON, 2024a).

Châteauneuf-du-Pape, located in the Rhône Valley, has a rich history dating back to the 14th century. Known for its unique terroir and Grenache grapes, the region blends traditional and modern winemaking techniques (Karlson & Karlson, 2016).

The analyzed winery Les Caves Saint Charles offers exclusive tastings in 13th-century cellars, led by Maître Sommelier Guy Bremond, providing a unique sensory experience (Les Caves Saint Charles, 2024).

Although studies such as those by Bruwer and Lesschaeve (2012) and Wu et al. (2024) acknowledge the critical role of online reviews in wine tourism reputation management, a significant gap remains: determining whether sentiment patterns expressed in reviews vary systematically across distinct national contexts. Given that Portugal, Italy, and France possess unique winemaking traditions, heterogeneous visitor profiles, and differentiated tourism integration models (Lima & Pereira, 2024; Wine News, 2023), it is plausible that cultural and operational factors modulate tourists' emotional expression. This possibility necessitates a robust comparative analysis that contrasts not only ratings, but also the underlying sentiment nuances of experiences across the three countries – particularly considering that sociolinguistic norms may distort objective quality perception (Hu & Liu, 2004). In light of this evidence, we propose the following hypothesis:

H_1: Tourist satisfaction (measured by sentiment scores) differs significantly among wineries in Portugal, Italy, and France.

Furthermore, the relationship between the textual length of reviews and assigned ratings lacks conclusive evidence within the wine tourism context. Studies such as those by Liu et al. (2017) suggest that longer reviews may indicate heightened emotional engagement (whether positive or negative); however, this dynamic remains untested in viticultural settings where the sensory complexity of experiences may generate idiosyncratic discursive patterns. The hypothesis that detailed reviews

(>100 words) predict extreme ratings (5 or 1 star) requires empirical verification, particularly given that sharing motivations (Fu et al., 2022) and cultural expectations (Amaral et al., 2024) may differentially modulate this relationship across the analyzed destinations. Against this background, we propose the following research hypothesis:

H_2: Longer reviews (>100 words) exhibit significantly higher ratings than concise reviews (≤100 words)."

METHODOLOGIES FOR ANALYZING ONLINE REVIEWS

Online reviews, as a form of electronic Word of Mouth (eWOM), have become essential for understanding tourist behavior and satisfaction, offering valuable insights for strategic planning and decision-making (Vu et al., 2019). With the growing availability of user-generated content, researchers have increasingly turned to tools and techniques from the field of Artificial Intelligence (AI) to analyse these data sources.

Several studies have applied text mining and machine learning methods to extract patterns and trends from online reviews. For example, Li et al. (2021) and Luo et al. (2020) used TripAdvisor data to explore hotel selection preferences and theme park perceptions. Vu et al. (2019) developed a framework to analyze dining preferences, while Liu et al. (2017) applied data mining techniques to identify key drivers of hotel satisfaction in Chinese reviews.

In the context of wine tourism, researchers such as Barbierato et al. (2022) and Fu et al. (2022) employed sentiment analysis and text mining to assess visitor emotions and satisfaction levels. Gao et al. (2022) introduced a deep learning approach using neural networks to detect satisfaction and dissatisfaction in reviews. Amaral et al. (2024) combined qualitative and quantitative methods, integrating content analysis with statistical techniques to study visitor experiences.

Wu et al. (2024) applied topic modelling and sentiment analysis to extract key attributes from reviews and conducted Importance-Performance Analyses (IPAs). These approaches, grounded in Business Intelligence and AI, enable a more nuanced understanding of tourist perceptions. Similarly, Bruwer and Lesschaeve (2012) adopted a holistic framework to assess wine region destination image, incorporating marketing theory and multi-dimensional analysis. Gaps Encountered in Previous Articles

While extensive, the existing research on online reviews in wine tourism has several limitations. A common issue across multiple studies is the sample size and selection method. For example, Treen et al. (2018) acknowledged that their sample of 45 winery websites was small and potentially unrepresentative due to the simple Google search rule used. Similarly, Fu et al. (2022) noted that their study of 58 top wine bars in Busan limits generalizability, and Barbierato et al. (2022) highlighted the limitation of using Tripadvisor data, which may not fully represent the broader tourist population due to its user base and hierarchical data structure.

Another major limitation is the reliance on a single data source, often Tripadvisor, which restricts the scope of findings. Amaral et al. (2024) pointed out that the data from Tripadvisor might not capture the full range of visitor experiences, especially as the platform's usage has declined. Wu et al. (2024) similarly acknowledged the limitation of using data from a single review platform, which might not represent a broader audience.

Furthermore, several studies noted methodological constraints. For instance, the sentiment analysis approaches used by Fu et al. (2022) and Barbierato et al. (2022) primarily focused on word frequencies, which can overlook the nuanced meanings of words and the deeper emotional context.

Lastly, geographical limitations were a concern. Barbierato et al. (2022) and Bruwer andLesschaeve (2012) recognized that their findings might be region-specific and not applicable to other wine tourism destinations, as different regions offer unique tourism products and experiences.

Research Objective

This study analyzes online reviews of wineries on Tripadvisor to understand wine tourists' demographics, sentiments, and emotions. Using Power BI and Orange tools, the authors aim to identify factors contributing to tourist satisfaction and dissatisfaction across regions. This analysis will help wineries leverage online sentiment and user-generated content for strategic planning in the wine tourism industry.

Furthermore, identifying factors influencing sentiments allows wineries to tailor their offerings to meet visitor expectations, enhancing the experience and encouraging repeat visits (Amaral et al., 2024; Gao et al., 2022). Moreover, leveraging insights from reviews can differentiate wineries in a competitive market, increasing customer loyalty and positive word-of-mouth promotion (Wu et al., 2024). In addition, analyzing reviews from different regions helps understand cultural nuances and tailor offerings for diverse tourist profiles, which is crucial for attracting international visitors (Bruwer & Lesschaeve, 2012). Further, using advanced data mining and business intelligence tools ensures decisions are based on robust evidence, leading to effective marketing and operational strategies (Li et al., 2021; Liu et al., 2017; Vu et al., 2019).

Simultaneously, we aim to investigate whether a country's affiliation influences sentiment expression and whether review length correlates with assigned ratings, through the research hypotheses outlined above.

Addressing Previous Limitations

This study addresses several limitations identified in previous research:

By collecting 432 reviews from three different countries, the study intends to portray a more representative and diverse sample. Moreover, utilising tools like Power BI and Orange allows for comprehensive data analysis, including sentiment analysis. This addresses the methodological constraints of more straightforward analytical techniques used in previous studies (Fu et al., 2022; Gao et al., 2022). By comparing reviews from wineries in different regions, the study also accounts for regional differences in wine tourism experiences, providing a more holistic understanding of global wine tourism trends, as Bruwer and Lesschaeve (2012) suggested.

By addressing these gaps and leveraging advanced data analytics, this study aims to contribute valuable insights to the wine tourism industry, helping wineries improve their services, enhance visitor satisfaction, and achieve a competitive advantage in the market.

Sentiment Analysis Methodology

The data for this work is based on Tripadvisor reviews. The authors of this paper collected 432 reviews in English from the three wineries introduced above. In addition to the full-text reviews, information like rating, published date, name, nationality, city of residence, and gender was gathered.

The methodology used in this paper focused on business analytics and text mining in accordance with the work of El-Masri et al. (2017). Business Intelligence was conducted utilizing the Microsoft program "Power BI". It allowed us to analyze the development of the reviews over the past years and the most frequent months. Furthermore, it provided insights into the author's demographics, using line charts, bar charts, matrices, and map charts to illustrate the findings. Orange Data Mining helped us understand the most-used words and word combinations using word clouds. Moreover, a scatterplot was used to show the spread of the ratings. Finally, sentiment analysis using Vader (Hutto & Gilbert, 2014) and SentiArt (Jacobs, 2019) algorithms, an artificial intelligence approach considering the text mining method, provided further insights into the reviews with the most positive or negative and the highest or lowest sentiment scores. The findings gathered by utilizing those programs are described in the following section, which is complemented with statistical inference to add a different approach to our research.

Statistical Inference Methodology

Considering the two hypotheses outlined above, statistical methods will be employed to draw inferential conclusions.

For the first hypothesis – that tourist satisfaction (measured by sentiment scores) differs significantly between wineries in Portugal, Italy, and France – the independent variable is country (a categorical variable), and the dependent variable is the sentiment score derived via the VADER algorithm (an artificial intelligence method). Given that the variables violate the assumption of normality, the non-parametric Kruskal-Wallis test will be applied.

For the second hypothesis – that longer reviews (>100 words) have significantly higher ratings than shorter reviews (≤100 words) – Pearson's correlation between sentiment score and rating, as well as between word count and rating, was analyzed. Here, the independent variable is review length, and the dependent variable is the rating assigned to the enotourism experience.

RESULTS AND ANALYSIS

Sentiment Analysis Results

Portugal

Data was collected from 100 reviews between 2016 and 2023. Analysis through Power BI provided the following insights, presented in Figure 1. Looking at the year of publication, most reviews (from the sample collected) were published in 2018. In that year, 38 reviews were written, and since 2020, less than ten reviews have been published each year (Figure 1).

Figure 1. Published reviews by year - Portugal

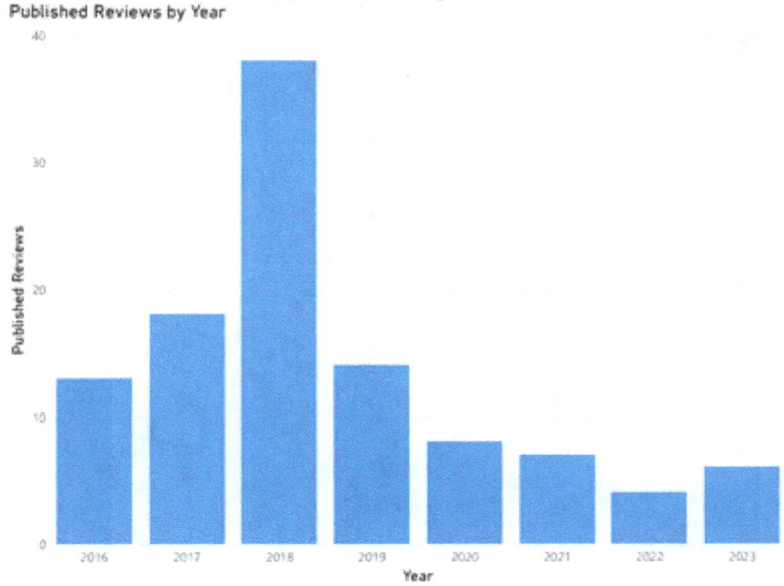

The months in which most reviews were published were July, followed by August and September, and the ones in which fewer reviews were published were January and December, with 2 reviews each (Figure 2).

Figure 2. Published reviews by month - Portugal

Analyzing the reviews by gender through Power BI, most reviews were written by people whose gender wasn't possible to verify. Sorting the reviews by author city and country, most reviews were written by citizens from the United Kingdom, more specifically London (10), followed by the United States of America and Portugal. Other nationalities writing reviews for Quinta dos Vales included Canada, France, Ireland, Sweden, the Netherlands, Angola, Australia, Estonia, Hungary, India, Norway, Spain, and Switzerland (Figure 3). The person's chosen username wasn't attributed to a gender.

Figure 3. Representation of reviews by author country - Portugal

Through the analysis conducted with Orange, it was possible to retrieve the following findings:

The word cloud analysis showed that the top 5 words were "wine", followed by "tour", "tasting", "great" and "wines" (Figure 4). Adjusting to word combinations, the ones with more weight were "wine tasting", "quinta dos" and "dos vales", among others (Figure 5).

Figure 4. Word cloud representation of Quinta dos Vales reviews - Portugal

Figure 5. Word cloud with most common phrases

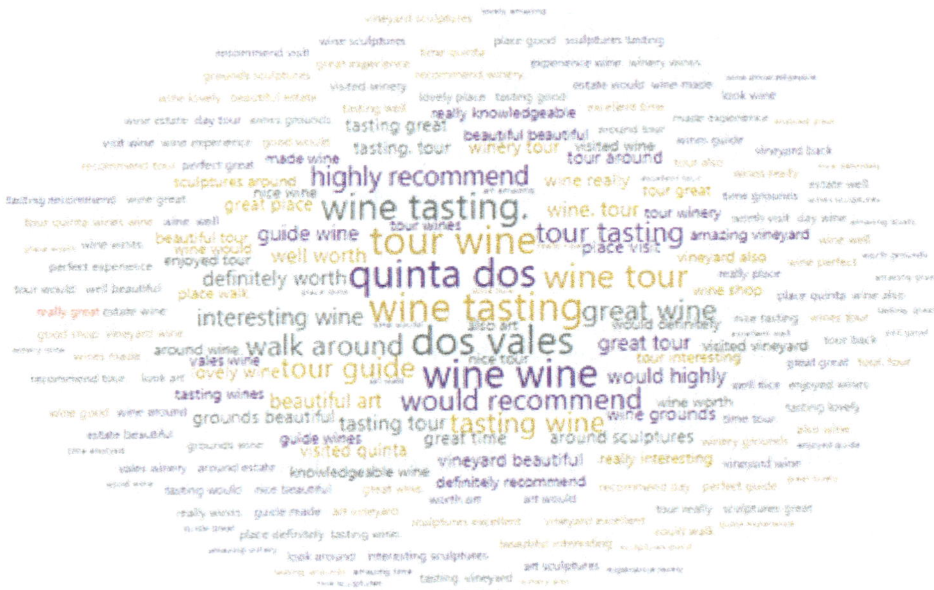

The scatterplot chart in Figure 6 shows that most reviews were classified with a 5 rating, therefore positive. The reviews were from multiple nationalities. For 1 and 2 ratings, there were only singular critiques from the UK and the USA, respectively.

Figure 6. Scatterplot of rating per author country - Portugal

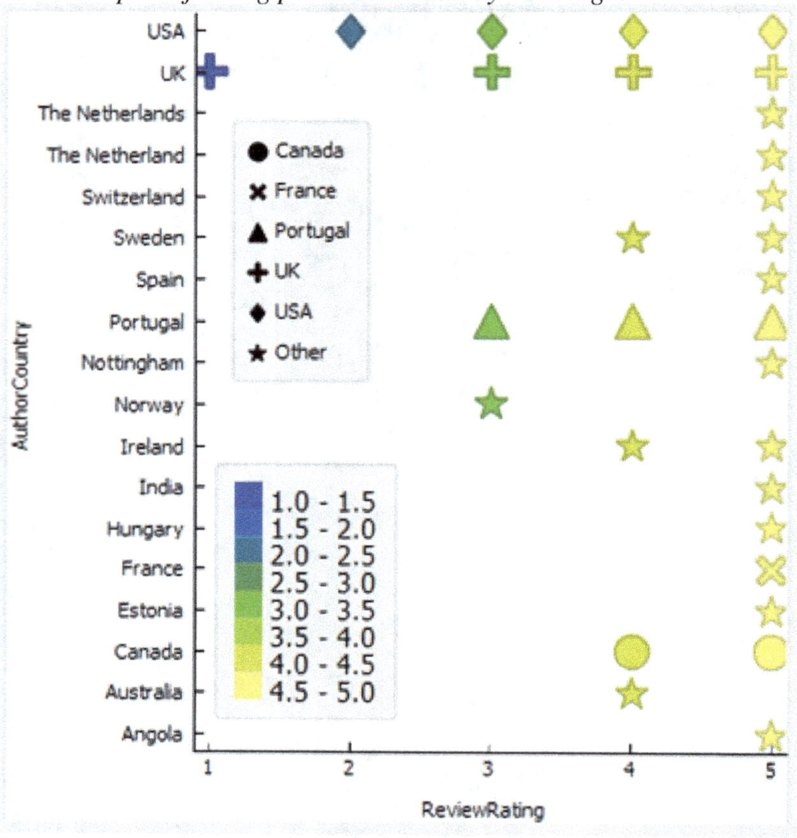

According to the sentiment analysis, the reviews with the most positive and most negative sentiment, as well as the critiques with the highest and lowest sentiment scores in general, could be identified:

Most Positive (2 most positive - descending order)

"Fantastic day, fantastic wine and a really knowledgeable and friendly tour guide. Definitely worth the visit " (Review #71).

"A nice winery! They had a beautiful garden with beautiful art. Thanks for the nice tour. Nice wine!" (Review #21).

Most Negative (2 most negative - descending order)

"Stayed in the apartment next to the BBC area. Could sound nice but awful in reality. One area for 4 apartments with the only washing machine in the same area. People always round there. Water not potable, causing quite a discomfort. Very rude lady at reception. Wine tasting people randomly strolling without caring about you privacy. Not recommended!" (Review #31).

"The grounds are beautiful we wanted to do a tour, but no one was around. We walked around for about 30 mins and then found the tasting room. We have drunk wine from around the world and for us this was some of the worse we just didn't like it but that's our personal taste." (Review #37).

Highest Sentiment Score (2 highest sentiment score - descending order)

"A nice winery! They had a beautiful garden with beautiful art. Thanks for the nice tour. Nice wine!" (Review #21).

"We (32&34) spend part of our honeymoon on this nice vineyard. What a beautiful place! We enjoyed the tranquility of the evenings, enjoying the sunset from our terrace and enjoying a nice glass of wine. The host was super friendly and took well care of us!" (Review #66).

Lowest Sentiment Score (2 lowest sentiment score - descending order)

"There are no words to explain how good it is. You need to go, to visit and try everything. 100 characters longer, hein? Ha, ha, ha!" (Review #25).

"A bit difficult to find but so worth it in the end! Coming from Albufeira direction, look for the Slide&Splash signs, then cross the railway tracks and turn right into Estombar. Once we got there, we had a great guided tour with Duarte, but you can also do just a wine tasting in their shop if you come without a reservation." (Review #63).

Italy

The Icario Winery is in the top position of the best Italian wineries, with more than 100 reviews, between 2017- 2024. Looking at the year of publication, most reviews (from the sample collected) were published in 2023, and some in 2024 (Figure 7). The months in which most reviews were published were May and June, followed by July and September, and the ones in which fewer reviews were published were January and April (Figure 8).

Figure 7. Published reviews by year - Italy

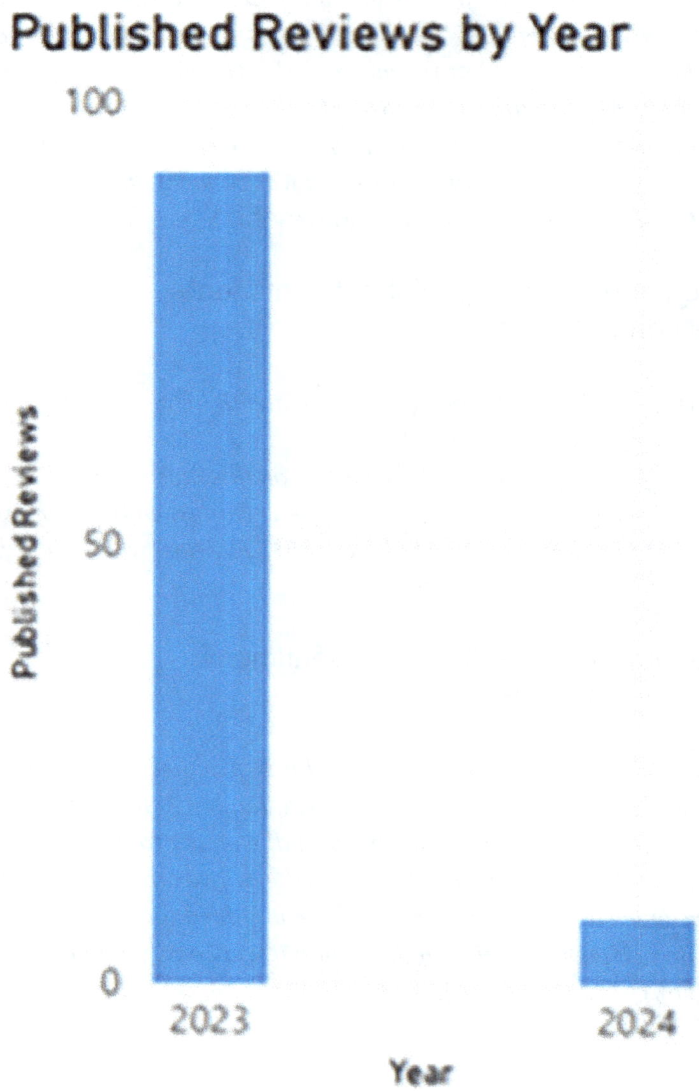

Figure 8. Published reviews by month - Italy

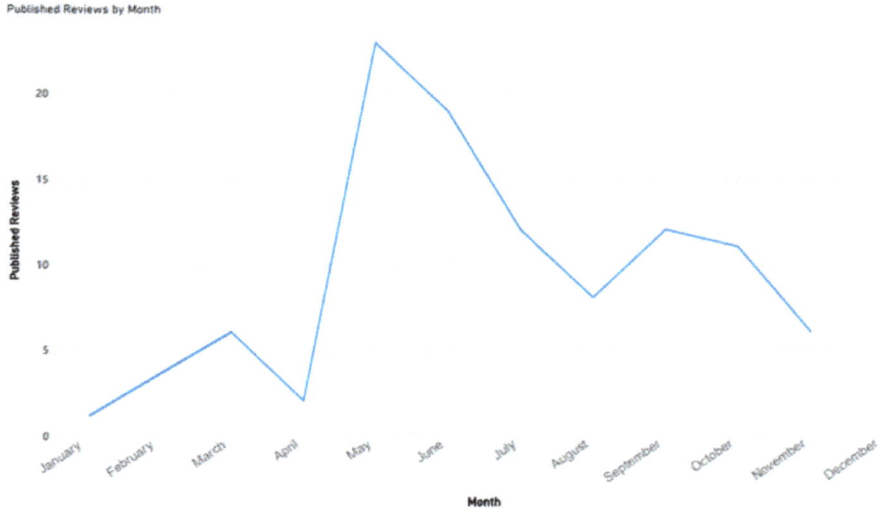

Analyzing the reviews by gender through Power BI shows that most reviews were written by women and by people whose gender wasn't possible to verify. When looking at the author's country, most reviews were written by citizens from the USA (48), followed by the UK (12) and Canada (10). Other nationalities included France, Ireland, Sweden, Netherlands, Vietnam, Australia, South Africa, Turkey, Perú, New Zealand, Spain, Germany, and Morocco (Figure 9).

After the Power BI analysis, Orange was used to create word clouds and conduct sentimental analysis. The two-word clouds below show the most common words (Figure 10) and the most frequent combination of words (Figure 11).

Figure 9. Representation of reviews by author country - Italy

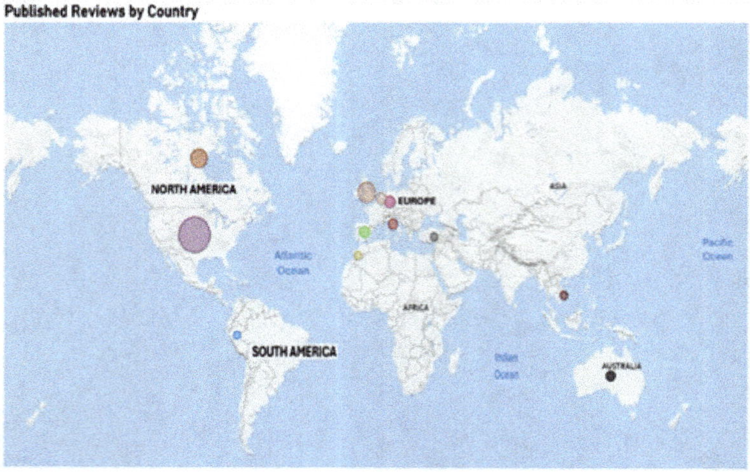

Country	Published Reviews
USA	48
UK	12
N/A	11
Canada	10
Germany	3
Netherlands	3
Spain	3
Australia	2
Italy	2
Morocco	1
New Zeland	1
Perù	1
SouthAfrica	1
Turkey	1
Vietnam	1
Total	**100**

The most common words were great, wine, winery, experience, tasting, visit, etc. (Figure 11). The most used combination of words was wine tasting, at Icario, wine tour, tasting experience, great experience, great wine, etc. (Figure 12).

Figure 10. Word cloud representation of Icario Winery - Italy

Figure 11. Word cloud with most common phrases - Italy

This scatterplot chart shows that most of the reviews were classified with a 5 rating, therefore positive. The reviews were from multiple nationalities. The US and an unidentified country had a 3 and a 4 rating, respectively (Figure 12).

Figure 12. Scatterplot of rating per author country - Italy

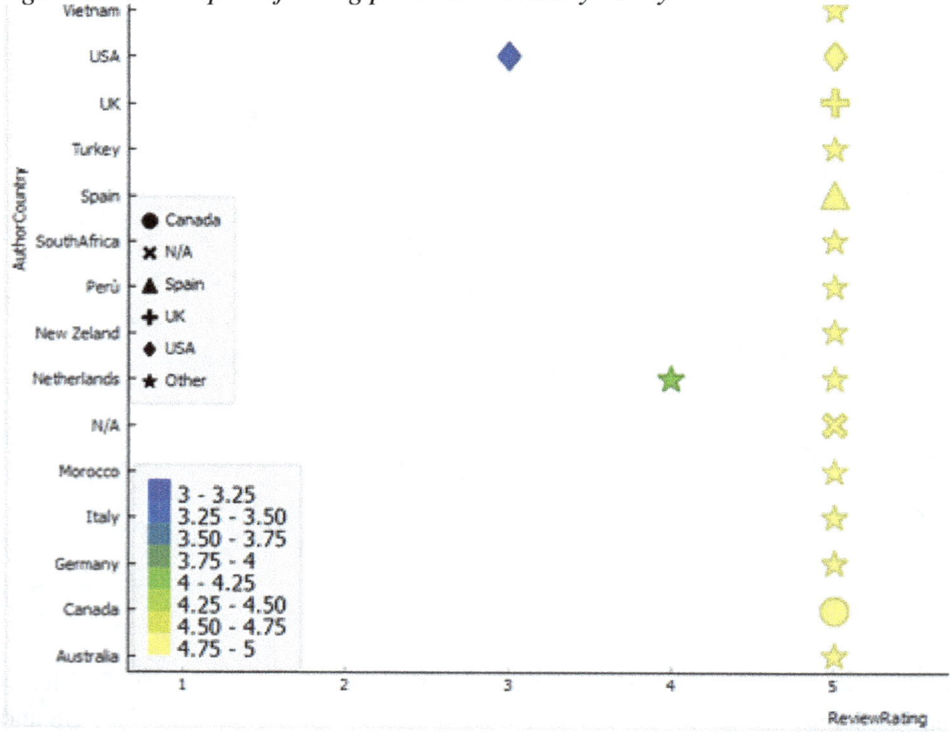

For sentiment analysis, most reviews were classified with a five-star rating, therefore positive. The reviews were from all nationalities illustrated above.

Most Positive (2 most positive - descending order)

"Fabulous time with Paula! Wines are lovely with a generous pour. Best vineyard/ winery tour I have ever done! I had a lot of questions and learnt so much. Olive oil also wonderful" (Review #92).
"His winery was absolutely perfect. A wonderful experience in Tuscany. The wine was fantastic and the story behind their wines was incredible. We are so happy to have stumbled upon this beautiful gem" (Review #87).

Most Negative (2 most negative - descending order)

"Our trip was last year, and it was amazing. The reason I am writing so late is that we ordered some wine to be shipped to us in the UK but sadly never arrived. I contacted the winery and spoke to the [...] who was fantastic in helping us

get our wine. Brilliantly helpful and understated of our situation and shipping the wine within 2 weeks. My dad was so happy. I seriously cannot thank him enough for all he has done for us. Give that man a Raise 1000% Thank you X" (Review #24).

"What an amazing setting! Simply fantastic. This was much more modern than the other wineries we visited in Tuscany. The Nysa rose with the single best rosé we have ever had, and we have drunk some wine in our lives! It was a very unique rose that you must taste for yourself. Our tasting guide was very knowledgeable and shifted between three different languages, easily, making everyone feel at home. We did the gold tasting but were very sad that we did not book the tour in advance. We did not expect this place to be as spectacular as it is, so that was a missed opportunity. This is a "must-do" for serious wine lovers. Very impressive" (Review #54).

Highest Sentiment Score (2 highest sentiment score - descending order)

"We had a lovely tour of the winery. They were running late so we began with a complimentary Chardonnay. After hearing about their breadth of wines, two different kinds of soil, and what makes the Nobile wine so special. We finished with a taste of 5 red wines and a sparkling, each better than the last and some of their delectable virgin olive oil and a meat and cheese plate. Beautiful views, incredibly knowledgeable and helpful staff, and delicious wines. I recommend this vineyard for all looking to the "noble" experience in the Montepulciano region" (Review #26).

"Beautiful contemporary winery in wonderful Montepulciano, fantastic friendly staff and great wines - must visit! " (Review #93).

Lowest Sentiment Score (2 lowest sentiment score - descending order)

"We had a great time; we came late in the day, and we thought it was closing but the host was so nice. There were a few people sitting in the back, so we joined them. No rush, very friendly. I'm not sure what's better, the wine or the views! Stop by and enjoy!" (Review #94).

"What a great winery! Completely stumbled across this on google maps and loved it! Really good wines and amazing views! Highly recommended!" (Review #60).

France

Data for the winery "Les Caves Saint Carles" was collected from 224 reviews posted between 2013 and 2023. In this sample, most reviews were published in 2014 and the least in 2021 (Figure 13).

The months with the most frequent publications were May, July, and August, and the ones with the least frequent publications were January, December, and March, with 3 reviews each (Figure 14).

Figure 13. Published reviews by year - France

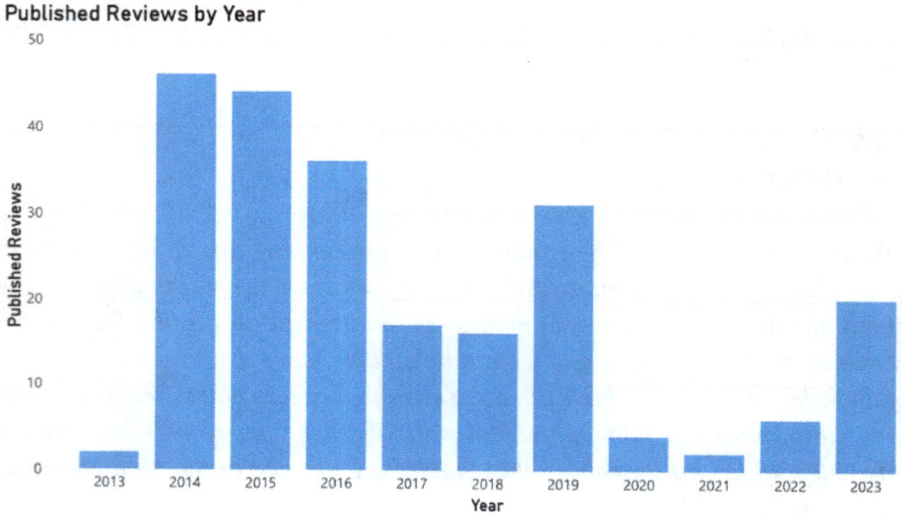

Figure 14. Published reviews by month - France

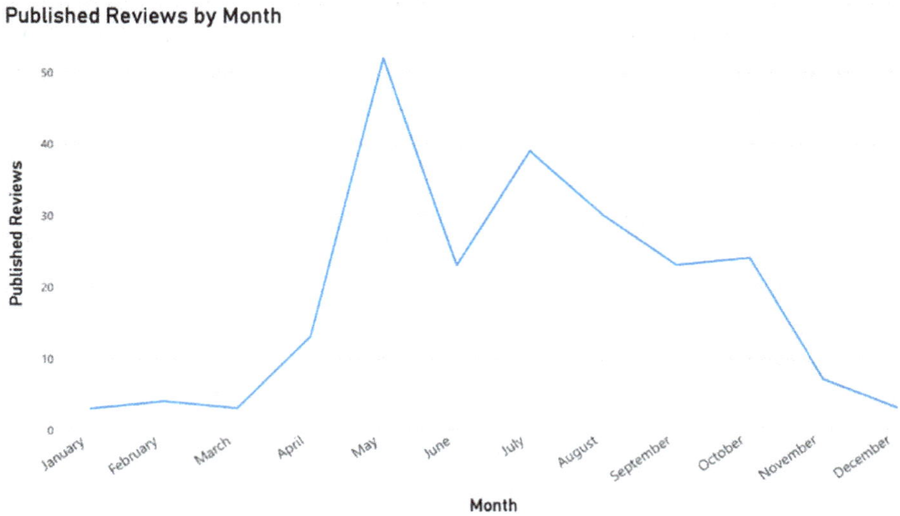

When looking at the origin of the authors of the reviews, nearly 60% of the reviews were published by Americans, followed by British and French citizens. Thereby, most reviews were posted by people based in New York City, London, and Boston (Figure 15). For the reviews sorted by country and city, the non-applicable ones (users without this information given) were removed to have a correct illustration of the graphs.

Figure 15. Representation of reviews by author country - France

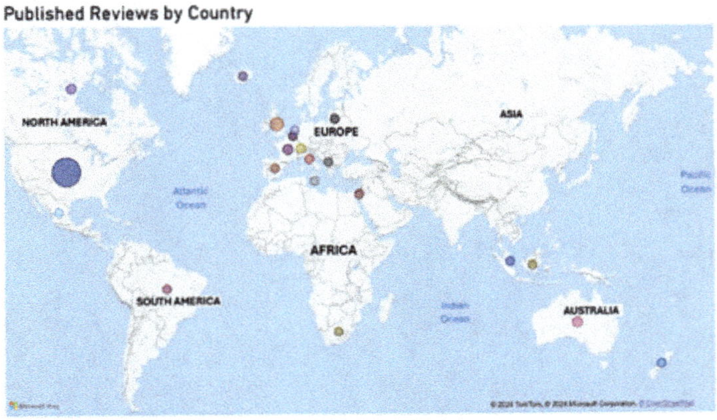

Country	Published Reviews
USA	115
UK	25
France	10
Australia	9
Canada	8
Switzerland	7
Italy	2
Malta	2
New Zealand	2
Singapore	2
Spain	2
Belgium	1
Brazil	1
Iceland	1
Indonesia	1
Israel	1
Kosovo	1
Lithuania	1
Total	**194**

In addition to the analysis in Power BI, Orange provided valuable insights from text mining, especially sentiment analysis. The two-word clouds below (Figures 16 and 17) show which words and word combinations were the most used in the reviews.

For words, those were "wine", "tasting", "wine", "guy" and "us" (Figure 16). The most frequent word combinations were "du pape", "wine tasting", "chateauneuf du", "saint charles" and "les caves" (Figure 17).

Figure 16. Word cloud representation of Les Caves Saint Charles reviews - France

Figure 17. Word cloud with most common phrases - France

Furthermore, a scatterplot chart illustrates the classification of the ratings by nationality. Most of the users gave a 5 rating. However, some Americans rated the experience with one, two, three, or four stars. Also, the UK, Italy, and Canada show a few reviews with varying ratings (Figure 18).

Figure 18. Scatterplot of rating per author country - France

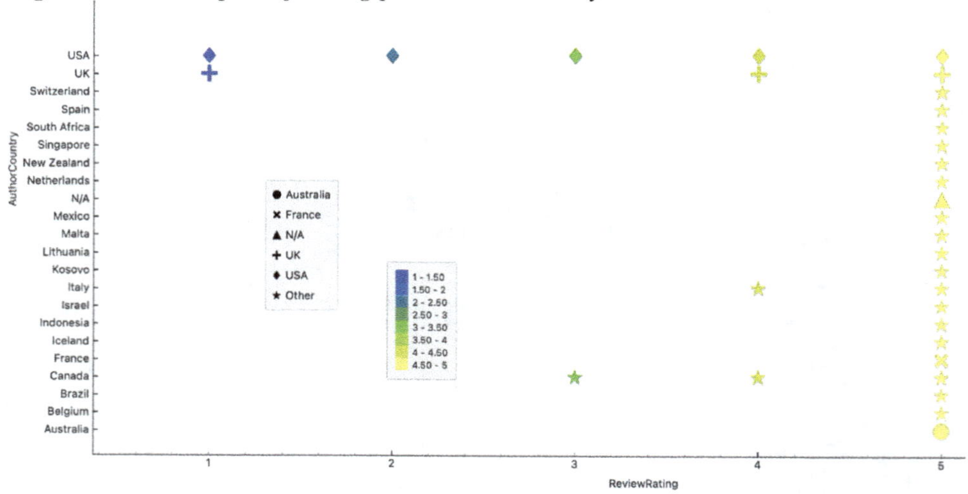

According to the sentiment analysis, the reviews with the most positive and most negative sentiment, as well as the critiques with the highest and lowest sentiment score, could be identified.

Most Positive (2 most positive - descending order)

"Perfection! What a fantastic journey through Provence and its wine history. Great setting, great value and a great and knowledgeable entertainer in Guy! Highly recommended!" (Review #20).

"Must do! Great wine tasting, great cellar and scenic views, great wines. A wine lovers delight. knowledgeable, great sample and selection of wines of the region, recommend it highly" (Review #141).

Most Negative (2 most negative - descending order)

"One of a kind wine experience Mathias was our sommelier while visiting Les Caves Saint Charles to purchase a case of wine. If you are a serious wine-lover, this experience will stick with you forever. You have a vast source of knowledge in [...] who is also perfectly friendly and accommodating. Not to mention, his English is perfect, and he still indulged my fiancé's basic French attempts. The wine selections are nothing but the best - it is appointment only and they don't just stick with the popular picks or Robert Parker selections. They offer you the best of the region without any bias. We can't say enough fabulous things

about our experience here and it would be an absolute shame to miss out on it!" (Review #185).

"First of all, I was hesitant to write this negative review about "Les Caves Saint Charles." My wife and I had a nice private wine tasting at the wine shop with Mathias on June 25th, 2015. The experience was so nice that euphoria swept me up and I decided to order two cases of wine totaling >$1200. Alas, fast forward to today (August 10th, 2015), and we still have not received anything. I would think that this type of professional business would have all the kinks worked out with their distribution to the United States. Unfortunately, this is not the case. According to the spotty tracking e-mails that we received, the wine was finally shipped out on July 22nd and has been sitting in some warehouse since then awaiting to be cleared by customs. Of course, there was a delay due to paperwork not being filled out on their end. I wouldn't care that much if I knew the wine was being stored in proper conditions. But I'm pretty sure the warehouse conditions are not suitable for fine wine storage. My advice: do the wine tasting and buy a few bottles to bring back home... but avoid buying anything to be shipped later. It has been a disaster and severe disappointment" (Review #155).

Highest Sentiment Score (2 highest sentiment score - descending order)

"Great wine tasting, great cellar and scenic views, great wines. A wine lovers delight. knowledgeable, great sample and selection of wines of the region, recommend it highly" (Review #141).

"Thank you [...] for an informative and thorough presentation of the CDP produce. Very nice presentation of four beautiful yet different wines. Also, special thanks for taking the young ones of our group into consideration - the seed for future master of wine may just have planted A highly recommended experience " (Review #50).

Lowest Sentiment Score (2 lowest sentiment score - descending order)

"We made an appointment for a private tasting over 6 months in advance. After driving 1 1/2 hours to Chateauneuf-du-Pape, we were told that the sommelier was running late, and the receptionist would start the tasting. Well, he never showed and never even sent an email of apology. Although the receptionist graciously tried her best, she simply didn't have the training. As we were here to stock our

wine cellar, it was a total waste of a half day of vacation. There are many, many cellars in this area. Don't risk wasting your time" (Review #223).

"This is one of the best wine tastings I've been to. Guy is incredibly knowledgeable and the cellar where the tastings take place is magical. It's a great thing to do if you want to expand not only your knowledge of Chateauneuf du pape wines but the village and history as well. We ordered all the wines we tried, and we're excited to hear about Guy's supper clubs that he is planning to bring to the UK" (Review# 61).

Reviews of Wineries in Europe

The analysis of the reviews of those three wineries showed various similarities and differences.

Although all three wineries had reviews on Tripadvisor for over ten years, the spread differs greatly. For Quinta dos Vales, most reviews from the sample collected were published in 2018, for the Icario Winery in 2023, and for Les Caves Saint Charles in 2014. Regarding the months with the most published reviews, they are quite unanimous. In all three wineries, most critiques were published in the (early) summer months and September.

When looking at the gender of the authors, more men have written in Portugal and more women in Italy and France. This is interesting, as more men are currently engaged in wine tourism, which is somewhat contradictory (Bibicioiu & Crețu, 2013; López-Guzmán et al., 2014; Santos et al., 2020).

The geographical distribution of wine tourists is mainly in English-speaking countries (USA, UK, Canada, and Australia). However, this is also due to the fact that only English-language reviews were considered for this analysis.

Focusing on the text mining results, and especially the word clouds, most reviews centered around the topics of "wine", "wines", "tour", "tasting", "experience", "winery" and the adjective "great". The word clouds illustrating the combination of words focus mainly on recommending the tour, the winery, and the experience.

An analysis illustrated by a scatterplot showed that most of the experiences at the wineries were rated positive (5 starts), with only a few deviations.

The reviews with the most positive/negative and the highest/lowest sentiment scores were analyzed for each winery. While for those with a positive/high score, it definitely makes sense from a reading perspective, there are also ratings with negative/low sentiment scores that sound positive regarding understanding and were not rated badly. This means those reviews must contain more negative words than positive ones to achieve those results (Hu & Liu, 2004).

Regarding the study, it can be stated that aspects such as wine quality, service, and ambience increased the positive sentiment in the reviews.

Statistical Inference Results

Regarding the statistical results for the first hypothesis (H1), the Kruskal-Wallis test indicated no statistically significant differences between groups [$\chi^2(34) = 29.495$, *p* = 0.688]. When comparing sentiment levels across countries, the test revealed no significant variations. These results suggest that sentiment expressed in reviews does not systematically vary by tourists' country of origin.

Although studies like Bruwer and Lesschaeve (2012) and Wu et al. (2024) established the influence of cultural and regional factors on enotourism perceptions, a significant empirical challenge remains: determining whether online satisfaction expressions vary systematically among traditional wine-producing nations. Given that Portugal, Italy, and France possess distinct hospitality traditions, visitor profiles, and landscape contexts (López, 2024; Pereira, 2023), it is plausible that sentiment patterns in TripAdvisor reviews reflect not only service quality but also cultural expectations and socio-linguistic norms of emotional expression (Hu & Liu, 2004). This gap necessitated a robust comparative analysis of cross-cultural nuances in enotourism experiences – yet the results show no discernible differences.

For the second hypothesis (H_2), examining whether longer reviews (>100 words) yield higher ratings or sentiment – Pearson's correlation was applied in two stages: between Review Rating and Word count and between VADER-derived sentiment (compound) and Review length. A statistically significant but extremely weak negative correlation emerged between ReviewRating and Word count ($r = -0.094$, $p = 0.050$, $n = 432$), indicating a marginal tendency for higher ratings to associate with shorter texts. The effect size was negligible ($R^2 = 0.88\%$), implying word count explains <1% of rating variance.

A non-significant weak positive correlation was observed between Word count and compound sentiment ($r = 0.066$, $p = 0.173$, $n = 432$), with only 0.44% of sentiment variance explained. Thus, review length demonstrates no substantive linear relationship with either ratings ($p = 0.050$, marginally significant but trivial) or sentiment polarity ($p = 0.173$, non-significant).

The relationship between review length and ratings/sentiment remains underexplored in enotourism. Pioneering work (e.g., Liu et al., 2017) suggests longer reviews correlate with memorable experiences, but this pattern remains unvalidated in wine-tourism contexts where sensory complexity may trigger unique discursive behaviors. This study found no empirical support for the hypothesis that longer reviews predict higher ratings or more extreme sentiments.

CONCLUSION

The analysis of three popular wineries in Europe highlighted the growing prominence of wine tourism and its increasing integration into broader travel experiences. The use of Business Intelligence techniques enabled the extraction of valuable insights from customer reviews, revealing demographic patterns, preferences, and visitor perceptions. Furthermore, the application of Machine Learning for sentiment analysis provided a deeper understanding of tourists' emotional responses and satisfaction levels.

The findings underscored the strategic importance of wine tourism as a driver of tourism growth in Portugal, France, and Italy, significantly contributing to the attractiveness and economic success of these destinations in the global wine tourism market. This study offered a more targeted analysis compared to broader state-of-the-art research by focusing on individual wineries and exploring unique aspects of customer feedback through Artificial Intelligence-based approaches.

It is thus concluded that, in the analyzed sample, neither geographical origin nor textual length account for measurable variations in satisfaction with enotourism experiences, challenging established notions in the literature

Limitations of this work include the reliance on a single data source (TripAdvisor), which may restrict the scope of the findings and not capture the full range of visitor experiences. Furthermore, while the results provide valuable insights, the study is based on a specific set of reviews from three wineries, limiting the generalizability of the findings to the broader wine tourism industry.

Future research could expand the geographical scope, incorporate multiple review platforms, apply advanced statistical methods, and conduct longitudinal studies to capture evolving trends in wine tourism. The adoption of Business Intelligence tools and Machine Learning algorithms can support wineries in strategically analyzing online reviews, enhancing service quality, personalizing offerings, optimizing marketing efforts, and investing in continuous staff training. These measures are essential for improving visitor satisfaction and securing a sustainable competitive advantage in the market.

ACKNOWLEDGMENT

This paper is financed by National Funds provided by FCT- Foundation for Science and Technology through project UID/04020: Research Centre for Tourism, Sustainability and Well-being.

REFERENCES

Alebaki, M., Psimouli, M., & Kladou, S. (2022). Social Media for Wine Tourism: The Digital Winescape of Cretan Wineries in the Era of COVID-19. In Global Strategic Management in the Service Industry: A Perspective of the New Era (pp. 81–98). Emerald Publishing Limited. https://doi.org/DOI: 10.1108/978-1-80117-081-920221006

Amaral, M. M., Kuhn, V. R., dos Anjos, S. J. G., & Flores, L. C. da S. (2024). Experiences in a wine tourism destination from the visitors' perspective. *International Journal of Wine Business Research, 36*(1), 85–102. DOI: 10.1108/IJWBR-05-2023-0028

Barbierato, E., Bernetti, I., & Capecchi, I. (2022). Analyzing TripAdvisor reviews of wine tours: An approach based on text mining and sentiment analysis. *International Journal of Wine Business Research, 34*(2), 212–236. DOI: 10.1108/IJWBR-04-2021-0025

Bibicioiu, S., & Creţu, R. C. (2013). ENOTOURISM: A NICHE TENDENCY WITHIN THE TOURISM MARKET. *Scientific Papers. Series Management, Economic, Engineering in Agriculture and Rural Development*, **•••**, 13.

Brown, G., & Getz, D. (2005). Linking Wine Preferences to the Choice of Wine Tourism Destinations. *Journal of Travel Research, 43*(3), 266–276. DOI: 10.1177/0047287504272027

Bruwer, J., & Lesschaeve, I. (2012). Wine Tourists' Destination Region Brand Image Perception and Antecedents: Conceptualization of a Winescape Framework. *Journal of Travel & Tourism Marketing, 29*(7), 611–628. DOI: 10.1080/10548408.2012.719819

dos Vales, Q. (2024). Quinta dos Valees Winery. About us. https://quintadosvales.pt/en/about-us/

El-Masri, M., Altrabsheh, N., & Mansour, H. (2017). Successes and challenges of Arabic sentiment analysis research: A literature review. *Social Network Analysis and Mining, 7*, 1–22. DOI: 10.1007/s13278-017-0474-x

Fu, W., Choi, E.-K., & Kim, H.-S. (2022). Text Mining with Network Analysis of Online Reviews and Consumers' Satisfaction: A Case Study in Busan Wine Bars. *Information (Basel)*, *13*(3), 127. DOI: 10.3390/info13030127

Gao, D., Xia, H., Deng, W., Muskat, B., Li, G., & Law, R. (2022). Value creation in wine tourism – an exploration through deep neural networks. *Journal of Vacation Marketing, 135676672211406*. Advance online publication. DOI: 10.1177/13567667221140605

Garibaldi, R. (2024). Enoturismo. https://www.robertagaribaldi.it/enoturismo-turismo-del-vino-trend-e-previsioni/

Hospitality, O. N. (2024a). The Development of Wine Tourism. https://hospitality-on.com/en/trends/development-wine-tourism

Hu, M., & Liu, B. (2004). Mining and summarizing customer reviews. *Proceedings of the Tenth ACM SIGKDD International Conference on Knowledge Discovery and Data Mining*, 168–177. https://doi.org/DOI: 10.1145/1014052.1014073

Hutto, C., & Gilbert, E. (2014, May). Vader: A parsimonious rule-based model for sentiment analysis of social media text. In *Proceedings of the international AAAI conference on web and social media* (Vol. 8, No. 1, pp. 216-225). https://doi.org/DOI: 10.1609/icwsm.v8i1.14550

Icario. (2018). Icario Winery. https://icario.it/reality/?lang=en

Jacobs, A. M. (2019). Sentiment analysis for words and fiction characters from the perspective of computational (neuro-) poetics. *Frontiers in Robotics and AI*, *6*, 53. DOI: 10.3389/frobt.2019.00053

Karlson, P., & Karlson, B. (2016). Châteauneuf-du-Pape: Rich and Spicy Wines With a Historical Heritage. Forbes. https://www.forbes.com/sites/karlsson/2016/07/07/chateauneuf-du-pape-rich-and-spicy-wines-with-a-rich-historical-heritage/?sh=35503a257590

Les Caves Saint Charles. (2024). Les Caves Saint Charles Winery. https://www.cave-saint-charles.fr/

Li, C., Gao, L., Liu, Y., & Li, H. (2021). Cluster analysis of China's inbound tourism market: A new multi-attribute approach based on association rule mining of tourist preferences at scenic spots. *Asia Pacific Journal of Tourism Research*, *26*(6), 654–667. DOI: 10.1080/10941665.2021.1887305

Lima, N., & Pereira, P. (2024). Inquérito Enoturismo. TravelBI. https://travelbi.turismodeportugal.pt/turismo-em-portugal/inquerito-enoturismo-2022/

Liu, Y., Teichert, T., Rossi, M., Li, H., & Hu, F. (2017). Big data for big insights: Investigating language-specific drivers of hotel satisfaction with 412,784 user-generated reviews. *Tourism Management, 59*, 554–563. DOI: 10.1016/j.tourman.2016.08.012

López, A. M. (2024). Wine Tourism in Europe. Statista. https://www.statista.com/topics/8997/wine-tourism-in-european-countries/#topicOverview

López-Guzmán, T., Vieira-Rodríguez, A., & Rodríguez-García, J. (2014). Profile and motivations of European tourists on the Sherry wine route of Spain. *Tourism Management Perspectives, 11*, 63–68. DOI: 10.1016/J.TMP.2014.04.003

Luo, J. M., Vu, H. Q., Li, G., & Law, R. (2020). Topic modelling for theme park online reviews: Analysis of Disneyland. *Journal of Travel & Tourism Marketing, 37*(2), 272–285. DOI: 10.1080/10548408.2020.1740138

Pereira, P. (2023). World's Best Vineyards 2023. TravelBI. https://travelbi.turismodeportugal.pt/turismo-em-portugal/world-s-best-vineyards-2023/

Santos, V., Ramos, P., Almeida, N., Marôco, J., & Santos-Pavón, E. (2020). Wine tourist profiling in the Porto wine cellars: Segmentation based on wine product involvement. *Anatolia, 31*(4), 577–590. DOI: 10.1080/13032917.2020.1802308

Statista Market Forecast. (2024). Wine - Worldwide. https://www.statista.com/outlook/cmo/alcoholic-drinks/wine/worldwide

Treen, E., Lord Ferguson, S., Pitt, C., & Vella, J. (2018). Exploring emotions on wine websites: Finding joy. *Journal of Wine Research, 29*(1), 64–70. DOI: 10.1080/09571264.2018.1433139

Vu, H. Q., Li, G., Law, R., & Zhang, Y. (2019). Exploring Tourist Dining Preferences Based on Restaurant Reviews. *Journal of Travel Research, 58*(1), 149–167. DOI: 10.1177/0047287517744672

Wine News. (2023). Positive sentiment in 2023 for Italian wine bigwigs. https://winenews.it/en/positive-sentiment-in-2023-for-italian-wine-bigwigs-mediobancas-sector-study_498374/

World Population Review. (2024). Wine Producing Countries 2024. https://worldpopulationreview.com/country-rankings/wine-producing-countries

Wu, D. C., Cao, C., Wu, J., & Hu, M. (2024). Wine tourism experiences of Chinese tourists: A tourist-centric perspective. *International Journal of Contemporary Hospitality Management*. Advance online publication. DOI: 10.1108/IJCHM-07-2023-1003

About the Contributors

Ankit Shukla: A Multifaceted Educator and Researcher in Hospitality Ankit Shukla is an accomplished educator and researcher with a robust background in the hospitality and tourism industry. His career spans various roles, from frontline food and beverage service to academic leadership and extensive research contributions. Ankit's journey exemplifies a blend of practical experience, academic acumen, and a passion for advancing knowledge in his field. Early Career and Industry Experience Ankit's career in the hospitality industry began at the Hyatt Regency Chandigarh in September 2015, where he worked in the Food & Beverage Service department. His responsibilities included taking room service orders, managing point-of-sale systems, and setting up banquets for events. This role provided Ankit with hands-on experience in the operational aspects of the hospitality industry. Following his time at the Hyatt Regency, Ankit transitioned into academia, joining Kumaun University Nainital as a teaching personnel in September 2016. He took on multiple responsibilities, including lecturing MBA Tourism students, overseeing practical training, and designing course curricula. This marked the beginning of his journey in education, where he contributed to shaping the next generation of hospitality professionals. Academic Leadership and Teaching Ankit's academic career continued to flourish as he assumed the role of Assistant Professor at SIHMCT & AN Tehri Garhwal in July 2017. He became the head of the Food & Beverage Service department, where he coordinated events, designed course curricula, and managed the department's overall functioning. From February 2018 to April 2022, Ankit served as an Assistant Professor at Shoolini University Solan. Here, he expanded his responsibilities to include training and placement coordination, NAAC coordination for the Hospitality Management department, and handling operations of a hotel with eight rooms. His leadership skills and ability to manage complex tasks were instrumental in the smooth functioning of the university's hospitality department. Contributions to Research and Academia Ankit's passion for research is evident through his significant contributions to academic literature.

He has presented research papers at various international conferences, focusing on topics such as ecology, local and regional food in destination marketing, and the influence of technological amenities on hotel guest satisfaction. His research has been published in reputable journals, with manuscripts addressing safety and security measures in hotels, tobacco consumption in India, and the role of Indian spices in Indian history. In addition to his research publications, Ankit has filed 40 patents, copyrights, and utility applications, with several already granted. His commitment to innovation and intellectual property rights underscores his dedication to advancing knowledge in his field. Current Role and Future Endeavors As of July 2023, Ankit is a lecturer at Raja Mangala University of Technology Thanyaburi in Thailand. In this role, he continues to inspire and educate students in the field of hospitality while pursuing his research interests. Ankit's extensive experience in both industry and academia, coupled with his passion for teaching and research, positions him as a key figure in the ongoing development of the hospitality and tourism sector. Ankit Shukla's journey from industry practitioner to esteemed academic is a testament to his dedication, expertise, and unwavering commitment to excellence in hospitality and education.

Nagendra Yadav is an 11th-year academia professional with a master's degree in hotel management from Kurukshetra University. With a passion for culinary arts and management, Nagendra has dedicated his career to enhancing guests' culinary dining and hospitality experience worldwide. His expertise spans various facets of hotel operations, including culinary innovation, guest services, and management practices. As an educator, Nagendra shares his knowledge and experience with the next generation of hospitality professionals, inspiring them to pursue excellence and creativity in their careers. He continues to explore new trends and innovations in the industry, aiming to elevate the standards of hospitality and enrich the experiences of guests and clients. He extensively researched promoting local and authentic culinary offerings in destinations. His work has contributed to a broader understanding of how culinary experiences can drive tourism and support local economies.

Partho Pratim Seal is Associate Professor at Welcomgroup Graduate School of Hotel Administration (WGSHA), a unit of Manipal Academy of Higher Education (MAHE), Manipal. He is involved in designing and developing educational curricula as per the academic standards for both undergraduate and postgraduate courses. He guides and assists faculty with instructional strategies. He is also actively involved in guiding culinary students in undergraduate and postgraduate research projects As an avid wordsmith who believes that writing is the most influential and resourceful medium to disseminate information has authored, four books Computers in Hotels: Concepts and Application, and Food & Beverage Management from Oxford

University Press- India, How to Succeed in Hotel Management Job Interviews from JAICO India, and Food Anthropology in India from Routledge, London and authored several book chapters in international publication. His research interests are in Human Resource Management, Gastronomy, and Tourism. He has also published various research papers in Scopus-Indexed journals and is a reviewer of various international journals.

Abhishek Tiwari is an accomplished academic with More than six years of teaching experience in Thailand as a university lecturer. His dedication to education and research is evident through his extensive work in various academic fields. Abhishek has published numerous research papers in prestigious Scopus-indexed and international journals, contributing significantly to his areas of expertise. In addition to his research, Abhishek has actively participated in and presented at several international conferences, showcasing his commitment to sharing knowledge and fostering academic discussions. His academic journey is further enriched by his participation in numerous Faculty Development Programs (FDPs), where he has both gained and imparted knowledge, enhancing the educational experience for himself and his peers. Beyond the classroom and research, Abhishek has also organized various community service projects, demonstrating his dedication to giving back to society. His efforts in arranging industrial visits have provided students with valuable practical insights, bridging the gap between theoretical knowledge and real-world applications.

***

Ajay Chauhan is a hospitality professional and academic with seven years of combined industry and teaching experience. He has worked two years in the hospitality industry and contributed five years to higher education, where he has mentored students with both practical insights and academic knowledge. He currently serves as an Assistant Professor and Culinary Expert at the School of Hospitality & Tourism, Bahra University, while also pursuing as Research Scholar for Ph.D. in Hotel Management at Bahra University, Waknaghat., focusing on research and innovation in hospitality and tourism studies.

Rohit Chauhan is an Assistant Professor at Chandigarh School of Business, Chandigarh Group of Colleges, Jhanjeri, Punjab. With over 10 years of experience in teaching, research, and academic administration, his expertise spans tourism, digital marketing, analytics, and sustainable development. A prolific researcher, Dr Chauhan has contributed significantly to transformative tourism, augmented reality in tourism, and rural entrepreneurship. He actively supervises PhD scholars and has organized numerous academic conferences and workshops.

Kaveri Samir Deosarkar is an Assistant Professor in the Department of Computer Science and Engineering at Indian Institute of Information Technology Nagpur, India. She holds M. Tech. and B. E. in Computer science and engineering. And has 5years of experience in academics.

Venkatesh Ganapathy, a seasoned academician with over two decades experience in the corporate world. Mentored thousands of students pursuing MBA in soft skills and pre-placement. Teaching courses in marketing, strategy and supply chain management.

Abhishek Ghai is an accomplished academic and hospitality professional with over 15 years of combined experience in teaching and the hospitality industry. He currently serves as an Associate Professor at Maharishi Markandeshwar (Deemed to be University), Mullana, Ambala. He has previously held esteemed academic positions at Panjab University, Chandigarh, and Kurukshetra University. Dr. Ghai holds a Ph.D. in Tourism and Hotel Management from Kurukshetra University, specializing in bar service operations within the Indian hotel industry. An alumnus of both Bangalore University and Kurukshetra University, he is a recognized expert in food and beverage service, alcoholic beverages, and molecular mixology. His hands-on training at international bars in the UAE and professional experience with luxury hotel brands such as One & Only Royal Mirage (Dubai), ITC Hotels, Taj Hotels, Radisson bring practical insights to his academic work. A prolific researcher and published author, Dr. Ghai has contributed numerous peer-reviewed articles, book chapters, and a co-edited book. His research interests span bar operations, food and beverage management and event management.

Megha Gupta is currently serving as Assistant Professor in the Department of Tourism and Hotel Management, Kurukshetra University. With over 18 years of experience, she holds a Ph.D. in Tourism and Hospitality. Her research interests include marketing, CRM, consumer behaviour, and gender studies. She has published widely and contributed to academic content development and training.

Nitin Gupta is a seasoned hospitality professional and academician with over two decades of rich experience in the hospitality industry and education. Currently serving as a Senior Associate Professor at Bhikaji Cama Subharti College of Hotel Management, Swami Vivekanand Subharti University, Meerut, India. Dr. Gupta has made significant contributions to both teaching and research in hotel management. He holds a Ph.D. in Management, with a research focus on motivational factors in the Indian hotel industry, and has published extensively in reputed national and international journals. His academic work includes authoring books, presenting papers at conferences, and supervising doctoral research. Beyond academia, he has professional experience with leading hospitality brands such as Jaypee Palace Hotel and The Suryaa, New Delhi. An award-winning educator, Dr. Gupta is known for his innovative teaching methods and leadership in hospitality education. His dedication to training future professionals and his passion for the industry are reflected in his numerous accolades, patents, and active involvement in academic forums and workshops.

Tejal Irkhede is a Professor in the Department of Computer Science and Engineering (Data Science) at St. Vincent Pallotti College of Engineering and Technology, Nagpur. She holds a Ph.D. in Computer Science and Engineering from Kalinga University (2024), with a passion for innovative teaching and applied research, she brings over 14 years of combined academic and industry experience to her role. She holds a Bachelor's degree in Computer Engineering (BE) and a Master's degree in Wireless Communication and Computing (ME), providing her with a strong interdisciplinary foundation in both theoretical and applied computing. Dr. Irkhede has published 17 research papers in reputed national and international articles and holds 3 patents in emerging technologies. Published book on IoT ecosystem. She also serves in academic leadership roles, mentors student projects, organizes technical events, and is a member of the International Association of Engineers (IAENG), ACM, ASI chapter.

Aruditya Jasrotia is working as an Assistant Professor in Amity Institute of Travel and Tourism, Amity University Uttar Pradesh. He completed his Ph.D. and M. Phil. from Central University of Jammu. He has published 13 Scopus Publications, 2 Patent and 23 Publications in total including research papers and book chapters. His broad research areas of interest are: Smart Tourism, Tourism Marketing, Sustainable Tourism, Tourism Policy and Planning, Tourism Products and Services. He is also associated as a reviewer for various Scopus Indexed and ABDC listed international journals like International Journal of Tourism Cities, Journal of Hospitality, Leisure, Sport & Tourism Education, International Journal of Tourism Policy and Kybernets. Associate Editor of ESTEEM Journal of Social Science and Humanities.

Sanket Joshi is an Assistant Professor in the Department of Computer Science and Business Systems at St. Vincent Pallotti College of Engineering and Technology, Nagpur, India. He has a Masters in Business Administration degree from USA and has over 15 years of experience spanning academia and the IT industry. His research focuses on artificial intelligence, machine learning, data analytics, and visualization. He serves in academic leadership roles, mentors student projects and organizes technical events.

Firoj Kabir is a researcher, Lecturer of the Department of "Tourism and Hospitality Management" at the Daffodil International University, Dhaka, Bangladesh, with a solid academic background in BBA and MBA majoring in Tourism and Hospitality Management from the University of Dhaka. His passion lies in exploring the intersection of travel and culture. His diverse areas of interest include archaeological tourism, sustainable tourism, and the unique characteristics of Generation Z travelers. Through extensive research and insightful writing, he sheds light on the intricate connections between historical sites, responsible tourism, and the evolving preferences of the next generation of travelers. His work not only enriches the field of tourism but also inspires readers to engage with the world around them in meaningful and responsible ways.

Vikesh Kashyap is an accomplished academic and professional in the field of Hospitality and Tourism. With a robust career spanning over a decade, he has held key positions, including Associate Professor and Head of Department at prestigious institutions like Bahra University and Chandigarh Group of Colleges. He holds a Ph.D. in Tourism Studies from Himachal Pradesh University and an MBA in International Hotel and Tourism Management from Vatel International School, France. A prolific researcher, he has substantial contributions to research, with publications in esteemed journals and with multiple book authorships. His work spans areas like creative tourism, community-based tourism, and sustainable practices in the Hospitality industry. He is an active member of "The Indian Society for Technical Education" and has chaired sessions and co-convened International Conferences on Sustainable Development and Innovation in Hospitality and Tourism. In addition to his academic achievements, He has industry expertise, having worked with prestigious organizations, including Boscolo Plaza Hotel, France. His dedication to advancing the field is reflected in his patents, conference contributions, and initiatives in promoting tourism's economic and cultural dimensions.

Rajashree S Kini currently serves as an Associate Professor and the Coordinator of the Department of Allied Hospitality Studies at WGSHA, MAHE, Manipal. In addition to her teaching role, she actively contributes to various academic and administrative responsibilities. She is a member of the Internal Quality Assurance Cell (IQAC), oversees MHM admissions, participates in the WGSHA Career Development Cell, and manages alumni relations. Before her current role, Dr. Kini held the position of Associate Professor in the Department of P.G. Studies and Research in Law at SDM Law College, Mangalore. Her extensive academic journey also included positions at MBA Department of MITE, Moodbidri, School of Management, Manipal, and NITK, Surathkal, Mangalore. During her tenure at SDM Law College, Mangalore, she assumed various roles, such as the Coordinator of the Department of P.G. Studies and Research in Law, member of the Finance and Audit Committee, IQAC member, member of the Board of Studies for Add-on Courses, and Head of the Counseling Cell. Dr. Kini boasts an impressive educational background, holding a Ph.D. in Organizational Ethics and Human Resource, an MBA in Finance, an LLM in International Business Law, a PGDIT in IT, and an MSc in Psychology. Her diverse professional experience includes working as an HR consultant and Project Manager for a private company for 1.5 years and practicing as an advocate for seven years in the Mangalore Bar under the guidance of Advocate Mr. Ranjan Rao. She has made significant contributions to academic literature, including a chapter in the book 'The Scanner,' which focuses on solving previous years' question papers on income tax of Mangalore University. Additionally, she co-authored a book titled 'Personal Growth and Interpersonal Effectiveness,' which serves as a textbook for MBA students under VTU. Dr. Kini has also published a chapter titled 'Analyzing the Challenges of Government Schemes/Initiatives on Skill Development in India' in the book 'Skill Development - A Key to Nation Building.' Her dedication to research is evident through her participation in over 20 research papers presented and published in national and international conferences and journals. Beyond academia, Dr. Rajashree is a passionate counselor for youth and a facilitator for 'Personality Development in Employability Skills.' She has conducted numerous training sessions for students pursuing MBA, CA, Engineering, and undergraduate degrees, as well as for faculty members through Faculty Development Programs (FDPs). She also serves as the Soft Skills Facilitator at the 'World Konkani Centre' in Mangalore, and her expertise has led her to deliver many guest lectures as a resource person on topics related to management, commerce, law, personality development, and psychology.

Suman Lata is working as an Assistant Professor in Amity Institute of Travel and Tourism, Amity University Uttar Pradesh. She completed his Ph.D. and M. Phil. from Central University of Jammu. She has published 07 Scopus Publications, 2 Patent and 15 Publications in total including research papers and book chapters. Her broad research areas of interest are: Social Media, Tourism Marketing, Sustainable Tourism, Consumer Behaviour, Tourism Products and Services. She is also associated as a reviewer for various Scopus Indexed and ABDC listed international journals like International Journal of Tourism Cities, Journal of Hospitality, and Tourism Technology, Current Issues in Tourism.

Subhadip Majumder He is Professor and Corporate Chef at Silver Mountain School of Hotel Management in Kathmandu, Nepal. He is also Academic Head for Queen Margaret University's "Bachelor of Arts in International Culinary Arts" curriculum. He holds an Honorary Doctorate (PhD) in Hotel Management with a Food Production Specialisation from Washington University in Seattle, USA. His efforts are further recognised by 9 patents and 2 design patents. He wrote 12 books and 3 chapters. He has 12 academic research papers and there are upcoming. He lectures and researches on food production, hospitality trends, food journalism, food and architecture, and food nutrition. He was Head of School for Lovely Professional University's School of Hotel Management and Tourism before this. HACCP Auditor, Government of India-recognized hotel andragogy trainer, and AHLEI-certified hotel educator. He is a lifelong member of WACS, IFCA, CAFR, ISHE, CAWB, and WCCF.

Pranoti Mane is a prominent educator and researcher specializing in Computer Science and Engineering, boasting over nine years of teaching experience. She is dedicated to delivering technical knowledge and promoting innovation among her students. Currently, she is engaged in doctoral studies at Visvesvaraya National Institute of Technology (VNIT) and possesses both an M.Tech and a B.E. in Computer Science and Engineering. Throughout her professional journey, Pranoti has significantly contributed to technological progress through her research and teaching endeavors. Focusing on areas such as Cloud Computing, and Data Science. Her research interests lie in Data Mining and Deep Learning, and she has authored multiple publications in esteemed conferences and journals, including ICAIQSA-2024 and ICCICA-2021. In addition, she has secured copyrights for several research articles and holds NPTEL certifications across various technical fields. Notably, she has been awarded a patent for her innovative creation, the "Circuit Maker Machine." In acknowledgment of her outstanding teaching and student engagement.

Iram Nausheen is a dedicated academician, researcher, and technology advocate with over 13 years of teaching experience. She holds a Ph.D. in Electronics and Communication Engineering and has been passionately contributing to the fields of Artificial Intelligence and Data Science. Alongside her academic career, Dr. Nausheen actively engages in initiatives to empower women in technology. She serves as a Google Women Techmakers Ambassador and is the Vice President of the PSF Women Wing, where she works towards raising awareness about technology, digital literacy, and women's representation in STEM. Beyond the classroom, she is a prolific content creator, blogger, and columnist, regularly writing insightful articles for newspapers and digital platforms. Her writings aim to bridge the gap between complex technological concepts and public understanding, inspiring individuals to embrace the potential of AI and emerging technologies. Dr. Nausheen's journey reflects her commitment to education, innovation, and community service. She continues to explore, research, in the domain of AI and Data Science, striving to inspire the next generation of learners, especially women,

Célia M. Q. Ramos graduated in Computer Engineering from the University of Coimbra, obtained her Master in Electrical and Computers Engineering from the Higher Technical Institute, Lisbon University, and the PhD in Econometrics in the University of the Algarve (UALG), Faculty of Economics, Portugal. She is Coordinator Professor with Habilitation at School for Management, Hospitality and Tourism, also in the UALG, where she lectures computer science. Areas of research and special interest include conception and development of information systems, tourism information systems, big data, etourism, econometric modeling and panel-data models. Célia Ramos has published in the fields of information systems and tourism, namely, she has authored a book, six book chapters, conference papers and journal articles. At the level of applied research, she has participated in several funded projects.

Bharathi S Gopal Associate Dean & Dy. Academic Coordinator Head-Case Research Centre/Area Coordinator - Marketing & Strategy/Coordinator-Digital Assistantship, ICFAI Business School, ICFAI Foundation for Higher Education, Off-campus Centre Bangalore

Ramadevi Vitthal Salunkhe is an Assistant Professor in the Department of Artificial Intelligence at G.H. Raisoni College of Engineering, Nagpur, and a PhD research scholar at Visvesvaraya National Institute of Technology (VNIT), Nagpur. With over 11 years of experience across academia, government institutions, and industry, her expertise spans artificial intelligence, software development, and technical education. She has authored multiple books and book chapters, presented research at international conferences, and actively contributes to AI-driven solutions in education and sustainability. Her interests include AI personalization, human-computer interaction, and ethical innovation in smart systems.

Rakesh Sharma is an experienced academician and researcher specializing in hospitality and tourism management. With over 14 years of combined academic and industry experience, Dr. Sharma is currently serving as an Associate Professor at MMICT&BM (HM), MMDU, Mullana, Amabala. His expertise spans research coordination, curriculum development, experiential learning, and implementing outcome-based education (OBE) and National Education Policy (NEP) 2020. An active researcher, Dr. Sharma has published extensively in Scopus, Web of Science, and UGC Care-listed journals, contributing to books and conference proceedings on hospitality technology, service automation, and consumer behavior. He has also registered multiple copyrights related to hospitality operations and research methodologies. Dr. Sharma is passionate about leveraging technology for innovation in hospitality education and research, with a vision to bridge industry-academia gaps and enhance learning experiences in the field.

Vikas Sharma is an alumnus of IHM Gwalior. He has 9 years of experience, including teaching and industry experience in luxury brands like ITC grand Bharat Gurugram, The Leela Ambience Hotel and Residences Gurugram and Le Meridian Gurugram. Currently He is associated with Swami Vivekanand Subharti University, Subharti college of Hotel Management as Assistant Professor. He is currently pursuing PhD in Hotel Management from Swami Vivekanand Subharti University and has completed Masters in Travel and Tourism Management. He has 7 patents published, 16 Chapters Published and also published 12 research papers in various UGC care List, Springer Nature and Scopus Indexed Journal. He has attended 7 FDPs and workshops. He has also presented research papers in many national and international conferences.

Laura Siegl is a Master's student in Tourism Organizations Management at the Faculty of Economics of the University of Algarve in Faro, Portugal. Her research focuses on culinary, gastronomy, sustainable tourism and climate-conscious mobility, with a particular interest in behavioral responses to aviation emissions. She holds a Bachelor's degree in Tourism Management from Munich University of Applied Sciences and completed an exchange semester at the University of Wisconsin-Stout. Prior to her academic studies, she graduated from the Higher Institute of Tourism in Bludenz, Austria, with a diploma in Hospitality Management. Laura also brings over eight years of hands-on experience in the event and hospitality industries across Europe, including roles in catering, project management, and content marketing. Her international education and professional background have shaped a holistic and interdisciplinary perspective on tourism, sustainability, and innovation.

Amrik Singh is working as Professor in the School of Hotel Management and Tourism at Lovely Professional University, Punjab, India. He obtained his Ph.D. degree in Hotel Management from Kurukshetra University, Kurukshetra. He started his academic career at Lovely Professional University, Punjab, India in the year 2007. He has published more than 40 research papers in UGC and peer-reviewed and Scopus/Web of Science) journals. He has published 12 patents and 01 patent has been granted in the inter-disciplinary domain. Dr. Amrik Singh participated and acted as a resource person in various national and international conferences, seminars, research workshops, and industry talks. His area of research interest is accommodation management, ergonomics, green practices, human resource management in hospitality, waste management, AR VR in hospitality, etc. He is currently guiding 8Ph.D. scholars and 2 Ph.D. scholars have been awarded Ph.D.

Gurucharan Singh is an accomplished academic and hospitality professional with over a decade of experience spanning both the education and industry sectors. Currently serving as an Assistant Professor at MMICT&BM (Hotel Management), MMDU, Mullana, Ambala, he specializes in hospitality operations. His professional journey includes managerial roles in training and development within the hospitality industry and operational stints at renowned establishments like Marriott International and Country Inn & Suites by Carlson. An active researcher and contributor to academic literature, he has published numerous articles in reputable journals, focusing on topics such as Artificial Intelligence, guest experience, sustainability, technology, and operational management in the hospitality industry.

Dil Swarna, Independent researcher Dil Afrin Swarna is a renowned expert in education and research. She holds a research-based MBA in Human Resource Management (HRM) from the National University of Bangladesh. Swarna's passion for HRM drives her to explore various themes, showcasing her understanding of business dynamics and interconnections. Her research benefits both academia and industry. Swarna has delved into organizational behavior, employee development, and strategic management, demonstrating the critical role of HRM in corporate performance and offering practical solutions to current managerial challenges. She actively participates in conferences, publishes articles, and collaborates with other specialists in education and research. Her innovative approaches to HRM challenges are highly regarded and push the boundaries of existing knowledge. Through her dedication to research and teaching, Dil Afrin Swarna continues to make significant contributions to academia and her industry. Her work exemplifies how committed and insightful research can enhance organizational practices and deepen our understanding of human resource management.

Mohammad Badruddoza Talukder is an Associate Professor at the College of Tourism and Hospitality Management at the International University of Business Agriculture and Technology (IUBAT) in Dhaka, Bangladesh. Holding a PhD from Lovely Professional University, India, he is renowned for his innovative integration of cutting-edge technologies such as Artificial Intelligence (AI), the Internet of Things (IoT), and Blockchain into tourism and hospitality management. His research explores the transformative impact of these technologies on smart tourism, cloud kitchens, and the broader industry landscape. As a prolific author, Dr. Talukder has published numerous scholarly articles and books, with forthcoming titles that address

AI innovations in hospitality, the evolution of cloud kitchens, and the enhancement of digital literacy within hospitality education. His work shapes contemporary tourism and hospitality practices, skillfully marrying technological advancements with practical industry applications. Since 2008, Dr. Talukder has imparted his knowledge through a diverse array of courses in tourism and hospitality across various universities in Bangladesh.

Gautam Tanwar is a hospitality professional, culinary expert, and academician with more than 15 years of experience across the hospitality industry and collegiate education. He currently serves as an Assistant Professor and Culinary Expert at the School of Hospitality & Tourism, Bahra University, while also pursuing as Research Scholar for Ph.D. in Hotel Management at Bahra University, Waknaghat. His professional foundation was laid with The Oberoi Group, where he was trained under the prestigious Systematic Training & Education Programme (STEP) of the Oberoi Centre of Learning & Development. This rigorous training instilled in him the values of discipline, service excellence, and a deep understanding of the culinary and hospitality domains. Over the years, Gautam has held key roles that allowed him to bridge operational expertise with culinary innovation. In academia, he has carved a niche by blending tradition with modernity, by teaching classical culinary practices while embracing contemporary techniques and global hospitality trends. His teaching philosophy emphasizes not just skill, but also adaptability, professionalism, and creativity. Passionate about traditional Indian cuisine and committed to advancing research in hotel management and food safety, His career reflects a seamless union of professional integrity, academic rigor, and the art of hospitality.

Yashwant Thakur is a hospitality professional, culinary expert, and academician with over seven years of experience spanning the hospitality industry and higher education. He currently serves as an Assistant Professor and Culinary Expert at the School of Hospitality & Tourism, Bahra University, while also pursuing a Ph.D. in Hotel Management at Bahra University, Waknaghat. His research interests focus on innovation, culinary sciences, and academic advancement in the hospitality and tourism sector.

Monika Baltoo Verma is an accomplished professional with over 12 years of experience spanning both industry and academia. A deep understanding of both practical and theoretical aspects of Management. Currently, Dr. Monika Baltoo Verma serves as the Head of the Department of Management & Commerce at A P Goyal Shimla University, where they lead academic initiatives, foster research, and guide students in their professional development.

Vaibhav Verma is an accomplished academician, hospitality professional, and innovator with over 15 years of experience in academia and the hotel industry. He currently serves as Associate Professor and Head of the School of Hospitality & Tourism at Bahra University, where he provides visionary leadership, fosters academic excellence, and drives innovation in hospitality education. Holding a Ph.D., MTTM, and BHM, Dr. Verma specializes in hospitality and tourism management, with research interests in sustainability, technology integration, and destination marketing. He has published research papers, authored books, and registered patents and designs, reflecting his strong commitment to research and innovation. Dr. Verma's professional career includes roles with leading hotel brands such as The Oberoi (Wildflower Hall) and The Radisson Hotel, where he gained invaluable operational and managerial expertise. He is also the Secretary of Him Anchal Chef Association (HIMCA) and an active member of the Association of Hospitality Professionals. Dedicated to blending industry insights with academic rigor, Dr. Verma continues to shape the future of hospitality and tourism education.

Index

I

K

L

M

N

O

P

R

S

T

U

W

www.ingramcontent.com/pod-product-compliance
Lightning Source LLC
LaVergne TN
LVHW062338300926
849161LV00007B/1742
* 9 7 9 8 3 3 7 3 4 0 4 2 5 *